住 哪？[3]

Where to Stay ?

一个设计师的“找窝”之路 | 区伟勤 著

中国建筑工业出版社

图书在版编目(CIP)数据

住哪？3/ 区伟勤 著.—北京：中国建筑工业出版社，2017.11
ISBN 978-7-112-21439-6

Ⅰ.①住… Ⅱ.①区… Ⅲ.①室内装饰设计-中国-图集
②散文集-中国-当代 Ⅳ.①TU238-64②I267

中国版本图书馆CIP数据核字（2017）第262656号

责任编辑：杨　晓　唐　旭
责任校对：王　烨

《住哪？3》是作者继《住哪？》、《住哪？2》“再”住过的六十余个酒店的室内设计记录绘本，每张图都配有作者感性而又带有敏锐、细腻艺术感受的散文。从本书中旅馆房间的各处细节，可以窥见设计者的巧思，甚至各地的地域风情、历史等，不但适于高等院校室内设计、艺术设计等相关专业学生、从业者参考阅读，还可当作选择下榻旅馆的参考，更是建筑或空间设计的简易读本，适于大众读者。

住哪？3
区伟勤 著
*
中国建筑工业出版社出版、发行（北京海淀三里河路9号）
各地新华书店、建筑书店经销
恒美印务（广州）有限公司印刷
*
开本：889×1194 毫米　1/20　印张：17　字数：340千字
2018年1月第一版　2018年1月第一次印刷
定价：135.00元
ISBN 978-7-112-21439-6
（31121）

出版说明

本书作者区伟勤先生在常年的四处差旅中，出于设计师的敏感，养成了一个独特的习惯：每到一处酒店的房间，都用纸笔将房间的平面图和细部装饰等内容以手绘的形式记录下来，并写下当时的有感体会。由于是即兴所作，手稿中难免存在字迹潦草、语句不通等不当之处。为保留作者原汁原味的推敲与记录，编辑仅对手稿部分进行了细微的调整，而在“文章注释”部分中，在原稿文字的基础上进行了加工整理。因此，本书文字内容以“文章注释”为准，所有照片除标注外，均来源自区伟勤先生在入住过程中的原创拍摄。特此说明。

韦玮 · *Wilson Wei*

凯世酒店集团 · *Cachet Hotel Group*
资深项目总监 · 建筑设计、施工及项目管理
Senior Project Director
Architectural Design Construction and Project Management

区伟勤让我给他的作品写序的时候，我是有几分惶恐的。一口气读完了这本书，“手不释卷”大致就是如此吧。

I was somewhat frightened when Ou Weiqin asked me to write preface for his book. After reading the book in one breath, I feel that I cannot take my hands off it.

想起当初坐在书店的角落里捧着日本建筑师浦一也的《旅行从客房开始》，也是同样的情景，当时常常想中国的浦一会是谁呢?

I remembered when I sat in the corner of the bookstore, holding Travel From the Guest Room written by the Japanese architect Kazuya Ura.In the same scene, I often think who will be China's Kazuya Ura?

一个走遍世界、热爱酒店的人；
A person who travels around the world and loves hotel;

一个对建筑和设计充满激情和敏锐观察力的人；
A person who is passionate and a keen observer on architecture and design;

一个既有很高的专业水平又有手绘和文笔功力的人。
A person who has a high professional level and hand-painted and writing skills.

最重要的是，可以为了这份热忱，坚持十数年如一日的人……直到两个月前，我与大学同班同学伟勤在上海重逢，他拿出《住哪?》递到我手里之时，我心中才有了答案。

The most important thing is that he must persevere for more than ten years for this enthusiasm... until two months ago, when I met with my college classmate Weiqin in Shanghai and he took out the book Where to Stay? I had the answer in my mind.

这无疑是一本专业书，作者分别从住客和设计师两个角度分析了多年来他在世界各地入住过的各个酒店客房。作为行业从业者之一，公正地说，书中的内容无疑对于酒店投资方、设计师和管理者都是宝贵的经验财富。

This is undoubtedly a professional book. The author analyzes hotel rooms he has stayed all over the world in many years from perspectives of guests and designers. As one of the industry practitioners, I think this book contains valuable experiences for hotel investors, designers and managers.

有意思的是，即便是作为一个普通的酒店爱好者，抑或只是个爱旅游的人，这本札记也可以让你品得津津有味。因为作者恰好也是一个背着行囊，揣着热情，用心去追求“诗和远方”的旅人。

Interestingly, even as an ordinary hotel enthusiast or just a travel enthusiast, this note can also make you feel entertaining. Because the author happens to be a traveler pursuing "poetry and distance"by carrying luggage and keeping enthusiasm.

韦玮 2017年8月1日于上海

Wilson Wei, August 1, 2017 in Shanghai

丁江涌 · *Alex Ding*

雅高酒店集团 · *Accor Hotel Group*

大中华区多品牌酒店设计及技术服务总监

Director of Design and Technical Services for Multi-Brand Hotel in Greater China

人在旅途，到每间旅店打开房门前都有种种期待，开门后或多或少都会给旅者惊喜，也是一种小确幸。

For people on the journey, every time before opening the door of hotel room, they are with a variety of expectations; after opening the door, there is more or less a surprise. This is a kind of small happiness.

客房，旅者逗留最久之处，须经得起细细玩味和挑剔，客房设计，最讲究设计者功力。

Guest room, a place where travelers stay for the most time, should must bear delicate pondering and selection. The room design asks for the most exquisite design skills.

走遍世界，酒店设计各有特色。欧洲的矜持、日本的精致、美利坚的“豪”、东南亚的“闲”……为旅人提供心仪的休憩处，是客房设计的终极目标。

Throughout the world, hotel design has its own features. The modesty of Europe, the delicacy of Japan, the pride of America, the leisure of Southeast Asia...To provide a thoughtful resting place for travelers is the ultimate goal of room design.

有幸释读伟勤先生的《住哪？》。作为资深酒店设计人，作者在行走世界时，仍笔耕不辍，商务旅游之余，拿起酒店信笺，随手勾勒客房平面，写下感悟，并集结成书，今已经是其成第三本矣。

It's a pleasure for me to read Where to Stay? by Mr. Wei qin. As a senior hotel designer, the author is still writing when he walks around the world. When business travel is over, he picks up the hotel letterhead to draw the room plan and write down his feelings, and collects into a book. Now this is the third one.

在纷繁浮躁的当下，能静心品味产品，字里行间，那一份幽默洒脱，那一份坚守与淡然，令人感动。

In the numerous and impetuous world, he can still calm down to taste hotels. Between the lines, we can feel a sense of humor and freedom and a sense of persistence and indifference, which is moving.

平面是设计的基础，建筑形态千差万别，设计者需准确把握细节，精当安排。

Plan is the basis of design. For different architectural forms, designers need to accurately grasp the details and arrange delicately.

感悟需浸淫多年方能有所得，挥洒创意又要艰难妥协，悟道之路痛并快乐着。

Feelings can be got only after many years of immersion.Being creative should also comprise sometimes. The road of enlightenment is with both pain and happiness.

此书对酒店设计初入行者是指引，对资深设计人的“悟道”是启迪。

This book is a guide for beginners on hotel design, as well as an inspiration for the "enlightenment" of senior designers.

行千里路，须有容身处。

Travel one thousand miles, you need a place to shelter.

住哪?

Where to Stay?

心安之，住哪都行。

With a peaceful heart, everywhere is perfect for staying.

丁江涌 *Alex Ding*
7/31/2017
July 31, 2017

戴昆 · *Dai Kun*

北京居其美业住宅技术开发有限公司

EASY-HOME Technology Development Co., Ltd.

设计是一种病。

Design is a kind of disease.

2016年10月中的一天，美国展前我自己在纽约悠闲几天，早上去中央公园跑步，白天又去看了中城的几个新的房地产项目。记得有扎哈・哈迪德设计的520西28街，还有MiMA豪华公寓和哈德逊广场 。等走回上城天色已暗，我几乎是习惯性地走去了常去的牛排馆，然后径直地往自己常坐的位子（一个角落）走去。却发现那里坐了一个人，只好悻悻然坐在旁边，不甘心地再扭头看一眼旁边的这位，正好对方也抬头看我，于是都是惊讶——居然是区伟勤！

One day in October 2016, I spent a few days in New York before the American exhibition. I jogged in Central Park in the morning and went to see several new real estate projects in midtown during the day . I remember 520West 28th designed by Zaha , MiMA and Hudson Yards. When I got back to the uptown, it was dark and I was almost accustomed to the usual steak house, and then headed straight to my usual seat (in a corner). I found there sat a man, I had to sit next, and then I turned to see who sat next to me, at the moment, that guy just looked up at me too, and we were surprised. It is actually Ou Wei qin.

于是坐在一桌边吃边聊，同样的建筑系背景做室内设计，共同语言不少，加之他才从芝加哥来，也有不少观感。饭后伟勤兄盛情邀请我去他住的巴卡拉酒店喝一杯，自然，出于设计病患者的习惯，我们先品了酒店的公共部分，然后是客房。进他房间，桌上已经摊着本书240~243页那一堆草图，伟勤解释他有“拆”酒店的习惯，入住一个酒店就要“拆解”客房一番，还专门集结成册出版。记得我当时还感叹自己烂笔头多年，却始终没有鼓起勇气来整理成册……

So we sat at a table to eat and chat. With the same architectural background in interior design, we had lots of common language. He just travelled from Chicago and also had lots of impression. After dinner, Weiqin invited me to have a drink at Hotel Baccara where he stayed. Naturally, from the habit of designers, we first tasted the public parts of the hotel and then the guest rooms. Entering his room, I could see piles of paper, which was page 240-243 of this book on the table. Weiqin explained that he has the habit of disassembling hotel, which means that once he stays in a hotel, he would analyze the guest room and collect all drawings and hotels to a book for publishing. I remember that I lamented that I had written for many years, but didn't have the courage to order to a book.

转眼半年多过去，一日伟勤兄来约文，原来又是一本“拆”酒店的帖记完成了！收到书稿，我着实有些感叹于伟勤兄的“博爱”，原来他不只是“拆”四季、瑰丽这些大牌，连沧州的假日酒店也是要“拆”的——于是有了由衷的感叹，设计是一种病！

Then half a year past, one day Weiqin asked me to write something for him. I know that he has finished another book of disassembling hotel.When I received the manuscript, I was amazed at Weiqin's "universal love". he disassembles not only big brands such Four Seasons and Rosewood, but also Holiday Inn in Cangzhou. Therefore, I had a heartfelt sigh: design is a kind of disease.

磨笔头于我等，大概是如钢琴家的练琴和舞蹈家的练功。为了能够流畅地用手表达思维，我们日复一日地写写画画，让自己的手能够跟得上脑子并不容易，只有当笔尖的流动像呼吸一样自然的时候，思绪才能无休止地不断演进下去。当下的年轻学子们，更多地把“手绘”当作一个速成练习的目标，临摹几张范本让自己有些招式，却不知手是脑最好的合作伙伴，设计更多靠的是人脑而非电脑，你的手给你不插电的思考机会！

Practicing writing for us is like practicing piano for pianist and practicing dancing for dancer. In order to express one's thinking fluently, we draw and write day in and day out.It's not easy to make the hand keep pace with brain, and only when the tip flows naturally like breathing, our thought could endlessly evolve. Young students today regard freehand sketching as an intensive training target. They copy some templates to grasp some moves, but they don't know that hand is the best partner for the brain, and design depends more on the brain than computer, and the hand gives you unplugged chance to think.

期待第三本！

Look forward to the third book!

“拆光”天下酒店！

Disassemble hotels all over the world !

戴昆 二〇一七年八月一日. 北京. 东直门.

戴昆，二零一七年八月一日，北京 东直门

Dai kun, August 1, 2007, Beijing, Dongzhimen

叶茂中 · *Ye Mao zhong*

叶茂中策划机构

YeMaozhong Marketing Inc.

住哪?

Where to stay?

这是抬脚出门前，头一个要解决的问题。这算不算是一个冲突? 对于那些常年四处奔走的人，的确如此。酒店的位置、价格、服务、品牌、硬件、配套，都是要考虑的要件，而就算万事俱备，最后的结局往往是没有空房，哈。

This is the first problem to be solved before I set up to a trip.Isn't that a conflict? For those who travel around all the year round, that's true. The location, price, service, brand, hardware and accessories, all elements of a hotel should be considered. And even if everything is ready, the final outcome is always that the hotel is full booking, Haha.

找到一家真正称心如意的酒店，绝不是一件容易的事。

Trying to find a really good hotel is by no means an easy thing.

而到了今天，人们又对"什么是一家好酒店"提出了新的要求，"一张床、一个厕所、一张写字桌"的千篇一律式设计已成过去，各种精品酒店、设计酒店层出不穷。这是好事，给行路的人们更多的选择，但也增加了烦恼，因为挑选时更加眼花缭乱。

And today, people propose new requirements of "what is a good hotel", which means the same design of "a bed, a toilet and a desk" has become the past. A variety of boutique inns and design hotels emerge in an endless stream. This is a good thing for people on trip. They have more choices, but it also adds trouble to them because the selection is more dazzling.

伟勤兄弟是路上的人，更要命的是，他又是眼光很毒的设计师，对于"住哪"这个问题，肯定比常人会更挑剌些，也肯定比很多人都更有发言权。因为伟勤兄弟不仅阅店无数，更是个懂美的人，好比一个懂画的人，不是什么画都能入眼的。

Weiqin is a person on trip, what is more, he is also a designer with keen vision. For the problem of "where to stay", he certainly will be more critical than ordinary people, which he also has more right to say more than a lot of people. Weiqin not only live in countless hotels, he is also a person knowing the beauty. For a person who understand paintings will only select good painting.

在这本书里的每家酒店，都有自己的味道，加之伟勤的解读，更是妙趣横生。

Every hotel in this book has its own flavor, coupled with the interpretation of Weiqin, it is more interesting.

在这本书里的每家酒店，都仿佛有着自己的灵魂，它们通过每一张手绘的平面图，通过每一篇手写的心得，逐渐变得生动而真实起来。

Every hotel in this book seems to have its own soul, which gradually becomes vivid and true through each hand drawn floor plan and each handwritten experience.

在这本书里的每家酒店，都已经不仅仅是一家酒店，而是一件件作品。从一个设计师的角度做的解读，能让我们好好体会创意是如何变成现实的过程。

Every hotel in this book is not just a hotel but a piece of art work. From a designer's point of view of interpretation, we can have a good understanding of how creativity becomes reality.

想找到一家好酒店？想找到一家有趣的酒店？这本书是一个好选择。诸多酒店的优点与缺点、风格与细节、惊喜与瑕疵，无所不有。

Do you want to find a good hotel? Do you want to find an interesting hotel? This book is a good choice. Advantages and disadvantages, style and details, as well as surprise and flaws of different hotels can be found in this book.

你是个学设计的？或者家里正在装修？这本书也是一个好选择，因为全是名家精品，如假包换，伟勤认证，童叟无欺。

Are you a student of design? Are you decorating your home? This book is also a good choice, because the content is boutique and certificated by Weiqin.

如果你是个对平面图有着特殊迷恋的“怪人”…… 这本书更是你的上上之选，别无分店。

If you're a "geek" with a special crush on floor plan, this book is your last choice, without any other choices.

这是一本酒店攻略，这是一本设计图册，这是一本日记流水，这是一本户型大全，这是一本给那些同样懂美的人准备的精品指南。

This is a hotel guide, this is a design chart, this is a diary, this is a house type collection, this is a quality guide for those who understand the same beauty.

这是一本很特别的书。

This is a special book.

希望你喜欢。

I wish you like it.

Contents

目 录

序

前言

宿

Contents

目 录

成果

路上的朋友们、轻功铁汉：

20/6，2017

"轻功铁汉"而"轻重相适"，这些小道理大家都懂，也要说，但坚持自我的还是不容易的。坚持每一次在路途都要写一下自己看到的景色、做的选择、吃的东西，好的细节都会不被忘记地记下来；一旦遇到困难的、甚至是丑陋的，那就懂了。而记录我们走过、也知道都是谁、可以懂了！

同样，开始整理下我的资料，写了旁边的，再整理出来的东西、修改、重写，这一些都是在这过程的时候。所以多数利用在路上的时间，高铁上、住宿里、甚至汽车上也不断地作出的修改工作。真是佩服"以文字为生"的作家们的毅力。希望我也能了这个的级别了。总是坚持在路上，写写画画；画画，写写在路上的一些故事、美人、一些风情、再加一些所见的体会、感动。

轻功铁汉，不敢妄称这个名字。

[illegible]

Writing on the Road, Accumulates
路上写写字，积少成多

“积少成多”，这个小道理大家都懂，真要我、你坚持这样做还是不容易的。坚持每一次去住酒店都画一下平面，遇到好的设计、好的装饰、好的家具、好的细节都会小激动地画下来，一旦遇到平庸的，甚至是丑陋的，那就惨了，但为了记录我的足迹，也就勉为其难。可纠结了！

"Every little makes a mickle". Everyone knows it. Yet it is hard for one to stick to it. Every time I put myself up in a hotel I will habitually draw a plan sketch of it. When fascinated with some good design, decoration, furniture or detail, I will draw it down. When it is mediocre or ugly, it will turn out to be an awful experience. Yet, in order to record my footprints, I will settle for it with great reluctance!

同样，开始整理原始资料，写写草稿文字，再慢慢地看看，修改、重写，这一些都是点点滴滴的时间，而且多数还是利用在路上的时间，高铁上、住店里，甚至汽车上也可以做小小的修改工作，真是佩服“以文字为生”的作家及供稿人，看来我是混不了文字这口饭的了，只能坚持在路上写写画画、画画写写，写写在路上的一些故事、一些人、一些风情，再有一些自己的体会，足矣。

Likewise, I have begun to sort out original materials, write drafts, spend time poring over, revising and rewriting them. All this is done with little bits of time, mostly while I am on the road, on the highway, in hotels, or even in cars, when I will do some small revision. I really admire writers and contributors who make a living out of writing as it seems that I don't have it in me to be one of them. I can only keep writing and drawing while on the road, narrating certain stories, certain people, certain customs and some of my feelings. And I will be satisfied with that.

积少成多，不积就永远多不了。

Many a little makes a mickle. Without accumulation there will never be so much.

恋上你的床

你会睡陌生的床吗？不要想歪了，是指若果住店的人，有很多会睡不安甚至失眠于陌生的床。

真的是“金窝不如狗窝吗？”

这几年酒店在“睡”上面确实下了不少的功夫，硬件上从床上用品的手感，枕头的芯的用料细节、床垫的品质及其厚度，再到床背的形式；软性上从睡眠区的灯光气氛、冷风量，再到窗帘的位置及隔音研究、香薰、音乐的引入等，努力让你睡得安心。

床有两种：好的与不好的。

所以是好的床，“柔软而有韧力”；厚重而有分量，能够让你在酒店能睡到比你自家的更好的床，那你就会定期或不定期“入住”酒店的各式各样的床，那酒店就成功了。

但那都要有床，有好的床，就能让我——恋上你。

恋上你的床

你会睡陌生的床吗？不要想歪了，是指差旅住店的人，有很多会睡不安稳甚至失眠于陌生的床。

Can you ever sleep on a bed other than your own? Don't get amorous ideas. I'm talking about people who stay awake all night on a hotel bed.

真的是“龙床不如狗窝吗？”

Is it true that "even an emperor's bed can not compare with my homey bed"?

这几年酒店在“睡”上面确实下了不少功夫，硬件上从床上用品的手感、枕头内芯的用材细节、床垫的品质及其厚度，再到床背的形式，软性上从睡眠区的灯光气氛、新风量，再到空调的位置及低噪声研究、香薰系统的引入等等，努力让你睡得安心。

Over the past few years hotels have really been making efforts to make guests sleep well, paying attention to the feel of bedding, the details of the pillow padding, the quality and thickness of the mattress and the form of the bed back, as well to the the lighting atmosphere and fresh air amount in the sleeping zone, the study of the location of the air conditioner and low noise, and the introduction of the incensation system, and etc.

床有两种：好的与不好的。

Beds come in two kinds: the good ones and bad ones.

什么是好的床，柔软“有韧劲”，厚度够分量，够宽够大。在酒店能睡到比你自家更好的床，那你就会定期或不定期“白住”酒店各式各样的床，那酒店就成功了。

What are good beds? They are soft and tough, thick and weighty enough, wide and large enough. When a hotel's beds feel better than yours at home, you'll spend nights on the various beds of that hotel regularly or irregularly "free of charge". And then the hotel will succeed.

住哪都要有床，有好的床，就能让我——恋上你。

Whenever you go, you need a bed. Whenever there is a good bed, I will fall in love with it.

退堂鼓

《住哪》的筹备期间，心里也曾打过退堂鼓，一方面是手写难度，二来担忧，考虑到网络时代，谁还看这种书？这么传统甚至显得落后的表达方式。大家都在讨论"元宇宙"，讨论AI（人工智能）了。

手写书，如"保罗们的思考"，也许短期内更加珍贵，过了十几年，连泛黄的纸质书也变得稀缺如钻石，当然这本书更重要是继续记录了我的"足迹"。

世界各地的酒店建筑是否可以写得更加广泛一些，不再局限于讲述酒店里面的"看得见"空间、活动与事件，也还是这种"看不见"的表达层面，让我打起了另一面的"退堂鼓"，大家会不会觉得没有新鲜感呢，毕竟是第三本了《住哪》？

~~世界各地~~的酒店还是各式不同，每住一间也会有不同的惊喜与感受，要是长篇大论去陈述几十甚至上百间酒店的印象、感受、批判、评价等等，以一己之力也难以胜任。

想想也罢了，就安安心心写写住过的，就当是小小写写客房，让我再来也同你说一下"住在哪"的一点点趣事。

对，就一点点。

《住哪？3》筹备期间，心里也曾打过退堂鼓，一方面是手写耗时，二来想想，都互联网时代了，谁还看“这种书”，这么传统甚至是落后的表达方式。大家都在讨论“永生”，讨论AI（人工智能）了。

While preparing for Where to Stay? 3, I once thought of giving up halfway. For one thing, writing by hand is too much time consuming; for another, who will read "this kind of book" expressed in such a traditional and even behind-the-time way in this Internet age, when everyone is talking about "everlasting life" and the artificial intelligence?

手写书，如“侏罗纪的恐龙”，也许短期内更加珍贵，过了十几年，连活着的纸质书也变得稀缺如钻石，当然这本书更主要是继续记录了我的“足迹”。

Handwriting, like dinosaurs in the Jurassic period, will become valuable in the short term. In dozens of years, living paper books will also become as rare as diamonds. Of course, this book is mainly a continued record of my "footprints".

也有身边的朋友建议是否可以写得更加广泛一些，不单单局限于讲酒店房间里面的见闻、经历与事情，也正是这种“窄狭”的表达层面，让我打起了另一趟“退堂鼓”。大家会不会觉得没有了新鲜感呢，毕竟是第三本了，《住哪？3》。

Some friends close to me also ask if I can write something more abroad, something not confined to the accounts of what I saw, experienced, or happened in hotel rooms. It was just such "narrowness" of narration made me about to give up, too. I was wondering whether it would fail to inspire some sense of freshness in the readership, after all, it would be the third of the serial, or Where to Stay? 3.

世界各地的酒店还是层出不穷，每住一间也有不同的惊喜与感受，要长篇大论去陈述几十甚至上百间酒店的印象、感受、批判、评价等等，以一己之力也难以胜任。

Still, the world abounds with hotels, and each room gives different surprise and sensation. It is impossible for one to give lengthy accounts of the impression, feeling, criticism, comment of the dozens and even hundreds of hotels, single-handed.

想想也罢了，就安安心心写写住与床，就专专心心写写客房，让我乖乖地向你说一下“住在哪”的一点点趣事。

Then I have settled for just thinking it over, just writing about my staying-at-hotels experiences and the beds, the guestrooms, and the little bits of fun of "living there".

对，就一点点。

You are right, just little bits of it.

过客

过客，是丁乙怀信的眼科，排挤酒店里周的逢浮去世人士。他对酒店的要求，"舒适粗养"，能容身之处是其基本，而能让"心安之"的酒店，那才是值得回归之处。

正如叶芝中某某所说，要找到一间称心如意的酒店，绝不容易，当然不可能"像在他的剑桥记忆里去翻"的那样，但有一点点心思去用心，造一个快乐和温馨的"过客"去住。也许《何时？》而能给过客们手里得到的一本书。

有一天，当我收到戴某师兄的方案，写以我"指家"的解释，觉得他完全了。不论，那里了，住的、吃的、玩的、吃的，都在我。我甚至可以称作"家"，"这也是一种家"，也许这种温馨的"家"让我们住得更好。而且在这个世界里中，能否能让设计者们从"老年视角"，那只是能自己去记了，等着吧。

Passenger

过客

过客，是丁总微信的昵称，雅高酒店集团的资深专业人士，他对酒店的要求“简单粗暴 ”，能容身之处是最基本的，而能让客人“心安之”的酒店，那都应当有独到之处。

Passenger is the nickname of Mr. Ding in Wechat. As the senior professional of Accor Hotel Group, his requirements of hotel is simple and strict.To shelter is the most basic quality, and for hotel that can make guests“feel peace”, it should have some unique quality.

正如叶茂中先生所说，想找到一间称心如意的酒店，绝不容易，当然不至于“没有好创意就去死！”的地步，但花一点点心思去挑选，当一个快乐的三五天的“过客”，去住住，也许《住哪？》可以成为你手里特别的一本书。

As Mr. Ye Maozhong said, it's never easy to find a good hotel, and certainly it's far away from “dying without good ideas.” To spend a little thought to choose and be a happy “passenger” for several days, and to live, maybe this Where to Stay? can be a special book in your hand.

有趣，当我收到戴昆帅哥的序，冠以我“拆家”的称号，笑死宝宝了。确实，习惯了住住、写写、画画、吃吃、睡睡的我，或者可以称作“病”，“设计是一种病”，也许这种坚持的“病”让设计做得更好，而且可以乐在其中。能不能延缓或者阻止“老年痴呆”，那只能自己去印证了，等着呗。

It's funny that when I received the preface from Dai Kun, a handsome guy, he uses “separating home” to describe me. It's true that I'm used to living, writting, painting, eating and sleeping. You can also say that I get a “disease”.“Design” is a kind of disease, and adhering to this “disease”may help me to design better, and I enjoy this process. Whether it can delay or prevent Alzheimer's Disease? Let's wait and see. Time will prove.

也许我不一定会体悟老先生沛一世的教养，但所坚持

的精神，让你们在世上一种小心的，给了一个没有生命的人生

足迹，相信正如弟师在同学们的勤奋。正在出书是你

“手不释卷”的一本！

再次感谢杜老师们的“学子之书”，让这本《建筑》

得一重意义，给我多一份的鼓舞。

那我就随便的我的过客，谢谢了！

4/8, 2013

也许我不一定有日本老先生浦一也的执着，但可以坚持几十年，让你们通过这一系列的书，看到一个设计师的人生足迹，相信正如韦玮老同学所期待的，这应当是你“手不释卷”的一本！

Maybe I am not so persistent as Japanese Kazuya Ura, but I can persist for decades and let you see a designer's life footprint through a series of books. Just as my old classmate Wilson Wei's expectation, you can hardly tear yourself away from this book.

再次感谢有你，有你们的“举手之劳”，让这本《住哪？》多了一重意义，给了我多一份的鼓舞！

Thank you again. Your "hand's turn" gives Where to Stay? significance and encourages me a lot.

那我就继续做我的过客、睡客了！

hen I'll continue to be a passenger and sleeper.

August 4, 2017

凯宾酒店
KAIBIN HOTEL

包头凯宾酒店

1

★★★★

KAIBIN HOTEL BAOTOU CHINA

Address : No.12 Wenhua Road, Qingshan District, Baotou,Neimenggu,China
中国内蒙古包头市青山区文化路12号

Telephone : +(86 472) 3336660

Fax : +(86 472) 3334888

Http : //www.btkbjd.com

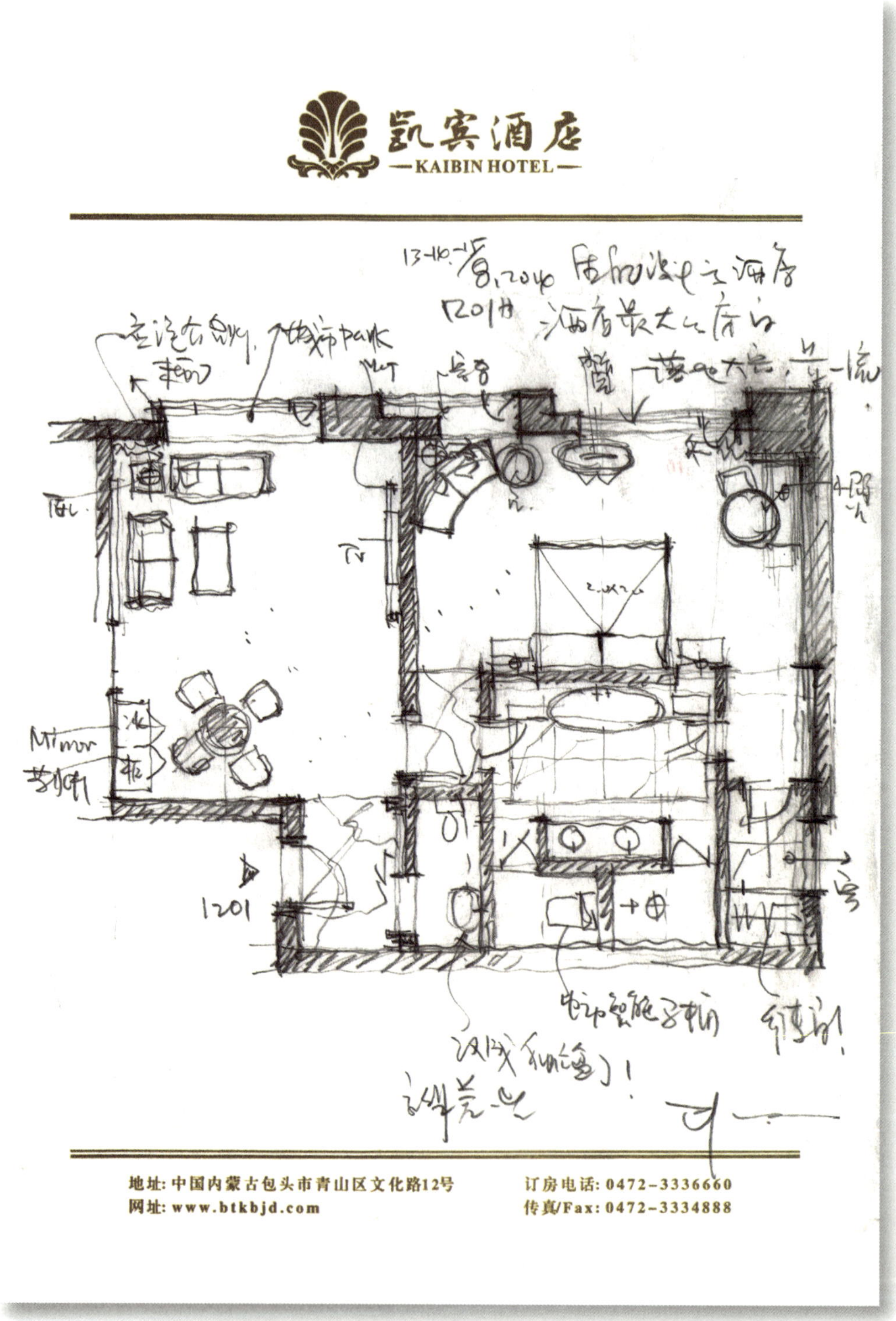

潘总你好：

住在自己公司设计的酒店，就像呵护小孩一样，深夜到达京宣华的，当然楼下人的热情也是我们感到幸运的原因。

前后两年多的时间，从第一次方册上会接洽，迅速deal签约，到开始的设计建筑（中间经历我们公司的劳作还是不少的）再到后来的工作有序进行，也让信任（对）是我们最大的恩惠，也是我们的责任。

做精做底的，再到大餐厅、材料、装饰、外墙、材料、色彩、公共空间、小的细节的这种酒店也开始适合我们去"打造"的！13000 m²，我们投资，这样下来也对设计施工成本控制也顺利，（成功）打造了后面运营的能力，这样的结合可让双方都满意！

"一分钱一分货！"将酒店布在包头的一些大气、豪华

凯宾酒店
—KAIBIN HOTEL—

我房的会所（未开出），大套房（就是我住了三天的）就够了，其余的是大堂、酒吧、餐厅、实用、高档到位，适合做设计的打分，可谓"上到顶上"。当然，酒店菜不太好也会少少打折扣，我发出的第一个酒店，很成功吧，相信你再接受下一个，就会更加成功了。

这次能做到舒适、漂亮、实用不易，我们这个酒店可谓小而精致，任务是这个大套房，给了东方高档酒店的背景（平衡）[保持]风格完全不同，大家对[illegible]市会同，可谓一流。

住三天，心满意足。

谢谢我们的团队、我们的客户！

中[illegible]
2014

地址：中国内蒙古包头市青山区文化路12号　　订房电话：0472-3336660
网址：www.btkbjd.com　　传真/Fax：0472-3334888

包头凯宾酒店
Kaibin Hotel,Baotou, China

Satisfied?
满意了吗?

住在自己公司设计的酒店，就像呵护小孩一样，深夜到达亦觉亲切，当然北方人的热情也是我们感到荣幸的原因。

Staying at a hotel designed by my own firm feels like taking care of my own child. It was in the depth of night when I arrived. Still, I felt warmth. Of course, the Northerners' hospitality also made us feel honored.

前后两年多的时间，从第一次在广州见面，投缘，迅速地签约，到开始优化建筑（事实证明我们的“功劳”还是不少的），再到设计工作的有序进行，业主的信任是对我们最大的恩惠，也是我们的幸运。

It took a period of more than two years, from our first meeting in Guangzhou when we hit it off and cliched a contract immediately, to the optimization of the building (facts proved that our "contributions" were remarkable), and to the orderly progress of the design process. The owner's trust was the greatest favor for us as well as our luck.

包头凯宾酒店
Kaibin Hotel,Baotou, China

做样板房间，再订大货产品，机电、装饰、外立面，样样包括，有时候觉得小规模的这种酒店也挺适合我们去“折腾”的！13000m^2，私人投资，这样可以达到双方共赢。成本控制也顺利，减少日后营运的压力，这样的配合可谓双方都满意！

Building the demonstration houses, ordering big items such as electronic-mechanic machinery, decorations, facades, the project was inclusive. Sometimes I think this small-scale hotel is worth our effort, the privately funded 13,000m^2 hotel was a win-win project.

The cost control was also smooth, reducing the pressure on the operation in future. Such cooperation was mutually satisfactory!

“一分钱，一分货！”将重要部位定位高一些：大堂、餐厅、私属会所（未开业）、大套房（就是我住了三天的），就够了，其余的空间尽量以标准、实用、易打理为主，结合灯光设计的技巧，可谓“点到即止”，当然工程的质量不太好也会使效果打点折扣，权当业主的第一间酒店，取取经验吧，相信如果再投资下一间，就会更加成熟了。

包头凯宾酒店 *Kaibin Hotel,Baotou, China*

“Quality is ensured by input”. Important parts should be highlighted: the hall, dining hall, private club (yet to open), big suite (as the one I had stayed in for three days). Then it suffices. As for other spaces, emphasis should be placed on being standard, practical and easy to manage, as well as the skill of lighting design. Everything should be “moderately to the point”. Of course, the project might have been comprised by a second-best quality of construction. It may well serve as the first pilot hotel of the owner. I believe that we will become more mature if the owner is to fund another hotel.

设计能做到多方满意实属不容易，我们这个酒店可谓小范例，你看看这个大套房，简单亦有高端酒店的骨架（平面），尽用景观资源，大窗对着城市公园，可谓一流。

It is really hard for a design to satisfy multiple sides. This hotel designed by us can serve as a small model. Look at this big suite: simple as it may look, it has the skeleton (plan) of a high-end hotel and optimizes landscape resources—the big windows offer a view of the city park. It can be class first class.

包头凯宾酒店 *Kaibin Hotel,Baotou, China*

住了三天，心满意足。

I stayed there for three days, content and satisfied.

谢谢我们的团队，我们的客户！

I am indebted to my team and my clients!

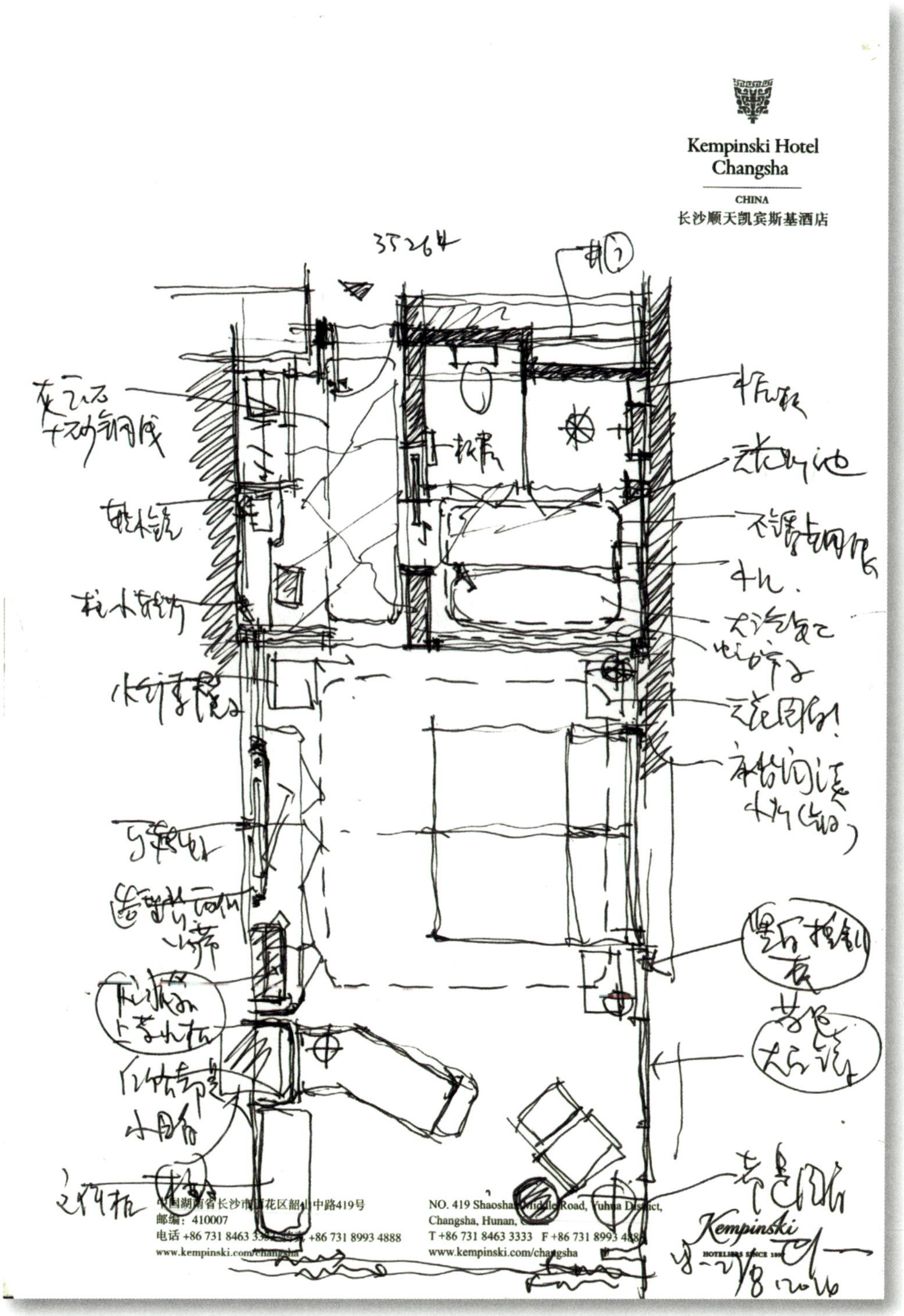

Kempinski Hotel
Changsha
CHINA
长沙凯宾斯基酒店
★★★★

2 KEMPINSKI HOTEL CHANGSHA CHINA

Address : No.419 Shaoshan Middle Road,Yunhua District, Changsha,Hunan,China
中国湖南省长沙市
雨花区韶山中路419号
Telephone : +(86 731) 8463 3333
Fax : +(86 731) 8993 4888
Http : //www.kempinski.com/changsha

走道上的洗手盆

记忆中，设计师与酒店管理公司一般都不轻易去挑战"规矩和习惯"，特别是不利于日后管理的"创新"。对于客房来说，洗手盆被放在公共区域的走廊区域，可说是一种"出界的行为"。大胆开放的洗手盆成为设计型酒店，如北京的柏悦酒店、香港的英迪格酒店（详见《住哪儿》）等等不同，象这一间长沙的凯宾斯基这么"粗鲁"的将洗手盆放在走道的一侧，不能不令我对设计师与设计公司的胆量和功夫佩服得不得了。

入住之后，体验一下还是有便利之处：不用关门，家人或情侣一起住的时候，一人如厕，另一半可以正常洗刷、化妆；不习惯之处只是在心理上，当我站在走道上刷牙、洗脸，动作要细致一些，温柔一些，免得水溅到地板上，毛巾、用品也小心翼翼的不能乱放，会不雅观。事后想想，也许这个也是乐趣。设计师就是让你大大方方的把这个小区域弄乱，体验自由的私居生活，这样我就不客气了。

你喜欢吗？在走道的洗脸！

Washbasin on the Passageway
走道上的洗脸盆

记忆中，设计公司与酒店管理公司一般都不轻易去挑战“规矩和习惯”，特别是不利于日后管理的“创新”，对于客房来说，洗手盆放在公共区域的走廊区域，可谓是一种“冒失的行为”，与全开放的洗浴区域设计型酒店，如北京的柏悦酒店、香港的英迪格酒店（详见《住哪？2》）等等不同，像这一间长沙的凯宾斯基这么“粗鲁”地将洗手台放在走道的一侧，确实令我对设计师、设计公司的嘴皮功夫佩服得不得了。

As far as I can remember, design firms and hotel management companies will not challenge “rules and customs” without provocation, by “being innovative” that does not favor management in future. As far as guests are concerned, placing the washbasin in the public passageway area is “rash act”. Unlike the fully open wash and bathing areas designed in hotels such as Park Hyatt Beijing and Hotel Indigo Hong Kong (see Where to Stay?2 for more), and etc, this Kempinski Changsha had the wash platform “rudely” placed on one side of the passageway, inspiring my heart-felt admiration for the eloquence of the design and the design firm.

入住之后，体验一下，还是有便利之处的：不用关门，单人或情侣一起住的时候，一个人如厕，另一半可以正常洗刷、化妆；不舒服之处只是在心理上，老是感觉在走道上刷牙、洗脸，动作要细致一些，温柔些，免得水溅到地板上，毛巾、用品，也小心翼翼地不能乱放，会不雅观。事后想想，也许这个也是乐趣，设计师就是让你大大方方地把这个小区域翻乱，体验自由的旅居生活，这样我就不客气了。

My stay here proved that such a design still has some advantage: the door need not be closed when you stay alone or with a lover. When one is using the toilet, the other can wash and dress up without interfering with each other; the discomfort is only psychological—always feeling that you have to be more careful and gentle when brushing your teeth and washing your face on the passageway lest the water should spill onto the floor; the towel and other articles cannot be misplaced lest it should look unseemly. On reflection, maybe it is also some kind of fun. The design of the designer is for you to make a mess of this small area feeling no guilt so as to experience a free travel life. If so, I will be free to use it.

你喜欢吗？在走道上洗脸！

Do you like it—washing your face on the passageway?

长沙凯宾斯基酒店
Kempinski Hotel, Changsha, China

INTERCONTINENTAL.
CHANGSHA
长沙北辰洲际®酒店

长沙北辰洲际酒店

★★★★

3

INTER CONTINENTAL HOTEL CHANGSHA CHINA

Address : 1500 Xiangjiang North Road,Kaifu District, Changsha,Hunan Province,China
中国湖南省长沙市开福区湘江北路三段1500号

Telephone : +(86 731) 8966 8866

Fax : +(86 731) 8966 8899

Http : //www.intercontinental.com

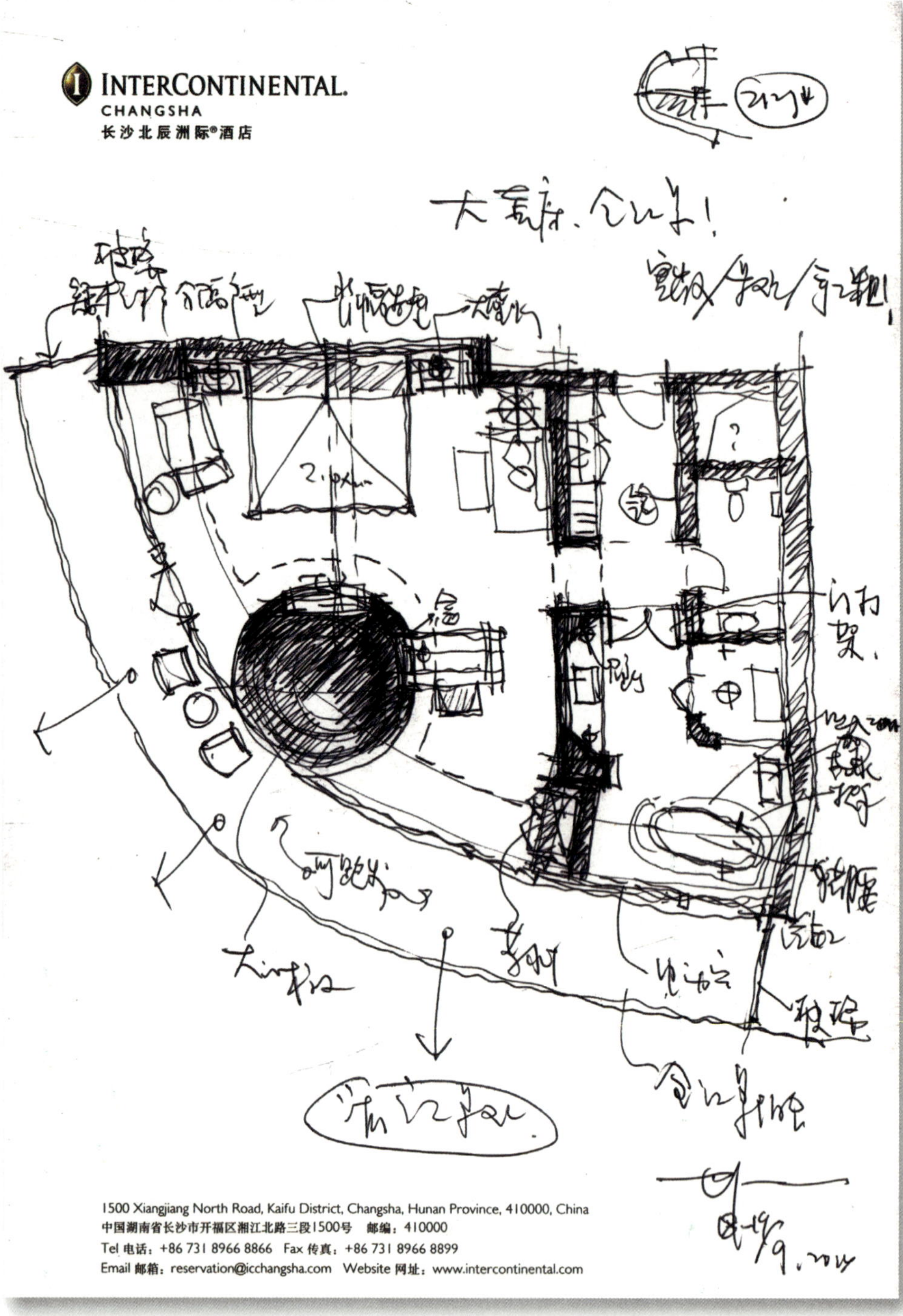

一个可以玩的酒店

长沙洲际酒店，地点很好，位于湘江与浏阳河的开阔之处交汇处，而建筑与室内设计由（HBA）负责，如出一辙，酒店一楼造型的设计得到了全面的考虑和建设，所以设计更是有明显的特征。

期待着这次正好在长沙的项目会议的第一日，入住洲际酒店的"试睡"地（可以随意体验的），挑选了一套位于端头的大床房体验一下。设计师还是"功力深厚"，不经意间的细节处处让人感到设计师的"多样"用心，每间客房都有其独特的体验，也更加深刻地"享受"设计师的思路。今天随处而有趣，不同的角落有不同的功能区域，让隐私也得到保护；设计富有人性化的细节，而为客户尽用心筹，路线给宾客指示，这样的客房也可以满足日常生活中默认的手法！设计者对入住者给一份安全感，同时又可以模拟出作为度假的"管制"！

睡眠区有一根大大的圆柱子，一侧是衣柜和储物空间，90°的侧面是卫浴间，皆以开放，分隔也靠透光玻璃，令睡房可一揽无遗的江景。

当然，最为精彩的还是倾斜的大大的全玻璃，可以进行不同视角的观景。只可惜还是中国式的工程毛病，整个房子的施工手工粗糙得不堪入目。

站在阳台，吹吹风，还是蛮惬意的好，在这样的阳台踱踱步，还是惬意，真的惬意之极！

长沙北辰洲际酒店 *Inter Continental, Changsha, China*

A Balcony You Can Run on
一个可以跑步的阳台

长沙洲际酒店，地点优，位处湘江与浏阳河开阔的交汇处，而建筑与室内设计（HBA）更是如出一辙，酒店一推出市场就得到充分的关注和热议，房间的设计更是有明显的特征。

Inter Continental Changsha is favorably located where Xiangjiang River meets Liuyang River. The architectural and interior design seemed to be the same work by HBA. As soon as it was marketed, the hotel received enough attention and debate. Its guestroom design especially has marked characteristics.

借着去北辰地产的项目会议就算当一回我们的甲方酒店的“试睡员”吧（可惜是自费的），精选了一套位于端头的大房型体验一下，设计师还是“功力深厚”，不规则的建筑条件也让我画的过程出现了好几次的叠笔“修正”，在我的“速画”体验中算是少见，也更加深刻地“寻觅”设计师的思路。分区自然而有趣，不同的角落有不同功能的区域，让你自然而然地停留。洗手间富有人性的细节，厕浴分离，尽用江景，浴缸临窗而设，这样的布局也可谓是时下高端豪华房型中默认的手法了，浴缸起水处设有扶手，多一份安全感，周边的水槽让水得到更好的“管制”！

Taking advantage of the project meeting held by Beichen Real Estate, I stayed at our Party A' hotel "sleeping test subject" (it was a shame that I had to pay out of my pocket). I carefully chose a big room at one end to experience. The designer was quite skilled. The irregular shape forced me to correct my drawing several times, something rare in my sketching, triggering a more in-depth search for the designer's line of thought. The division was quite natural and interesting, assigning each corner a different function, making you naturally pause. The bathroom abounded with human-friendly details:

the toilet was separated from the shower area; taking full advantage of the river view, the bathtub was installed near the window—such layout may be the default for high—end luxury rooms now; the tub was installed with hand rests for added security; the surrounding gutter had the water better "controlled" .

睡眠区有一根大大的圆柱子，一侧结合电视机的设置，90度角侧布置写字台，背江而坐，床设在靠近江景的位置，斜躺江景可一览无遗。

In the sleeping area there was a big round column. On one side of it, a TV set was arranged, and on another, at a right angle, a study—one would sit with back toward the river. The bed was so designed as to overlook the river view when lounging there.

当然，最为特别与吸引人的还是弧形的大大的阳台，可以让你不自觉地跑起来，只可惜还是中国式的工程老毛病，整个施工粗糙得不堪入目。

Of course, the most unique and appealing part was the arch-shaped balcony that ran the length of the room, on which you would find yourself running involuntarily. The only shame was that the whole construction and handwork was too crude for a look, an old problem with Chinese projects.

站在阳台，吹吹风，还是想想它的好，在逾十米的阳台踱踱，还是非常、真的非常的爽的！

Standing on the balcony to enjoy the breeze, I thought it better to look at its rosy side. After all, it was really quite pleasant to pace up and down the the balcony over ten meters long!

长沙北辰洲际酒店 *Inter Continental, Changsha, China*

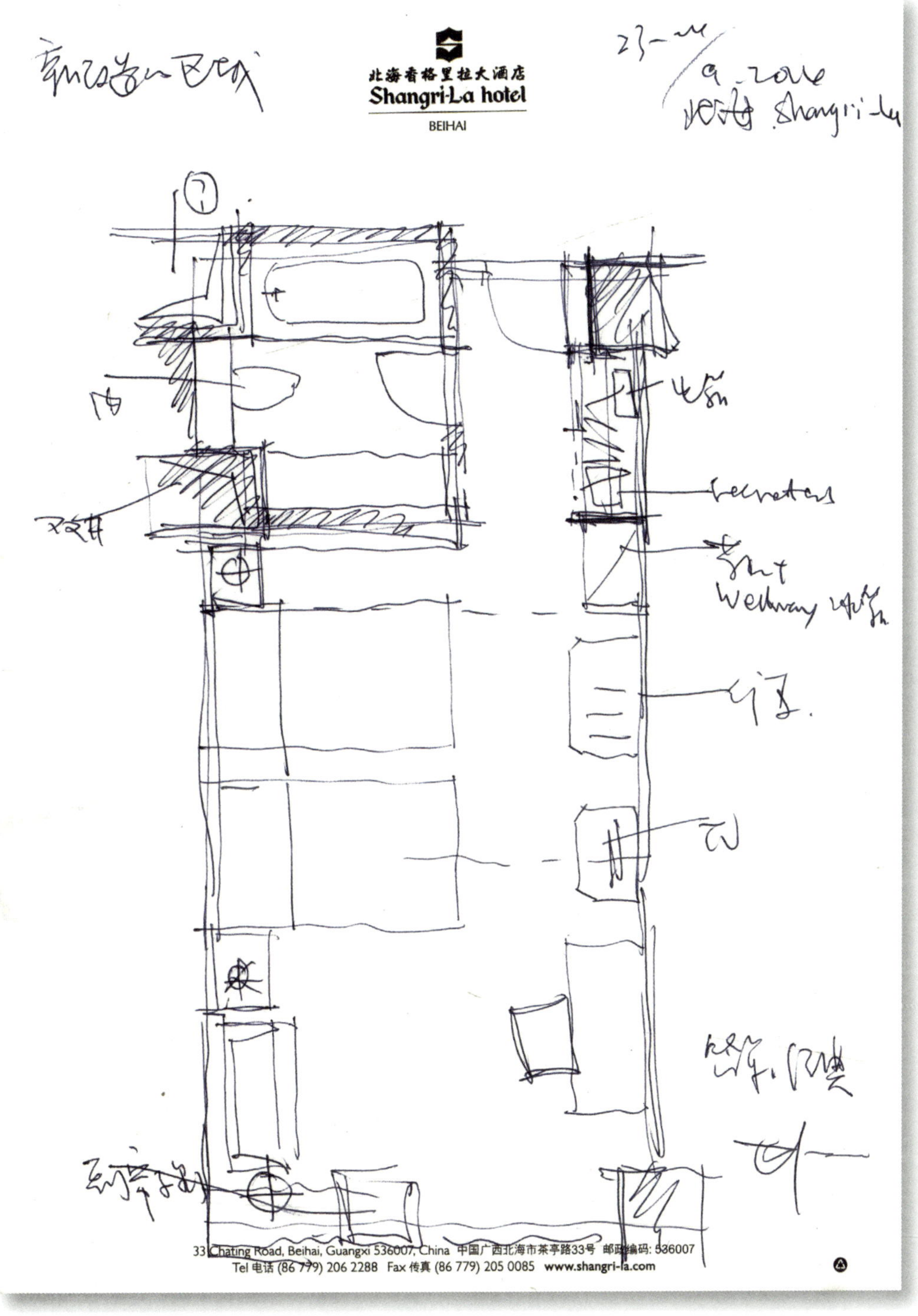

北海香格里拉大酒店
Shangri-La hotel
BEIHAI, CHINA

北海香格里拉酒店

★★★★★

4 SHANGRI-LA HOTEL BEIHAI CHINA

Address : 53600733 Chating Road, Beihai,Guangxi 536007, China
中国广西北海市茶亭路33号

Telephone : +(86 779) 206 2288

Fax : +(86 779) 205 0085

Http : //www.shangri-la.com

杰出的团队！

入住北海的Shangri-La，感受酒店、优质服务、体现公司之尊严和典范。

客房清新典雅和传统，做工中式之偏日本色，门窗的精致家具，温暖温馨的灯光，温馨中式的装饰，没有一点老旧的，与北海酒店的中国传统有一致的事物。我喜欢这里的风格和管理坚持！

回归、清洁、细腻、"完美"的杰出，和谐周到的服务、用心、身心舒畅、酒店的杰出细腻，这些都是酒店的细节都是令人佩服，18年的酒店老人，令我也会很多，我们和北海酒店带来的感动，永远不忘！

专注酒店的每一环，杰出的管理团队！

23-26/9.2016

33 Chating Road, Beihai, Guangxi 536007, China 中国广西北海市茶亭路33号 邮政编码: 536007
Tel 电话 (86 779) 206 2288 Fax 传真 (86 779) 205 0085 www.shangri-la.com

Warm Water is Very Important

热水很重要

入住北海的香格里拉,18年的酒店，佩服管理公司的先见和勇气。

The stay at the 18-year-old Shangri La Beihai inspired my admiration for the managing company's foresight and courage.

房间非常普通和传统，微微中式的偏红木色，订制的粗制家具，泛黄温馨的灯光，洗手间中规中矩，没有一点应有的与现在酒店设计潮流相一致的身影，或者这就是香格里拉的坚持!

The guestroom was quite commonplace and traditional. The mildly Chinese-style red wood, the customized crude furniture, the yellowing soft light, and the standard bathroom—there was no trace of the current trend of hotel design. Maybe it is just the insistence of Shangri-La!

睡前洗澡，迅速、"充沛"的热水，桂花香味的沐浴用品，身心舒畅，浴后浴缸去水迅速，这些配套的细节都令人佩服，18岁的酒店老人，有如此令很多新开业的酒店都望尘莫及的暗劲，确是不凡。

I took a shower before turning in. The gushing Warm water, the wash articles smelling of sweet osmanthus, quick drain of the used bath—all these details were admirable. It was really remarkable for an 18-year old hotel to have such hidden strength, putting many newly opened ones to shame.

考验酒店的好坏，热水确实很重要!

Warm water is really important to test the quality of a hotel!

北海香格里拉大酒店 *Shangri-La Hotel Beihai, China*

INTERCONTINENTAL
KUNMING
昆明洲际®酒店

昆明洲际酒店

5 ★★★★★

INTER CONTINENTAL HOTEL KUNMING CHINA

Address : No.5 Yijing Road,National Tourism Area of Dianchi, Kunming,650228 Yunnan Province,P.R.China
中国云南省昆明市滇池国家旅游度假区怡景路5号
Telephone : +(86 871) 6318 8888
Fax : +(86 871) 86318 6688
Http : //www.intercontinental.com

昆明洲际酒店
Inter Continental Hotel, Kunming, China

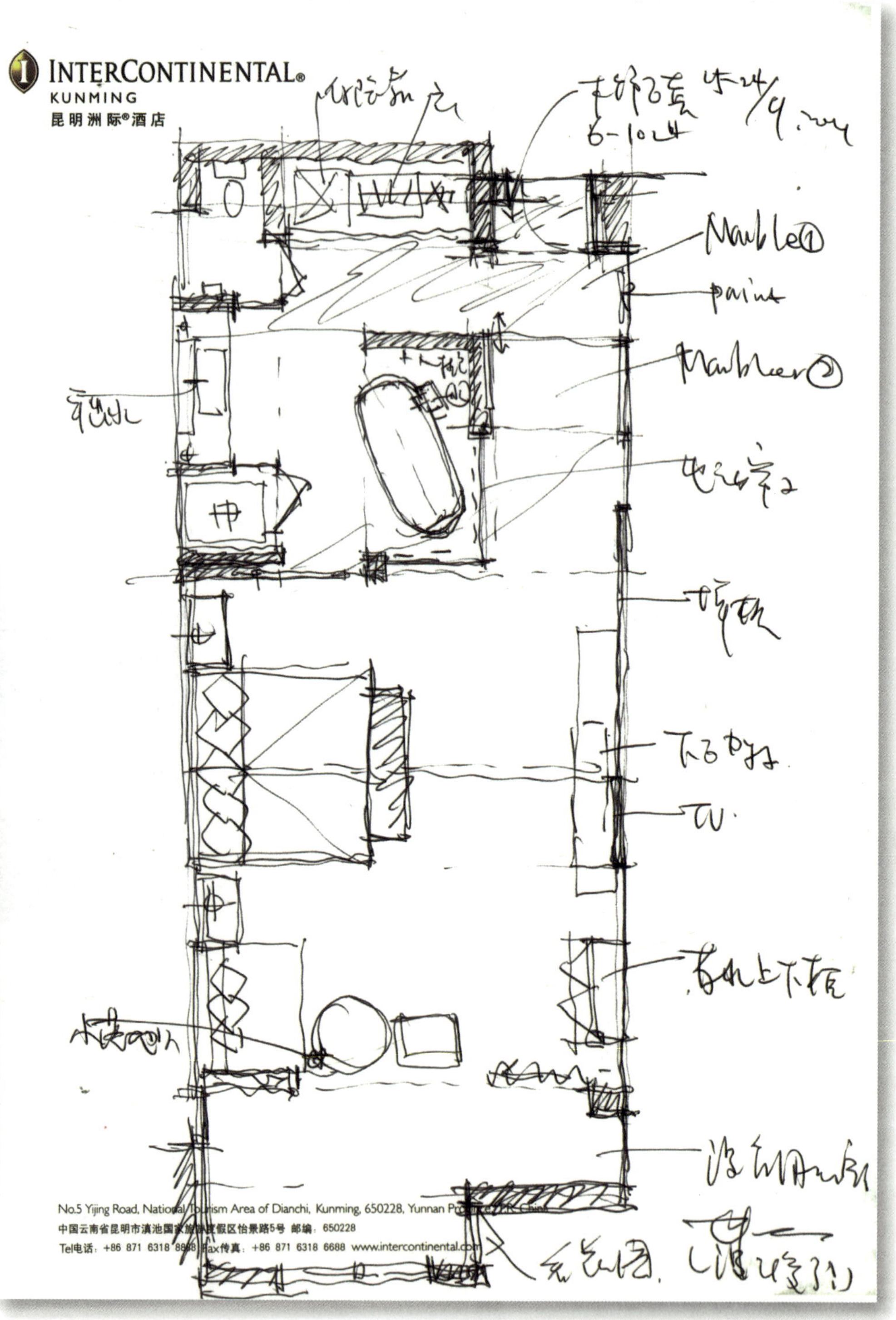

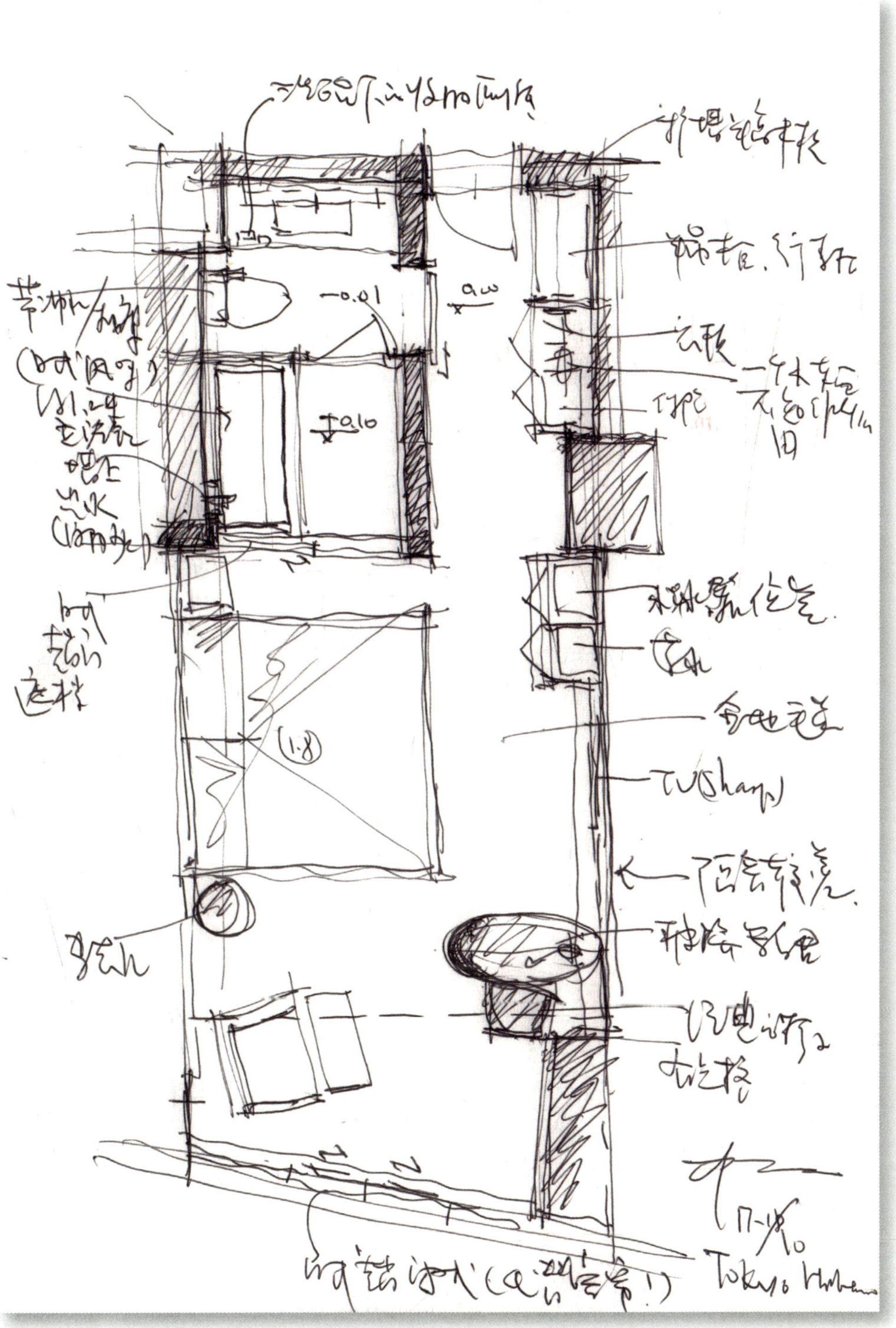

日本东京希尔顿酒店

6 ★★★★★

HILTON HOTEL TOKYO JAPAN

Address : 6-2 Nishi-Shinjuku6-chome,
Shinju-ku,Japan
日本东京都新宿区6-6-2

Telephone : +(81 3) 3344 5111

Http : //www.hilton.com.cn/

日本东京希尔顿酒店 *Hilton, Tokyo, Japan*

7

LOCKE ATAMI THE IZU JAPAN

日本伊豆热海洛克酒店
Locke, Atami The Izu, Japan

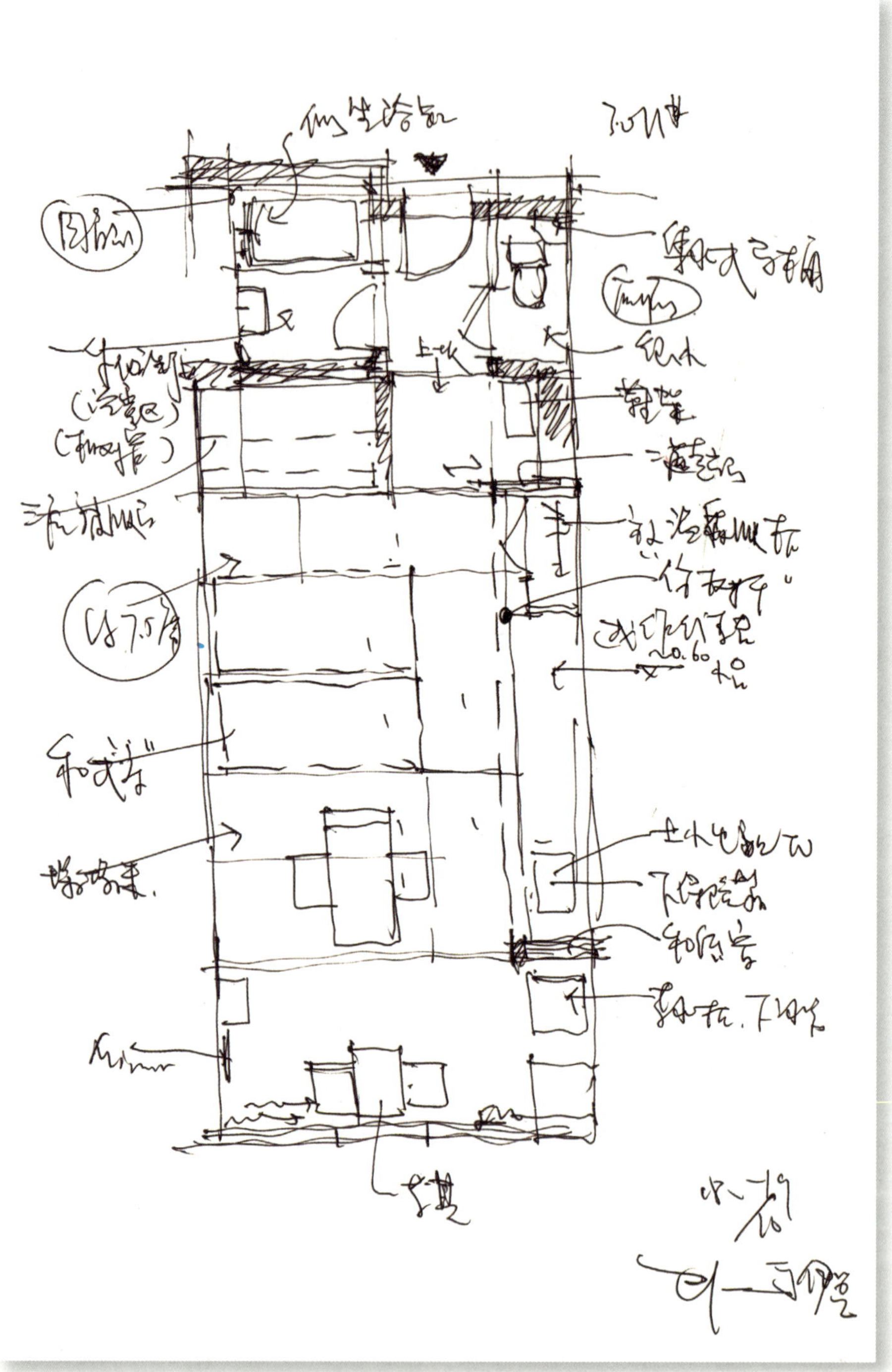

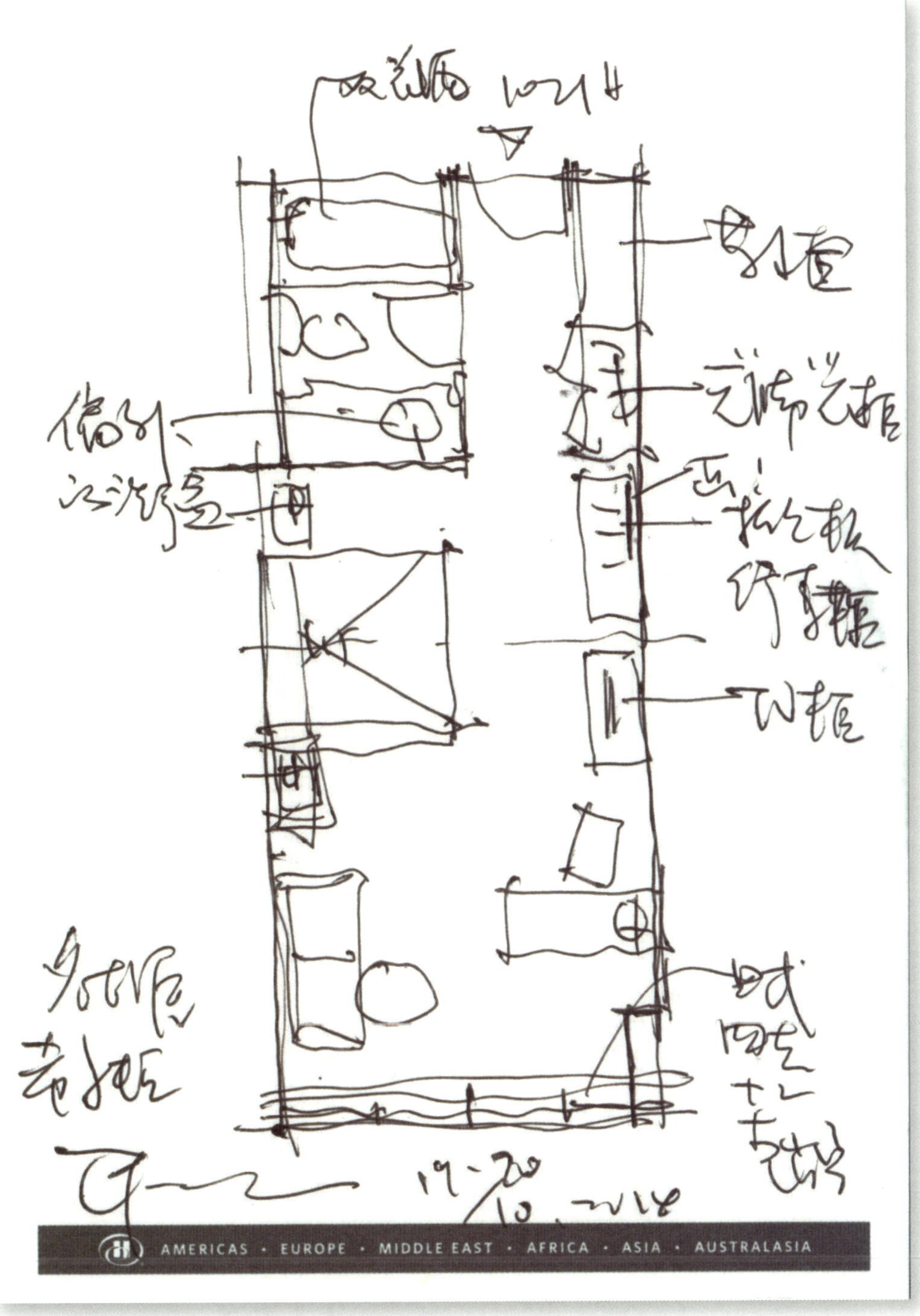

Hilton
HOTELS & RESORTS

日本名古屋希尔顿酒店
★★★★★

8

HILTON HOTEL NAGOYA JAPAN

Address : Nagoya,Japan
日本名古屋

Telephone : +(81 3)

Http : //www.hilton.com.cn/

日本名古屋希尔顿酒店 *Hilton, Nagoya, Japan*

R
RENAISSANCE®
SAPPORO HOTEL

日本札幌文艺复兴酒店

9 ★★★★★

RENAISSANCE SAPPORO HOTEL JAPAN

Address : 1-1,Toyohira 4-Jo 1-Chome, Toyohira-Ku,Sapporo,062-0904 Japan
日本062–0904札幌市丰平区丰平4条1丁目1–1
Telephone : +(81 11) 821 1111
Fax : +(81 11) 842 6191
Http : //www.rnsph.com/

日本札幌文艺复兴酒店
Renaissance Sapporo Hotel, Japan

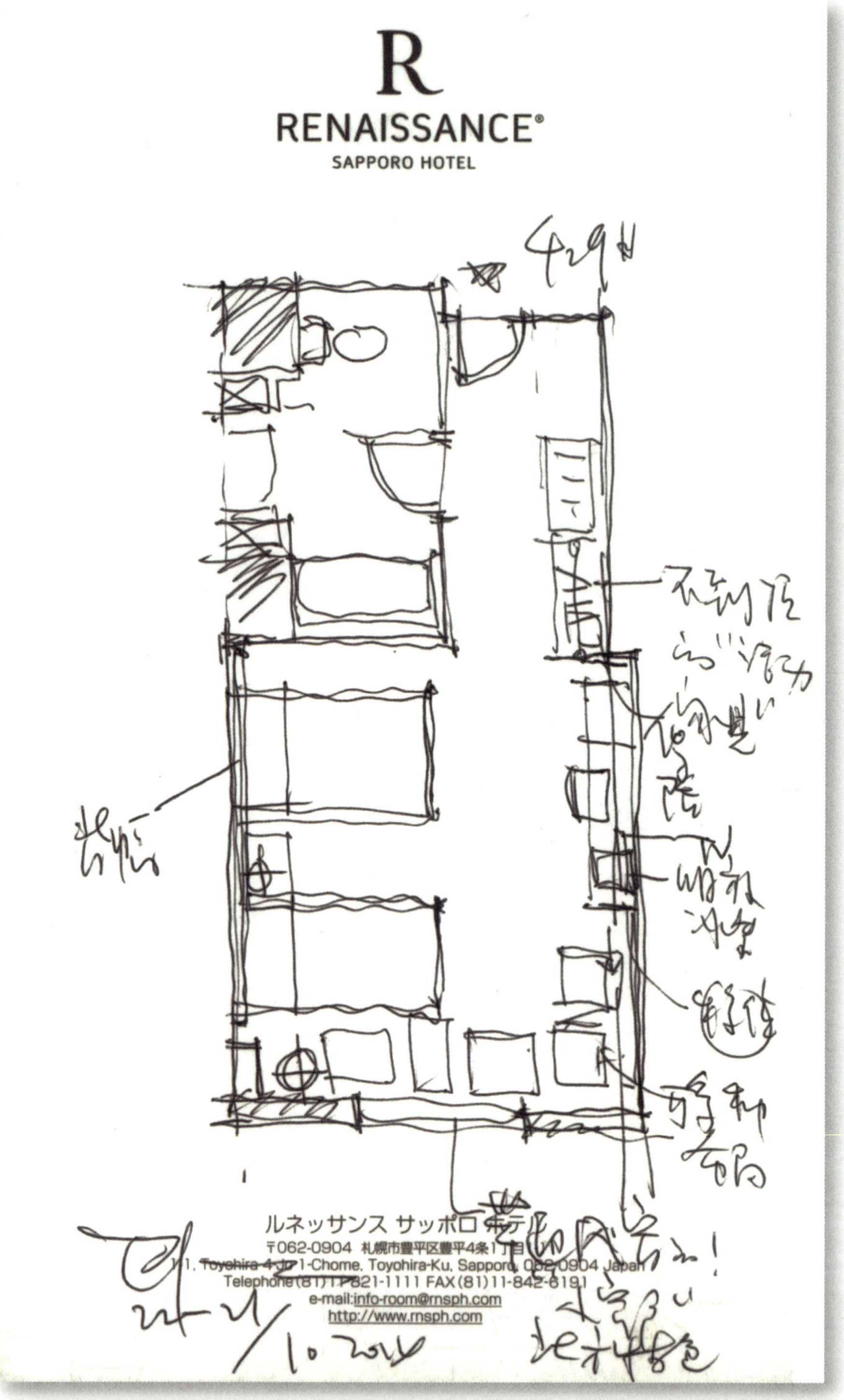

R
RENAISSANCE®
SAPPORO HOTEL
ルネッサンス サッポロ ホテル
〒062-0904 札幌市豊平区豊平4条1丁目1-1
1-1, Toyohira 4-Jo 1-Chome, Toyohira-Ku, Sapporo, 062-0904 Japan
Telephone (81) 11-821-1111 FAX (81) 11-842-6191
e-mail:info-room@rnsph.com
http://www.rnsph.com

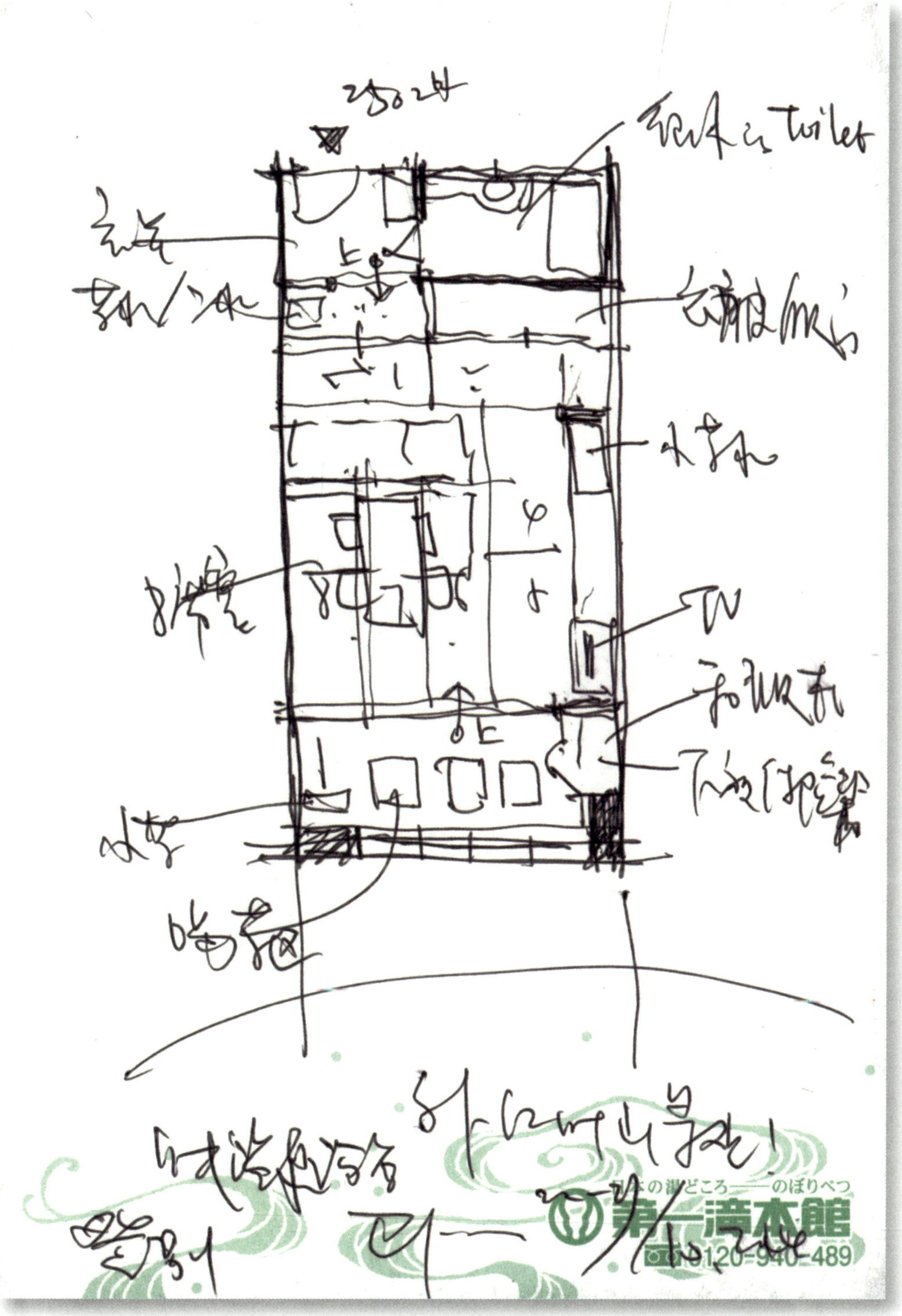

第一滝本館
Dai-ichi Takimotokan

日本登别第一龙本温泉酒店

10 ★★★★

DAI-ICHI TAKIMOTOKAN JAPAN

Address : Dai-ichi Takimotokan, Noboribetsu-onsen, Noboribetsu, Hokkaido, Japan 059-0595
日本国北海道登别市登别温泉町55番地

Telephone : +(81 143) 84 2111

Fax : +(81 143) 84 2202

E-mail : hotel_nidom@nidom.com

Http : //www.takimotokan.co.jp/

感受日式

去日本，一定会泡泡温泉，那自然而然就会住住榻榻米式的客房，穿个和服、袜子鞋，再加手握折扇装成一个日本人，晚上席地而睡，那就让我们看看这个房间有什么好玩的才能满足上面的要求呢？

实际入住后才知道什么是泡温泉的流程：进入房间，左右分别是厕所和浴室，两人分别可以如厕和洗脸化妆，换拖鞋，推开第二道的趟门，上台阶，换上专用的浴衣和服，轻轻松松的席地而座，慢慢泡一壶好茶，而私人露天温泉就存在小阳台内，欣赏着异国的风光。

去泡一个慢慢热热真真正正的温泉（既有男女分浴、赤条条的元一摩登物式的什么等）再吃一顿原汁原味的和式大餐。回到房间，服务员已经从那专门的被服柜取出两床被褥，铺在了与寝的榻榻米上，这样就是日式泡汤的感受。

泡温泉后，人很疲劳，很快人就进入梦乡了!!

Appreciate the Japanese Style
感受日式

日本登别第一龙本温泉酒店 *Daiichi Takimotokan Hotel, Japan*

去日本，一定要泡泡温泉，那自然而然就会住住榻榻米式的客房，穿和服和木屐，更加可以手握折扇装成一个东洋人，晚上席地而睡。那就让我们看看这个房间有什么好玩的才能满足上面的要求呢?

Once in Japan you must have a spa, then naturally you will find yourself in a tatami guestroom, wearing a kimono and a pair of getas, holding a folding fan as if you were posing as a Japanese. At night you will sleep on the floor. Then let us see what fun has to be there in the room to meet the above requirements?

实际入住后才知道什么才是泡温泉的流程：进入房间，左右分别是厕所和浴室，两个人分别可以如厕和洗脸、化妆，换拖鞋。推开第二道的趟门，上台阶，换上专用的温泉和服，轻轻松松地席地而坐，慢慢喝一壶日式茶，而私人贵重物品就存在小保险箱内，倚窗看看异国的风光。

Only after checking in did I know the process of taking a Japanese style spa: inside the room, you'll find a toilet and a bathroom on either side. Each of two guests can use the toilet, wash their face, put on makeup, change slippers, separately. Open the second sliding door, climb the steps, change into a special spa kimono. Then sit down on the floor, relaxed, and take your time drinking a pot of Japanese tea. Your private valuables are kept in a small safe. You can have a view of the exotic landscape outside the window.

日本登别第一龙本温泉酒店
Daiichi Takimotokan Hotel, Japan

去泡一个懒懒散散、真真正正的温泉（只有男女分浴，赤条条的无一障碍物式的才能算），再吃吃一顿原汁原味的和式大餐，回到房间，服务员已然从非常深的被服柜取出两床被褥，铺在7.5席的榻榻米上，这才是日式生活的感受。

Take a lazy and authentic Chinese spa (men and women separated, and stark naked), have a real big Japanese meal, and return to your room —by now the steward has taken from the very deep wardrobe two sheets and spread it on tatami. It is just what a Japanese life feels.

泡温泉后，人疲劳，很快就进入梦乡了！！

After the spa, so tired, you will go to sleep quite soon!

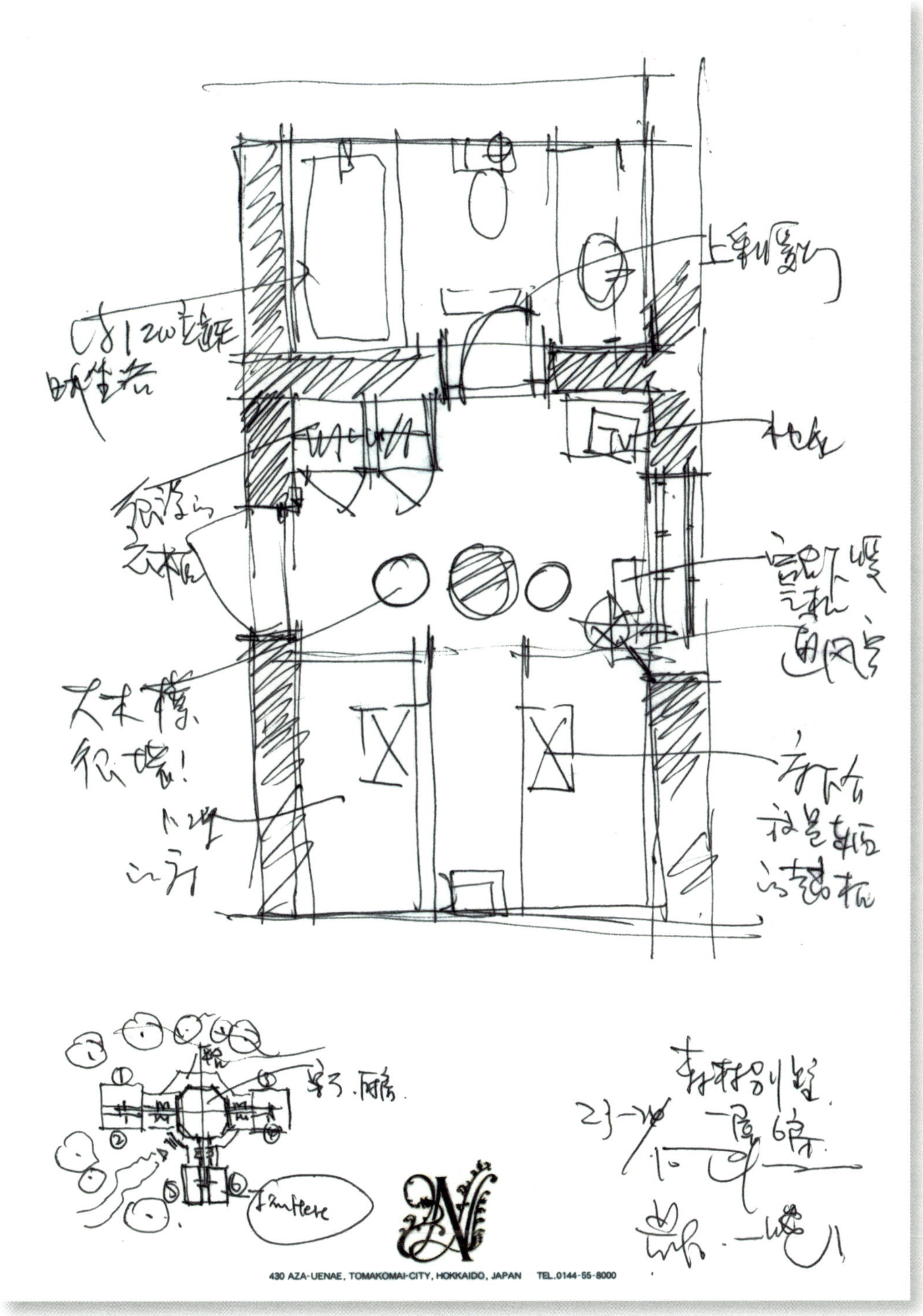

HOTEL-NIDOM

日本北海道NIDOM度假酒店

11

★★★★

NIDOM HOTEL HOKKAIDO JAPAN

Address : 430 AZA-Uenae, Tomakomai-city, Hokkaido,JAPAN 北海道苫小牧市字植苗430番地

Telephone : +(81 144) 55 8000

Fax : +(81 144) 55 8121

E-mail : hotel_nidom@nidom.com

Http : //www.hotelnidom.com/

非常可惜的短暂一夜

这是日本旅游的最后一个下榻的酒店，旅行团嘛，总是天黑透了才回到酒店，苫小牧NIDOM度假酒店，一座完全隐没在原始森林里的高级别墅酒店，佩服我们的导游，不知道是如何淘到这么不适合团队入住的高雅隐秘的地方的。

我们一行的中层领导们几乎包了两个大别墅，一个别墅分成四间房间，一个异常大的客厅、厨房、酒吧等等一应俱全的会客室的，当然还有KTV的设备（据说KTV是日本人发明的），安顿一下之后，就放肆地在会客大厅吃着喝着跳着唱着。别墅的外墙是就地取材的大大小小的原生圆木，房间里面非常温暖，倒是回到各层的睡房都是木制的狭小的“单间”，大家倒有种鼓励和亲爱的感觉，整体非常有北国雪域的感觉，非常尽兴地消磨到半夜，可惜，只有几个小时。

第二天早上还要赶早起来到处走走看看，哇！太美了，每一处都像幅画似的风光，红、绿、黄夹杂的树叶，大小弯弯、各自汇集的溪水，吸入的每一口是清新的空气都是甜的，小区域散步的片刻，享受了非常丰盛的早餐，西式的，依依不舍地离开了难得寻觅的一个世外桃源式的森林木屋。

可惜这么短暂的一夜。

日本北海道NIDOM度假酒店 *Nidom Hotel, Hokkaido Japan*

A Regrettably Short Night Stay
非常可惜的短暂一夜

这是日本游的最后一个下榻的酒店，旅行团嘛，总是天黑透了才回到酒店，苫小牧NIDOM度假酒店，一座完全淹没在原始森林里的高尔夫别墅酒店，佩服我们的同事，不知道是如何淘到这么不适合团队入住的高雅隐秘的地方的。

It was the last hotel we checked into during our Japan tour. As a group we did not return to the hotel until it was quite dark. The Tomakomai NIDOM Vacation Hotel was a golf villa hotel, in the depths of a primitive forest. I really admire our colleague who had somehow managed to find such an elegant and mysterious place so unsuitable for group travel.

我们一公司的中层领导们几乎包了两个大别墅，一个别墅有六间房，一个异常大的客厅、厨房、酒吧等等配套齐全的公共空间，当然还有KTV设备（据说是日本人首创的），安顿一下之后，就放肆地在公共大客厅吃吃喝喝、玩玩唱唱了。别墅的外墙是就地取材的，大大的原生圆木，房间里面非常温暖，倒是各自的睡房却是相当的狭小而“丰满”，大有侧重和鼓励社交的感觉，整体非常有北国雪域的感觉，非常尽兴地吵吵闹闹到半夜，可惜，只有几个小时。

The middle-level leaders of our firm occupied all of the two big villas. Each of them had six rooms, an extra big living room, kitchen, bar and fully equipped public spaces. Of course, there was also a KTV system (it is said to be invented by Japanese people). After some rest, we began eating, drinking, singing and playing with abandon. The outer wall of the villa was made

日本北海道NIDOM度假酒店 *Nidom Hotel, Hokkaido Japan*

of materials available at hand. The massive logs meant the the interior of the room was quite warm. Back in each bedroom it was rather cramped and "plump"—it gives the impression that it is intended to emphasize and encourage social activity. On the whole it felt quite characteristic of the snow-covered north of Japan. We reveled nosily till midnight. The shame was that it lasted only several hours.

第二天，早上还要趁早起来到处走走看看。哗！太美了，每一处忽微忽强的阳光，红、绿、黄夹杂的树叶，大小蜿蜒、高低汇集的溪水，吸入的每一口晨曦的空气都是甜的，在小区域散散步后，享受了非常尊贵的早餐，西式的，之后依依不舍地离开了难得寻觅到的这个世外桃源式的森林木屋。

The next day morning we had to rise early to have a look around. Wow, it was too beautiful: The omnipresent sunlight dimming and glaring now and then, the mosaic of tree leaves of red, green and yellow mixed, streams zigzagging and converging from high and low places. Each breath of the morning air was so sweet. After strolling around, we had a very noble Western breakfast before leaving this hard-to-find forest wooden house retreat, to which we found it hard to say goodbye.

可惜，这么短暂的一夜。

The night stay was so regrettably short.

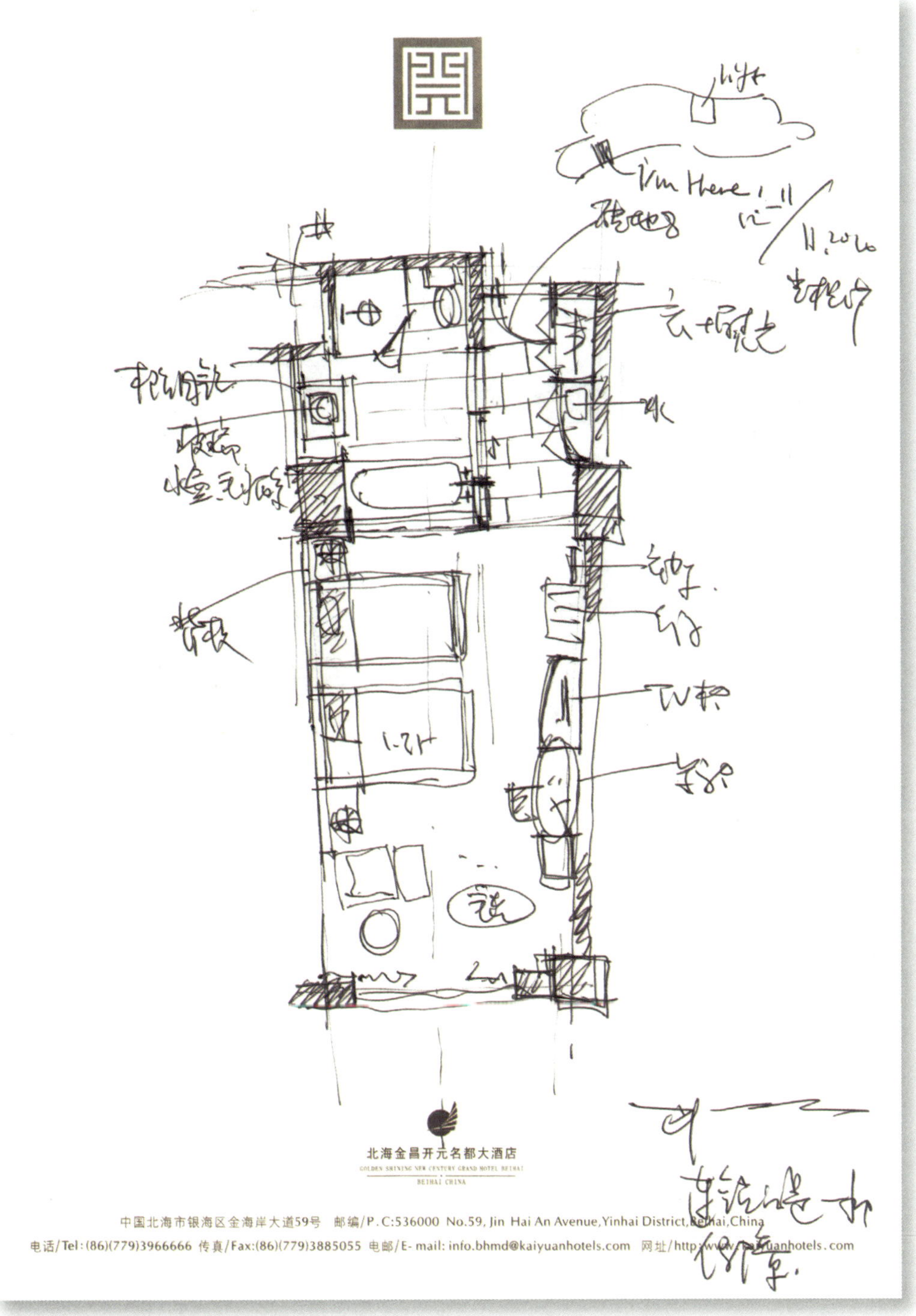

北海金昌开元名都大酒店
GOLDEN SHINING NEW CENTURY GRAND HOTEL BEIHAI
BEIHAI CHINA

北海金昌开元名都酒店

12

★★★★

GOLDEN SHINING NEW CENTURY GRAND HOTEL BEIHAI CHINA

Address : No.59,Jin hai An Avenue, Yinhai District,Beihai,China
中国北海市银海区
金海岸大道59号
Telephone : +(86 779) 3966666
Fax : +(86 779) 3885055
Http : //www.kaiyuanhotels.com

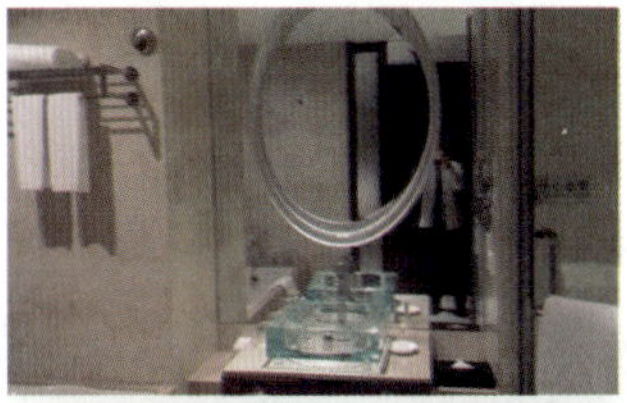

北海金昌开元名都酒店
Golden Shining New Century Grand Hotel, Beihai, China

静安香格里拉大酒店
Jing An Shangri-La
WEST SHANGHAI

上海静安香格里拉大酒店

★★★★★

13

JING AN SHANGRI-LA WEST SHANGHAI CHINA

Address : 1218 Middle Yan'an Road, Jing An Kerry Centre on West Nanjing Road, Shanghai 200040
中国上海市延安中路1218号南京西路静安嘉里中心，200040

Telephone : +(86 21)2203 8888

Fax : +(86 21)2203 8899

Http ://www.shangri-la.com

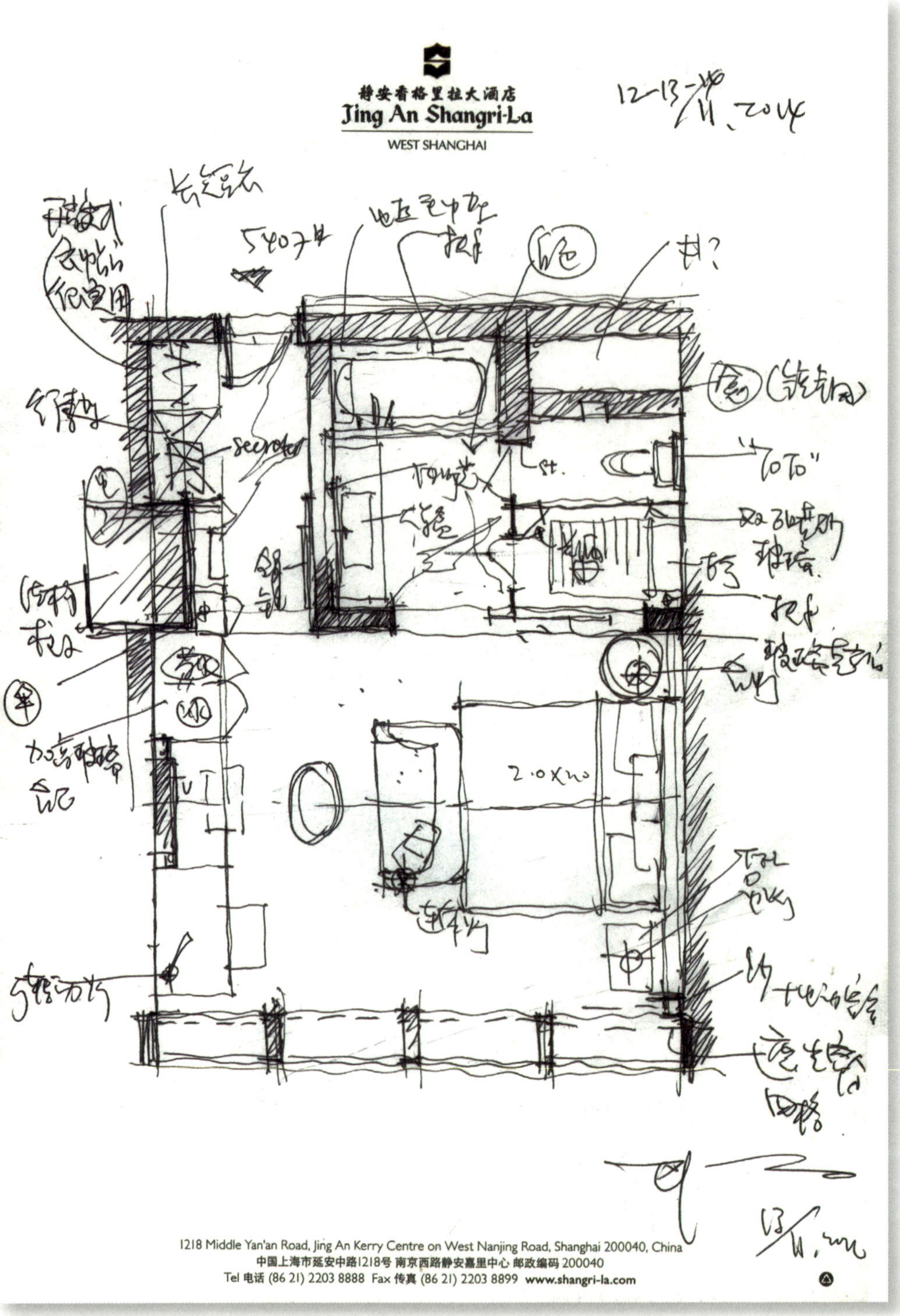

如果没有私人的不对称。

不论这是我去书店、超市，还是对着电视、去看他们的客厅和卧室，但对这个人人都可以看到（或者）的酒店，一直也反对没有私人的（包括卧室）同样的全都开放。也不知道酒店有没有哪个住过的没有私人的酒店，还是不理解房间里没有私人的"空间"也对着天井，而且包括这还是令人困惑，更是相矛盾的。

从我自己出差来想，酒店房间不过是会，只是情侣们以"城墙"来相隔开，而晚上睡觉、关灯（包括夜灯），还是可以看不见（情人之间）好的，有些人的卧房、如果开灯，会以为是干扰，冲水更加会放大大的嘈杂干扰，关门也不可避免有些人的这般声音，更令人整夜这样地如此干扰。

如果要对应开放，一要能够确保，二要不透明，三最好还是传统意义的墙的效果。

我也赞成部分开放，但还是传统地（没有你最不乐意听到的一句话）没有去考虑住，原来想到又被破坏的那些东西的设计，还是可以说，没有更大的空间。

什么道理呢？如果住酒店，如果住没有私人的，你问是没有私人的对称吗?!

Suppose the Bathroom Door Does Not Face the Bed

如果洗手间的门不对床

不知道是我“老土”、执着，还是对声音、光有独特的要求和理解。住过这么多的形形色色（式式）的酒店，一贯地反对洗手间的门（包括趟门）向床的方向开启。也不知道酒店设计师中有多少个住过自己设计的酒店，我还是不理解房间的洗手间门“光明正大”地对着床，而且这么近，让人深感不适，特别是玻璃门。

Either because of my “provincialism” and persistence, or my unique requirement and appreciation of sound and light, after staying at so many hotels of each and every description, I consistently oppose that the bathroom door (including the sliding door) opens onto the bed. There is no knowing how many designers have stayed in the hotels they designed. I still feel disgusting and cannot understand how come the bathroom door “boldly” faces the bed, so close apart—especially when it is a glass door.

从行为出发想想，白天用厕所还凑合，只是情侣们的“坦诚相对”而已，而晚上睡觉、关灯（有夜灯，通常还是非常不近人情的亮）如厕，有声音的干扰，如果开灯，有灯光的干扰，冲水更加有大大的噪声干扰；关门也不可避免有趟门的导轨声。总之，弊端远远比好处多。

From behaviorism, it is alright to use the toilet in the daytime and it is only a kind of “candidness” between lovers. However, at night, it is time to turn off the light to sleep. Using the toilet then involves the lights turned on, usually inhumanely sharp. There will be noise annoyance, of light annoyance, if the light is turned on, and too much noise annoyance from the flush toilet; there will be the sliding sound when the door slides open and shut. In a word, the disadvantages far outnumber the advantages.

如果要对床开门，一要距离较远，二要不透明，三最好还是用传统易用的掩门。

If you have to let a bed face the bathroom door, first, they must stand relatively far apart; second, the door cannot be transparent; third, the easy-to-use traditional hinged door is advisable.

我也尝试动手画画，门还是传统地（设计师最不乐意听到的一句话）设在走廊。原来趟门处设玻璃加电动窗帘的设计还是可以的。洗手盆没有现在气派。

I also try to design it. The door will still be traditionally(a word most disliked by designers) located in the hallway. It turns out that it is alright to have glass and an electric curtain installed on the sliding door. The washbasin will not be so stylish as this one.

你觉得呢？如果你是住客，如果你是设计师，你同意洗手间的门对床吗？

What do you think? If you are a guest, or a designer, would you agree that the bathroom door should face the bed?

上海静安香格里拉大酒店
Jing An Shangri-La, West Shanghai, China

THE PENINSULA

HONG KONG

香港半岛酒店

14 ★★★★★★

THE PENINSULA HOTEL HONGKONG CHINA

Address : No.59,Jin hai An Avenue, Yinhai District,Beihai,China
中国香港九龙尖沙咀
梳士巴利道22号

Telephone : +(852) 2920 2888

Fax : +(852) 2722 4170

Http : //www.peninsula.com

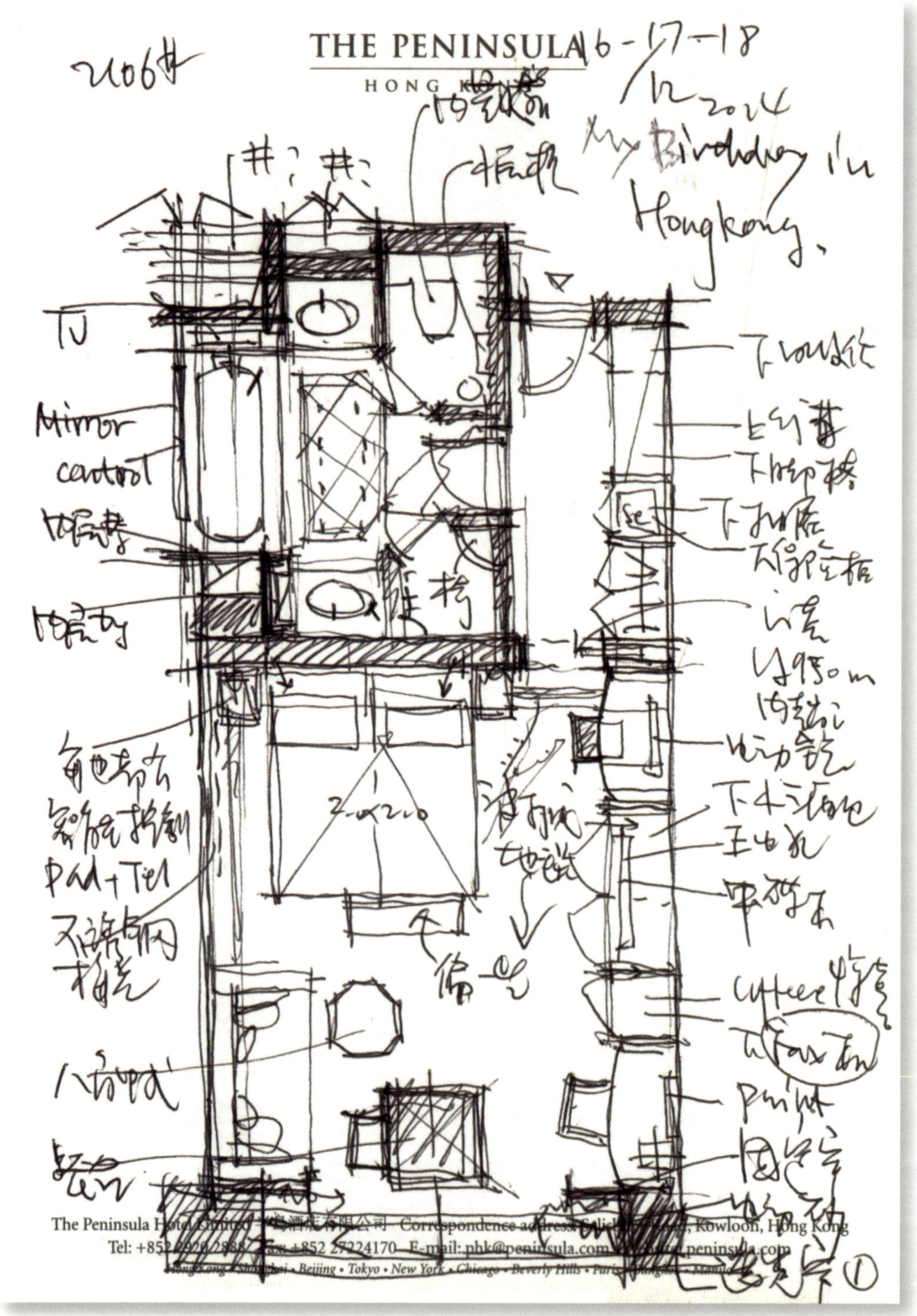

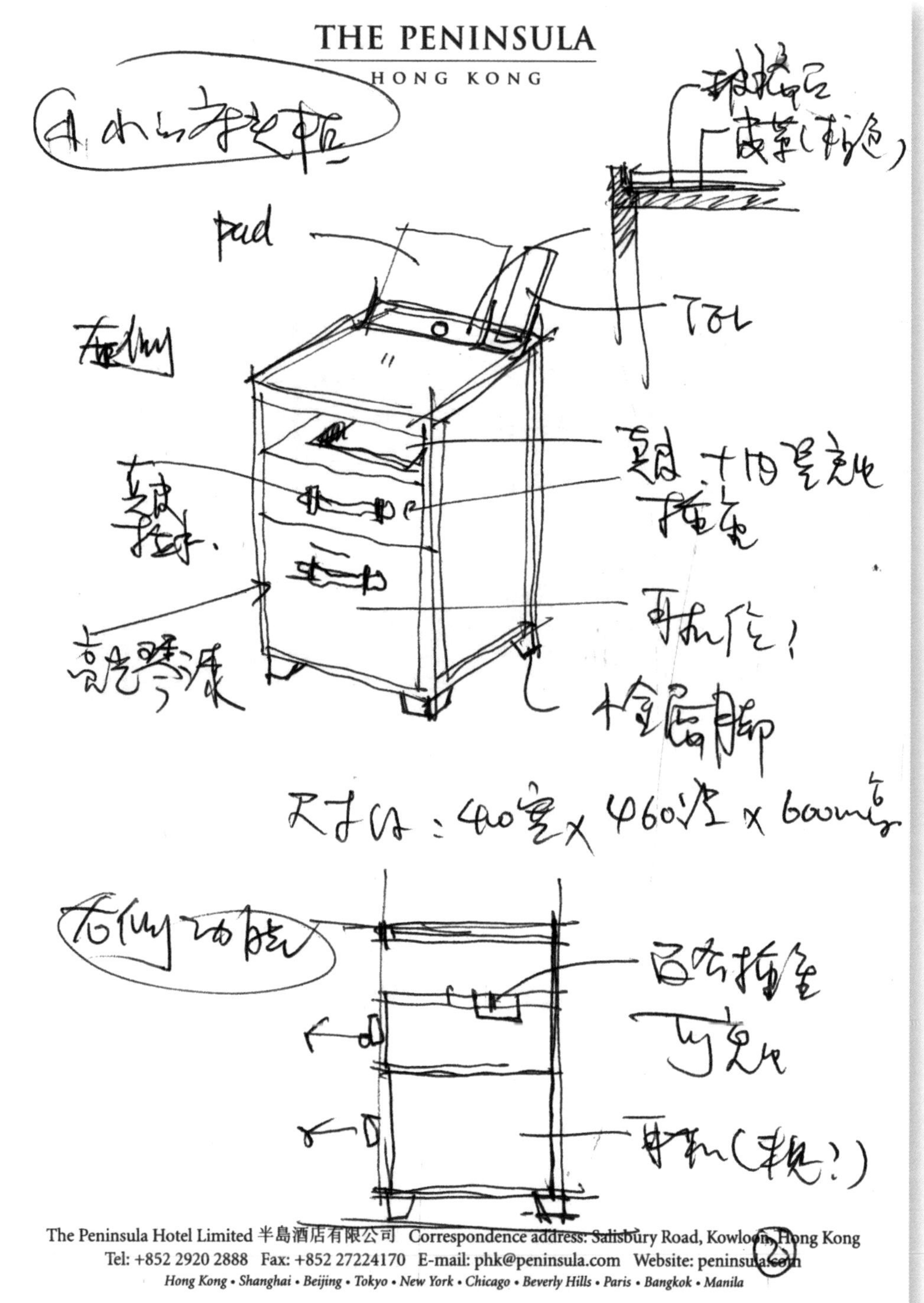
THE PENINSULA
HONG KONG
pad
The Peninsula Hotel Limited 半島酒店有限公司 Correspondence address: Salisbury Road, Kowloon, Hong Kong
Tel: +852 2920 2888 Fax: +852 27224170 E-mail: phk@peninsula.com Website: peninsula.com
Hong Kong • Shanghai • Beijing • Tokyo • New York • Chicago • Beverly Hills • Paris • Bangkok • Manila

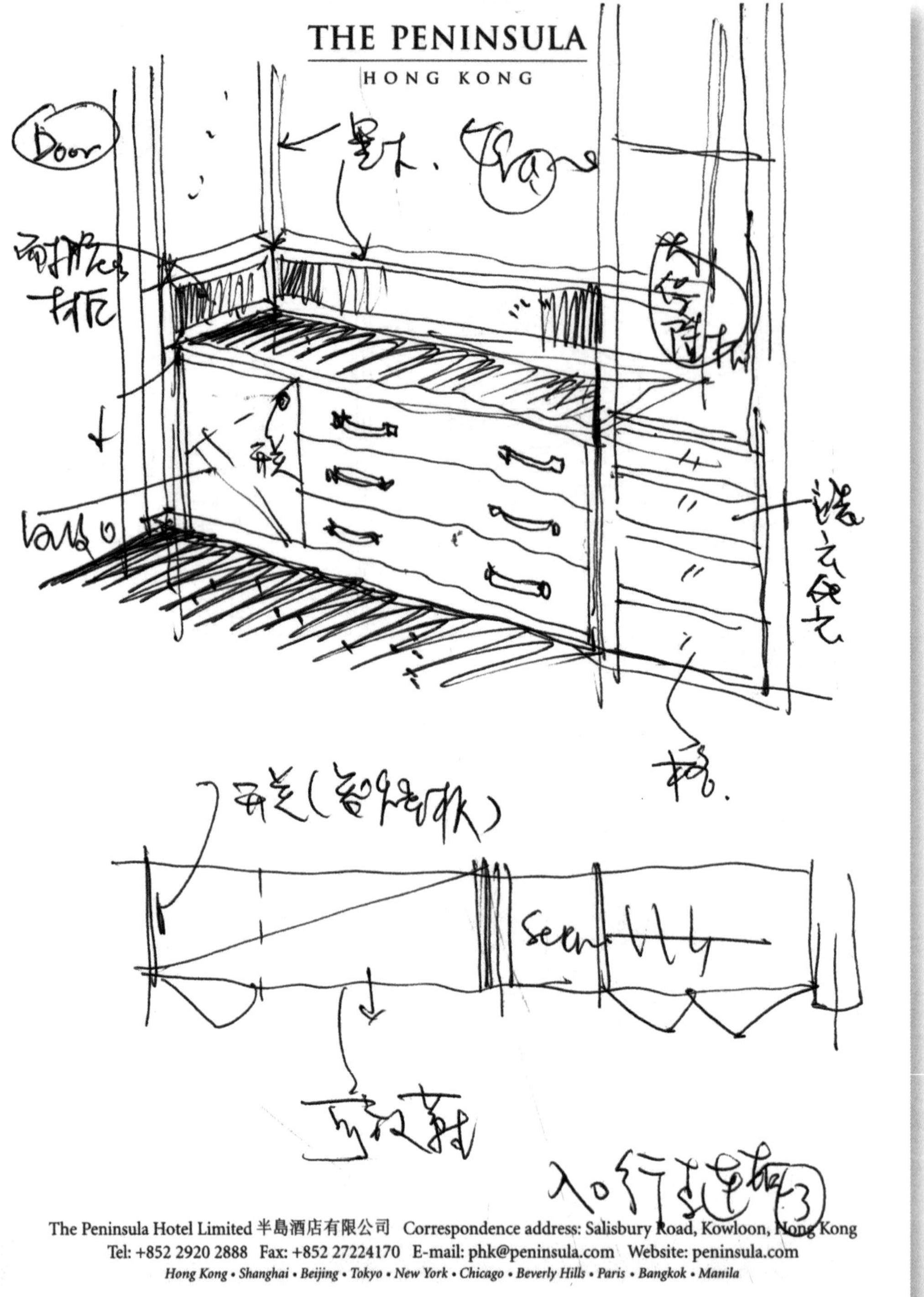
THE PENINSULA
HONG KONG
Door
The Peninsula Hotel Limited 半島酒店有限公司 Correspondence address: Salisbury Road, Kowloon, Hong Kong
Tel: +852 2920 2888 Fax: +852 27224170 E-mail: phk@peninsula.com Website: peninsula.com
Hong Kong • Shanghai • Beijing • Tokyo • New York • Chicago • Beverly Hills • Paris • Bangkok • Manila

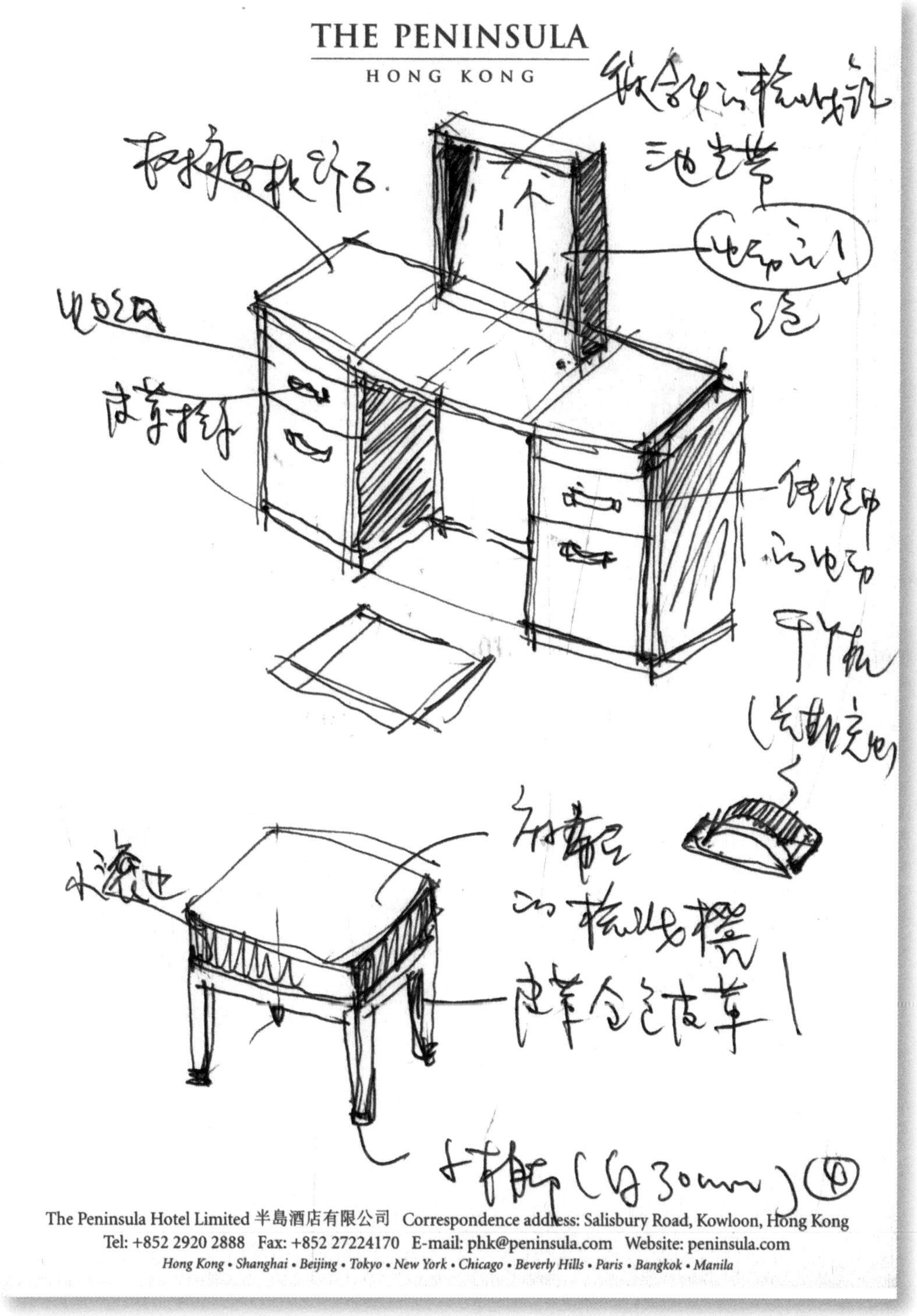
THE PENINSULA
HONG KONG
The Peninsula Hotel Limited 半島酒店有限公司 Correspondence address: Salisbury Road, Kowloon, Hong Kong
Tel: +852 2920 2888 Fax: +852 27224170 E-mail: phk@peninsula.com Website: peninsula.com
Hong Kong • Shanghai • Beijing • Tokyo • New York • Chicago • Beverly Hills • Paris • Bangkok • Manila

THE PENINSULA

HONG KONG

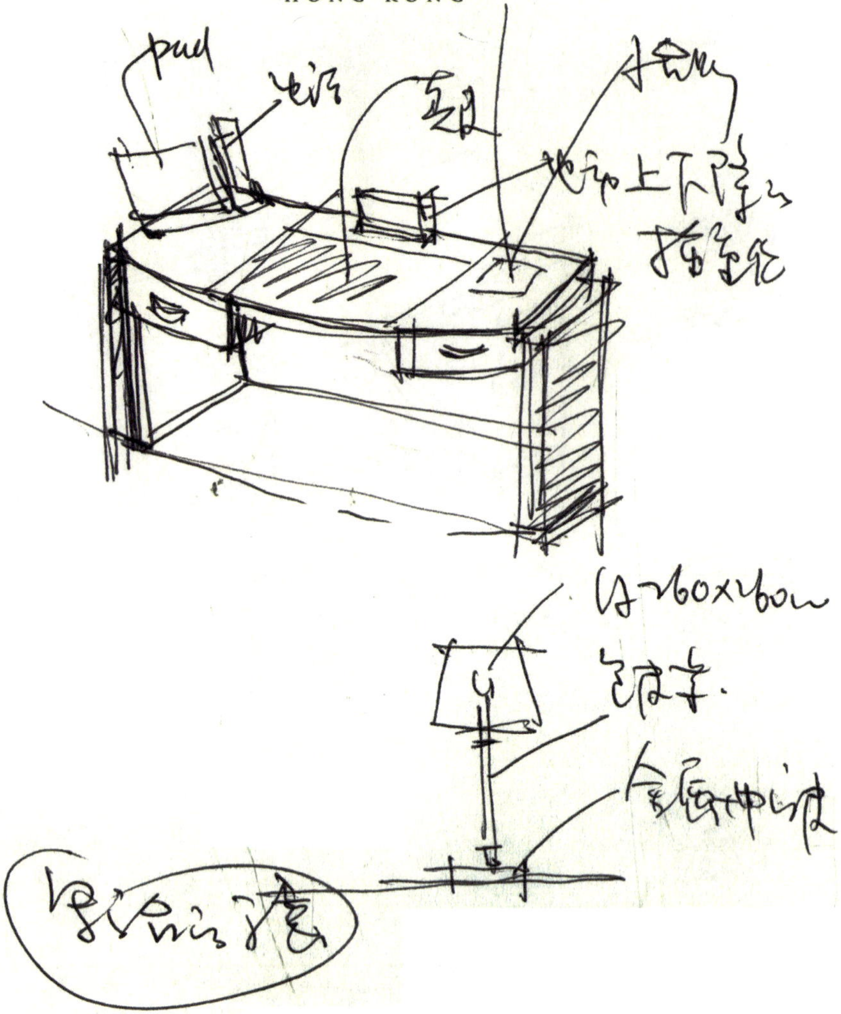

The Peninsula Hotel Limited 半島酒店有限公司 Correspondence address: Salisbury Road, Kowloon, Hong Kong
Tel: +852 2920 2888 Fax: +852 27224170 E-mail: phk@peninsula.com Website: peninsula.com
Hong Kong • Shanghai • Beijing • Tokyo • New York • Chicago • Beverly Hills • Paris • Bangkok • Manila

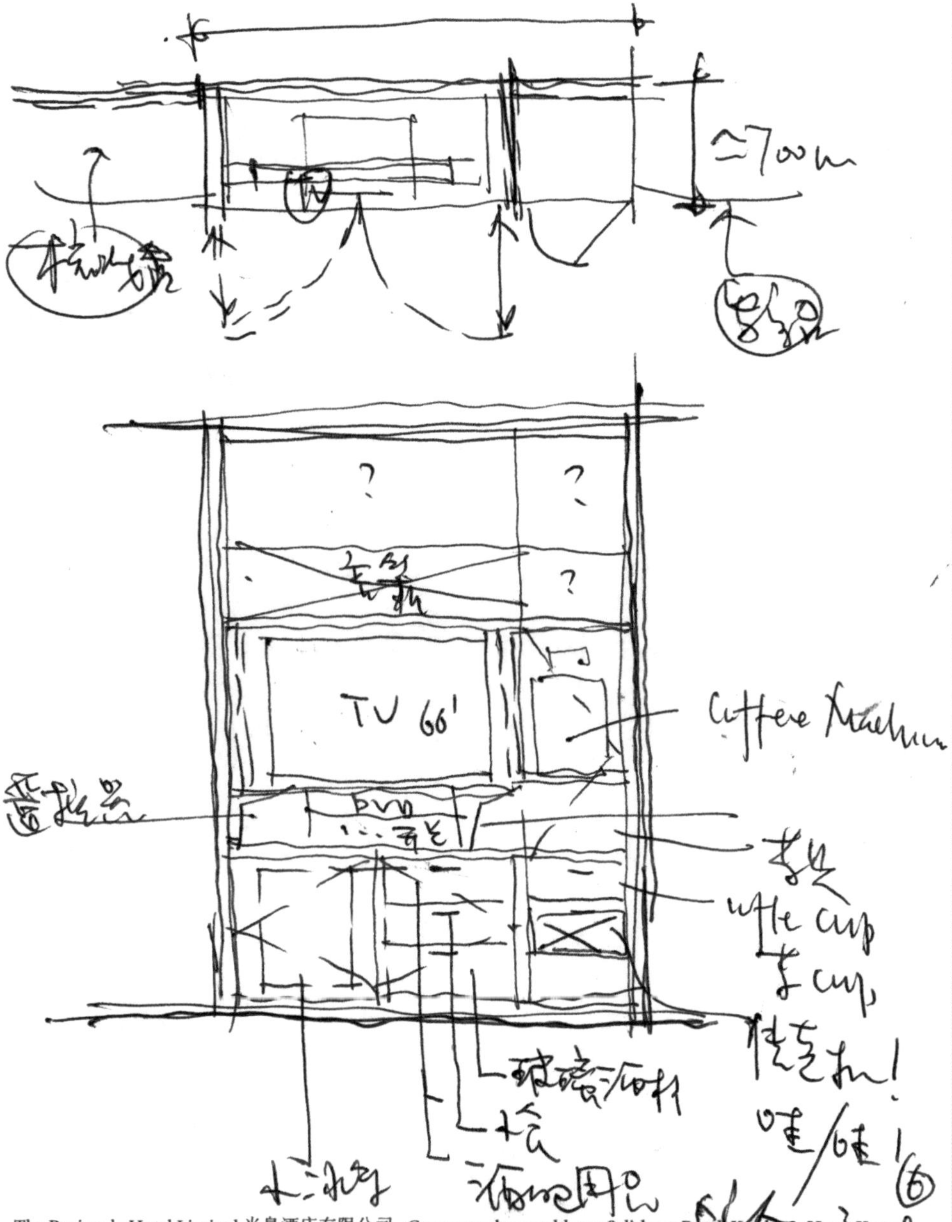
TV 66'
DVD

THE PENINSULA

HONG KONG

18/10, 2016

再住"半岛"。

再住设备的半岛酒店（香港，The Peninsula），这次是了六年前住过的"标准房型"，设备的布局基本不变，第一眼：洗手间台/浴缸，衣柜依旧！细细住下来一看，变化不少！

首先是全面智能化（如六年前住过的套房）（详见《住哪？》P155）改了一下，尤其三合一集化智能控制台：panel + 电话，书桌及双床头柜各一组；入门处，洗手间，梳妆台上方，各一组功能控制板，双床头控制开关及双边窗帘电动开关控制板……，当然还有全新设计的套餐家具。

①

这几面对的一点。每个人认为有细节的家具是带专业地写一篇。当然，不算难懂。但我以为上海名店的环境，家具与古董是大的比重。从布局、走位的合，到"诗在"的画面。这些与家具的，单独的水平和意了才成为好的。不然家什无色的魅力所在：

其中，入口走廊古老的行李台，下面的收口，又摆着花格时有大量的莲花在枝上的菖蒲。风水在枝干上"宝地"的大的恒温柜，挂在柜子上的雨伞架，整个气息的氛围，让好像都是地去旅行的书。

其中更显古老的是西洋风的电视柜，又如餐厅描述的一套，中国的大电视柜"，②

THE PENINSULA

HONG KONG

靠宗的书院，绝妙地方在山脚带上，核心地名

节地的核心镜——山脚风水偏后，不对称，湖

水连同前院隐藏，人性化，左侧也为风口，右

侧也有"绝招"——女士专用的移动式"厕所"

吸引美丽的地方！

书院与寨宫，既是传统建筑，而有其文化内

涵的，更有其深入的功能性和综合，其建的书院

型，如宫的转弯区域（与核心不同）

中国的"天地秩序"在此体现——

下节会论及书院水系统，小溪流/合流，影景器

四根柱，中轴大地风水（水系相对的），咖啡厅，

考生摆放处，不能不提的是，靠近书院

(3)

的"框如"，居然还可以在 Fax/打印机，更况

我是"股东"但这种世之有"Brother"等之中

每一间的坚持的合，我喜去出力去客人使用的

(除了全球性高级人士)是不出的所以才被这

接受股东的全球业务，我想到老板的手上签

Contract（合同），但取之教，那时是是个

以客人需求！

香港有一大处作用之地，自小年的接触

了细，让出了一大笔的中的建/作为设计，而

有可靠的，临车取之的区域，人情之世的

房的用名，不是建造地湖的有的电脑，总来

一切都是"出行如家"的生活所设计的 ⑩

THE PENINSULA
HONG KONG

好似不经意地被拍下来！

酒店的建筑说不出有什么惊喜，但每[illegible]下[illegible]了大力气，大量钱花在[illegible]忘记

设备更换，全面翻新，也入全面之智能化。

电视/窗帘/灯光/控制一一体化，控制是如此

便捷、大、稳、迅速！

小小一个房间，让我感受到为什么半岛是

全球数一数二的人[illegible]的酒店，但当

贴心，[illegible]之“本[illegible]”。

所付出成本之价格，才是[illegible]半岛之

味道独特之处！

The Peninsula Hotel Limited 半島酒店有限公司 Correspondence address: Salisbury Road, Kowloon, Hong Kong
Tel: +852 2920 2888 Fax: +852 27224170 E-mail: phk@peninsula.com Website: peninsula.com
Hong Kong • Shanghai • Beijing • Tokyo • New York • Chicago • Beverly Hills • Paris • Bangkok • Manila

香港半岛酒店　*The Peninsula Hotel, Hongkong, China*

One More Stay at the Peninsula

再住半岛

再住改造后的半岛酒店（香港The Peninsula），这次选了之前住过的“标准房型”。改造后的布局基本不变，第一眼：洗手间白/绿相衬，床向窗！细细住下来一看，变化不少！

Once more I checked into the Peninsula Hong Kong. This time I chose the "Standard Room" I had stayed in before. For all the remolding the basic layout remained the same at first glance: white and green set off each other in the bathroom and the bed faced the window! An examination, however, revealed many changes!

首先是全面智能升级（如之前住过的套房），（详见《住哪？2》P155），数了一下，共有三台一体化智能控制台：iPad+电话，书枱（台）及双床头柜各一组；入门处，洗手间、梳妆台上方各一个多功能控制板；双床头控光控制开关及双边窗帘电动开关控制板……当然还有精美的全新设计的奢华家具。

First, the overall smartness updating (just like suites I had stayed in before), (see Where to Stay2, P 155). I counted there was three sets of integrated smart controls: iPad+telephone, placed on the study desk and double night table each;a multifunction control board above the doorway, bathroom,and the dressing table each; the light-controlling switch at the bed head and the electric switch control board on the curtain ... Of course, there was also fine luxury furniture with completely new design.

这次耐心一点，每件认为有细节的家具都专心地画一遍，虽然不算难画，但我认为酒店房间的好坏，家具占有相当大的比重，从平面布局、高低组合到“内在”的配置，最为关键，半岛的水平和竞争力就在于此， 不妨我们看看它的魅力所在：

This time I was a little more patient, sketching each piece of furniture I found having details. Though it is not difficult to sketch, I think, to a big degree, the quality of a hotel is accounted for by its furniture. From the layout of planes, to combination of low and high, and to the “inner” equipment, there lie the key—where lies the Peninsula’s quality and competitiveness. We may well look at its glamour:

其中，入门走廊有长的行李台，下有回收口及考虑长住时可存放大量鞋的鞋柜，还有及时存放你的“宝贝”的大的保险柜，挂衣柜亦充分而便捷，整体走道约4米，让你“淋漓尽致”地去处理行李。

A long table on the side the indoor isle with an opening under it for recycling and a shoe cabinet for long-stay guests who might have lots of shoes to store there; a big safe for your “valuables”; a clothes-hanging wardrobe spacious and easy to use; a passageway, about 4 meter along, providing more than enough room for luggage.

香港半岛酒店 *The Peninsula Hotel, Hongkong, China*

亦是通常的单边组合电视柜。对称布局，梳妆台靠洗手间的一头，中间为“大电视柜”，靠窗为书写台。绝的地方在细节上，梳妆台带电动的梳妆镜，一则风水偏好，不对床，二则灯光同步隐藏，人性化，左侧电吹风，右侧就有“绝招”——女士专用的移动式“干甲机”，吸引美女的地方！

There was the usual one-way TV units. It was arranged symmetric: the dressing table near the bathroom, the big TV bench at the middle, the writing desk near the window.The ultimate lay in its details: the dressing table went with an electric mirror. First, it was geomantically placed, not facing the bed. Second, the light was synchronously concealed, quite user-friendly. On the left hand there was an electric fan, and on the right "unexpected tricky move"—ladies' nail drier, a place appealing to beauties!

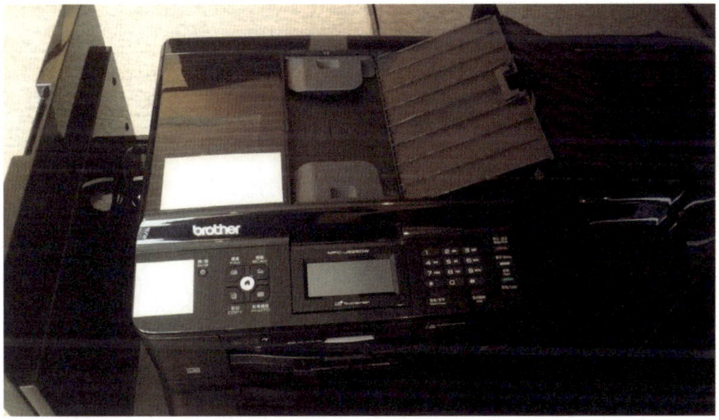

香港半岛酒店 *The Peninsula Hotel, Hongkong, China*

书写台靠窗，自然光线柔和，而角落处配有小台灯，更有升降的功能插座盒、真皮的书写垫、挑空的椅子区域（与梳妆台不同）。

The writing desk near the window received soft natural lighting. Placed at the desk corner was a small desk lamp, complete with a plug seat box that could move up and down, a real leather writing pad, with chair placed in an area whose ceiling was twice as high as other areas (different from the dressing table area).

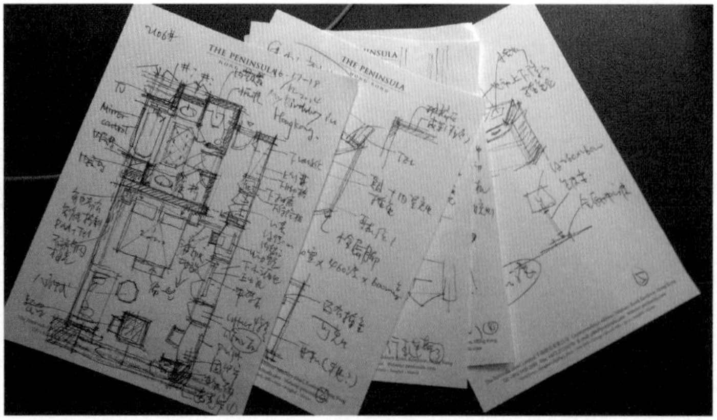

中间的“大电视柜”亦显心思——下部分综合小冰箱、小酒品/食物、杯具器皿放置，中部为大电视机（收褶双门）、咖啡机、茶具摆放处。不能不提的是，靠近书写台的柜内居然还配有FAX/打印机，虽然我是“脑盲”，但这稀奇之物“BROTHER”算是“半岛”一向的坚持配备，我想亦是少有客人使用的，（除了全球性商务人士）足不出户就可以方便连接电脑处理全球业务。我想到的是可以马上签Contract（合同）了，细致之极，哪怕是只有10%的客人需求！

The big TV cabinet also showed consideration—placed at the bottom were a small refrigerator, wine and foods, cups and wares; at the middle a big TV set (double doors), space for a coffee machine, tea set.I cannot help mentioning the Fax/printer placed in the cabinet near the writing table. Though I am computer illiterate, I know this rare Brother has always insisted on providing. I think guests seldom use them (except those who do global business). It enables people to handle global business simply by connecting their computer to the Internet. I was talking of signing a contract. It was consideration to the utmost—just to meet the needs of 10% guests just in case!

靠窗有一大处休闲之地，因为床的摆放原因，让出了一大空间作沙发/休闲之处，配有可看风景、喝茶聊天的区域，当然也可在房间用餐，确是其他酒店少有的想法，看来一切都是为你“如家”的生活所设计的。让你不经意地放松下来。

Close to the window was a big space for recreation. The bed was so placed that a big space could be used for a sofa/recreation. It was complete with an area for sightseeing, chatting over tea. Of course, you could have meals in the guestroom.It was an idea rarely allowed in other hotels. It seemed that all had been designed for "home-like" life, making you relax without knowing it.

香港半岛酒店 *The Peninsula Hotel, Hongkong, China*

卫生间区域虽说不出有什么惊喜，细看之下应亦是花了大力气、大金钱去升级的，石头部分更换，全面翻新过，加入全面智能化电视/音响/控制，一一俱全，特别是热水供应，大、猛、迅速！

Though I could not see any surprise in the bathroom area, scrutiny revealed upgrade through big efforts and spending—the stone replaced and fully revamped, and full smart TV/acoustics/controls added, especially the hot water supply—the water came gushing!

小小一个房间，让我感受到为什么半岛是全球数一数二、让人流连忘返的酒店，细节贴心，软硬兼施确是它的“撒手锏”。

The small guestroom made me realize why the Peninsula ranks first or second globally and makes guests deeply attached to it. Great care and consideration, complete with perfect hardware and software, is indeed its "secret weapon".

你住过或住多天、住多次，才能细细体味其独特之处！

You cannot appreciate its uniqueness unless you have stayed there, or you stay there for a few days or many times!

INTERCONTINENTAL
HOTELS & RESORTS

斐济洲际高尔夫酒店

15 ★★★★★

INTER CONTINENTAL FIJI GOLF RESORT & SPA

Address : Maro Road,Natadola Bay,Viti Levu,Fiji Islands (Private Mail Bag,Nadi Airport,Fiji Islands)

Telephone : +(679) 673 3300

Fax : +(679) 673 3499

Http : //www.fiji.intercontinental.com

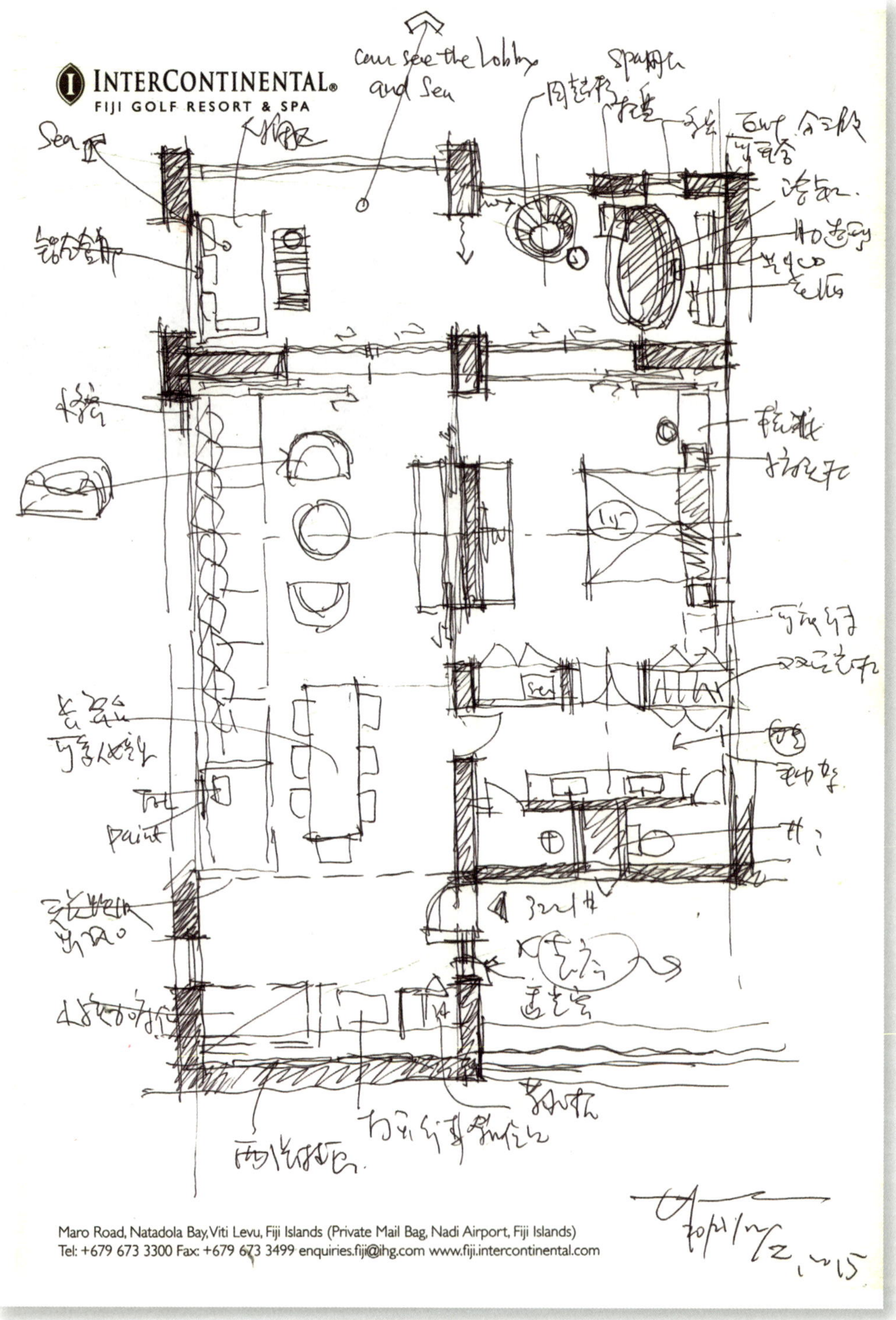

怎样的度假酒店的理想居住

这次前去太平洋岛国——斐济的度假的休闲是我们的目的，我就是元月份去的。经推荐选择了著名的度假酒店，两个酒店各住三四天。

一家三口，怎样住？

第一家的度假酒店，二楼，看到大海，哇！两间的大大的套房，行李进房各占一处，还有不少的位置让你轻松地打开大大的私行箱，小朋友的玩具也带了不少，准备占着藏入口的小天地（[illegible]）大人的组合，我们在这里乱吃东西，看书，坐坐画画，相信中国人的多人可能经过这样的[illegible]；[illegible]没问题，下午睡觉可以听着[illegible]入眠，而后特别选[illegible]藤椅相信是就地取材的材料，[illegible]太阳的大阳台合体有地上spa区，以当地特色的原住民文化的[illegible]，泳池（大浴缸）[illegible]半户外区域，[illegible]区与[illegible]的，[illegible]使用。[illegible]入口，[illegible]的游泳在度假中[illegible]。这里[illegible]，三个人轻松地[illegible]、洗澡脸，更衣（特别是大人的[illegible]回来后的一些小细节的清洗工作。

合口味、合习惯，是让你无法忘记的一种感觉，合川膳食在深圳地区的火爆也是让你依赖的另一个因素。

静静地、慢慢地，甚至不需要作诗，更是怡静的室内环境，阳光耀眼的灿烂，树影叶荫婆娑，盯着在熏风中摇摆的树影，十分钟八分钟你就会进入睡了。

那，你就来体验吧！

What Kind of House Encourage Long Stays?

怎样的房子可以让你长居

这次在太平洋岛国——斐济的行程以发呆为目的，或就是无目的的，纯粹躲躲春节的假期，两间酒店各住三、四天。

My trip to this Pacific island country Fiji was intended to kill time, or aimless —merely to shy away from the Spring Festival holiday by staying at hotels for three or four days.

一家三口，怎样住?

How to accommodate a family of three?

半山上的联排，二楼，看到大海，哗！两开间的大大的套房，行李进屋各占一处，还有不少的位置让你轻松地打开大大的旅行箱，小朋友的加床也事先准备，占着靠入口的小天地（应当是建筑阶段已经考虑到家庭旅游的需求）。长长的餐台，我们在这里乱吃东西、看看书、写写画画，相信中国人的多人打牌娱乐也是可以满足的。超过4米长沙发几个人平躺都没问题，下午睡觉可以听着电视入眠。两张特别造型的藤椅相信是就地取材的产物。晒着太阳的大阳台分为休闲区与spa区，以当地特色的麻质帘灵活分隔。泡池（大浴缸）处于半隐秘区域，方便连接睡眠区与主卫生间、衣帽间的使用。多出入口、多环路式的设计在度假产品中常见，这里更显老到，三个人完成刷牙、洗脸、更衣（特别是从沙滩回来后的一身小细沙的清理工作）。

斐济洲际高尔夫酒店
Inter Continental Fiji Golf Resort & Spa

The hillside Townhouse overlooked the sea on the second floor. It was a large two room suite. Inside, every piece of luggage had its own space, leaving adequate room for you to open your big suitcases with great ease. The kid's bed was prepared in advance, occupying the small space near the entrance (it seemed that the needs of family travel had been taken into consideration in the construction stage). There was a long dining-table, where we freely ate, read, wrote and sketch. I believe that it could satisfy Chinese people's need to play cards for the fun of it. The lounge, over 4 meters long, enabled several people to lie side by side. I could nod off on siesta with the TV still on. Two wicker chairs, specially shaped, were believed to be crafted from materials close at hand. The big sun-bathed terrace was divided into a recreation area and a spa area, flexibly partitioned by locally characteristic flax curtain. A big bathtub was located at a half-hidden area, conveniently connecting the sleeping area and facilitating the use of the master bathroom and cloakroom. It was common to see multiple entrances / exits and loops in the design of vocation hotel rooms. More mastery showed here. It was possible for the three of us to brush teeth, wash the face and change clothes (especially when we returned from the beach, we had to clean ourselves with the fine sand) at the same time.

斐济洲际高尔夫酒店
Inter Continental Fiji Golf Resort & Spa

全功能、全景观，是让你长住的一种前提，全日供应泳池边的大餐也是让你依赖的另一个因素。

Full function and full view, it is the prerequisite for your long stay. The rich dinner supplied all day beside the swimming poor is another factor that makes you attached to it.

舒舒服服，甚至不需要惊喜，只是恬静的室内环境，阳光耀眼的灿烂，绿树叶茂婆娑。盯着在热风中摇曳的树影，十分八分钟你就想入睡了。

It was comfortable. Dispensing surprise, it was more of the quietness of the interior ambiance, the glaring brilliance of the sun, the lush grace of the trees. Staring at the quivering tree shade, you would like to fall asleep.

那，你就长住吧！

Then, have a long stay there!

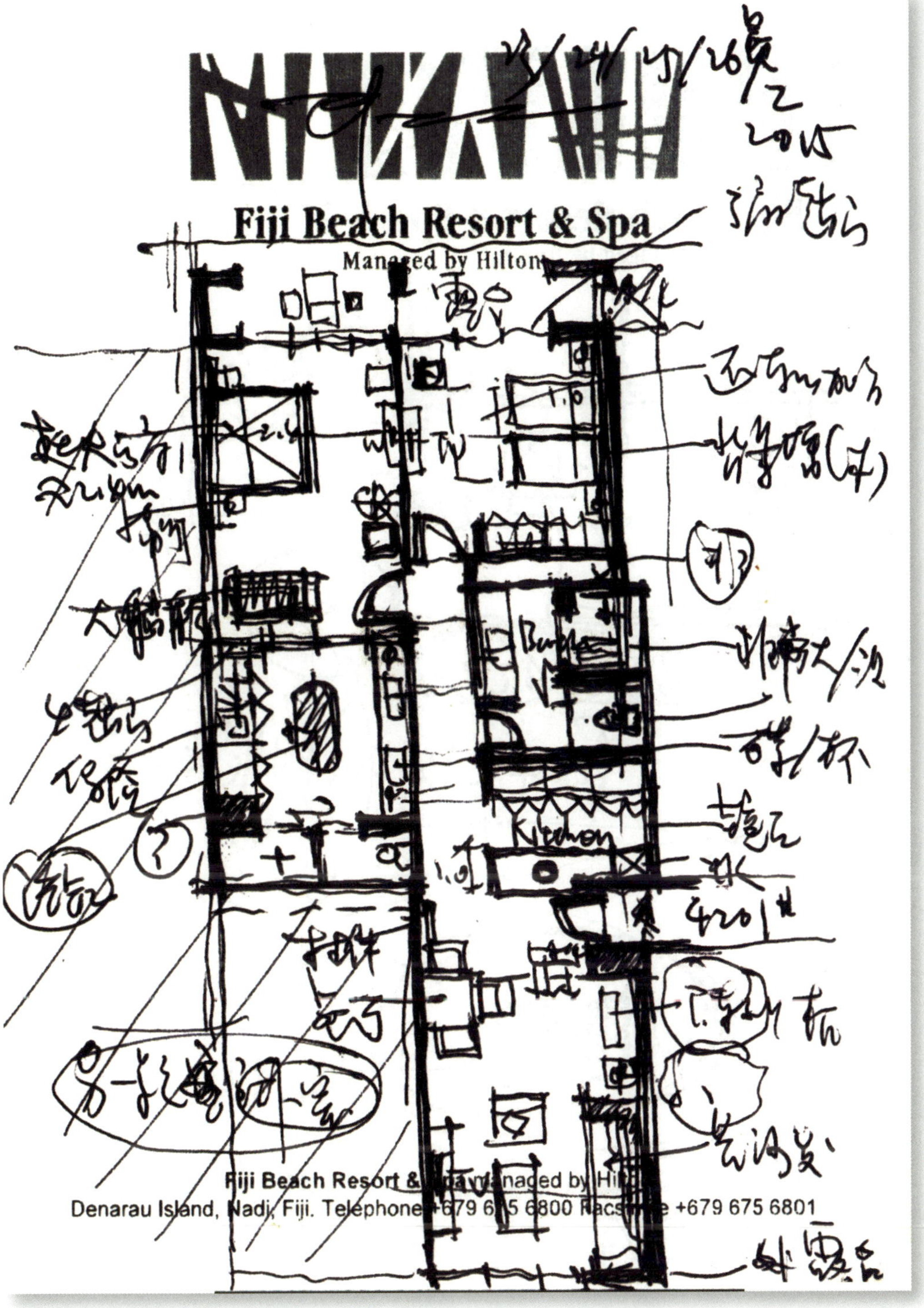

Hilton
HOTELS & RESORTS

斐济希尔顿酒店公寓

16 ★★★★★

FIJI BEACH RESORT & SPA HOTEL MANAGED BY HILTON

Address : Denarau Island,Nadi,Fiji.
Telephone : +(679) 675 6800
Fax : +(679) 675 6801

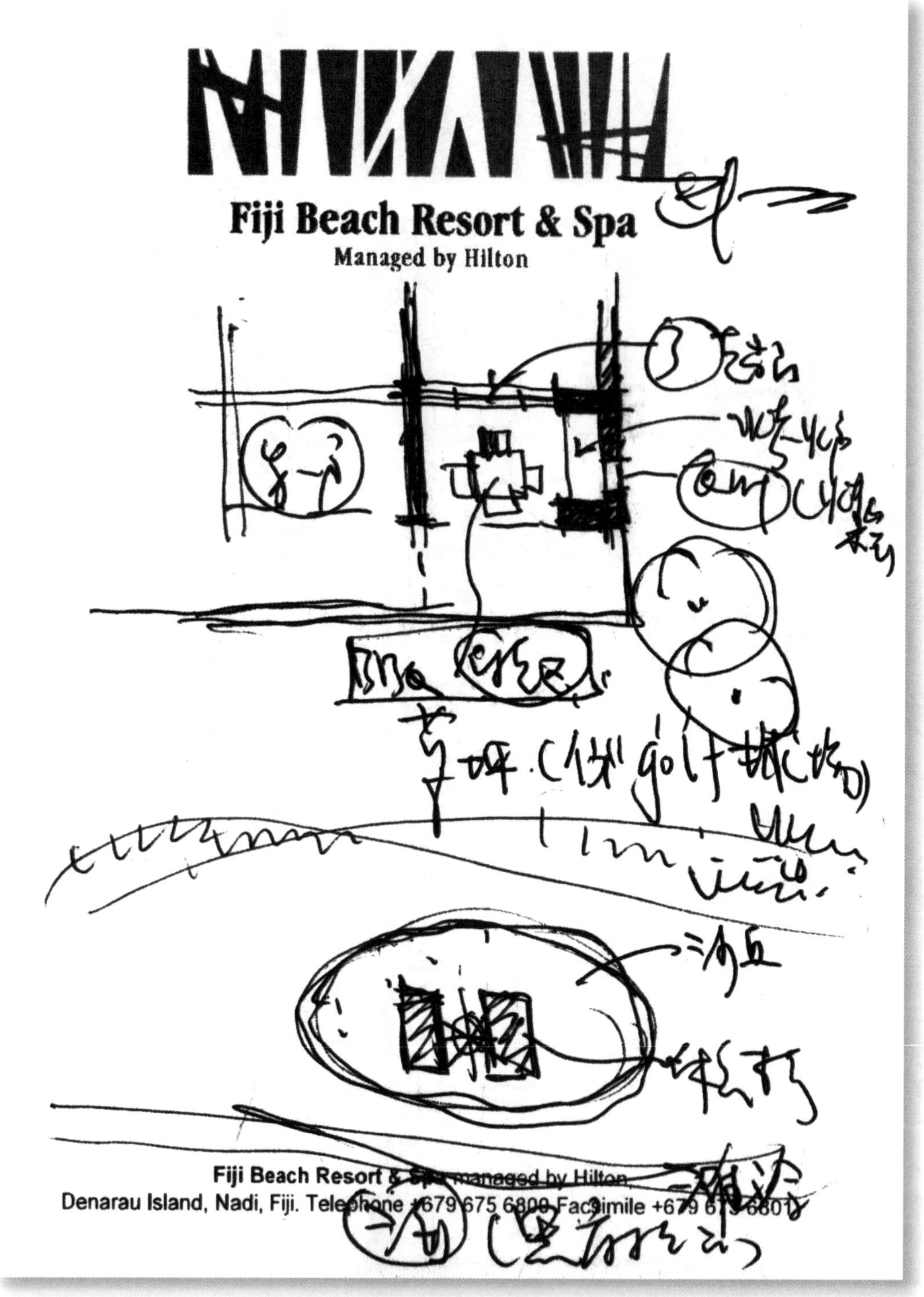

Fiji Beach Resort & Spa
Managed by Hilton
Fiji Beach Resort & Spa managed by Hilton
Denarau Island, Nadi, Fiji. Telephone +679 675 6800 Facsimile
golf

Fiji Beach Resort & Spa
Managed by Hilton

15/2, 2015

总经理：

没有想过来Fiji住到酒店，这次终于实现了，还是Hilton管理的。连住两天，让我又一次体验到Hilton的品牌和酒店的魅力（当然我平时也住在本公司的酒店管理公司管理的酒店）

对于我们这种旅游团不在意去[illegible]享受美食、建议以后，酒店[illegible]不多！

①

Fiji Beach Resort & Spa managed by Hilton,
Denarau Island, Nadi, Fiji. Telephone +679 675 6800 Facsimile +679 675 6801

Fiji Beach Resort & Spa
Managed by Hilton

厨房是用了几个锅子/炉子和冰箱。冰箱[illegible]水，还有一些剩余的剩菜，都吃了第二天早餐[illegible]。[illegible]可惜不会用，但是到外面海滩[illegible]一[illegible]，[illegible]。[illegible]（[illegible]）[illegible]。

②

Fiji Beach Resort & Spa managed by Hilton,
Denarau Island, Nadi, Fiji. Telephone +679 675 6800 Facsimile +679 675 6801

Fiji Beach Resort & Spa
Managed by Hilton

我们这个团9个人，不会点菜，用了[illegible]，留下的[illegible]给了当地的服务员。我在这里[illegible]用了，只是没有[illegible]。这100多平米的大套间，住得很舒服。[illegible]，让我不能不说是感受到了"以人为本"的管理理念。

我在这里住得很愉快，欢迎你们来参观我们的[illegible]！

Fiji Beach Resort & Spa managed by Hilton,
Denarau Island, Nadi, Fiji. Telephone +679 675 6800 Facsimile +679 675 6801

Fiji Beach Resort & Spa
Managed by Hilton

顺便提一下，这些家具都是木质的，一样的[illegible]，[illegible]。[illegible]，[illegible]，[illegible]一部分[illegible]部分，[illegible]。

[illegible]，[illegible]上，[illegible]！

总经理，再次感谢Hilton的[illegible]

[illegible]

Fiji Beach Resort & Spa managed by Hilton,
Denarau Island, Nadi, Fiji. Telephone +679 675 6800 Facsimile +679 675 6801

斐济希尔顿酒店公寓 *Fiji Beach Resort & Spa Hotel Managed By Hilton*

This Flat

这公寓

没有想过来斐济住到公寓，这次终于有幸住到了，还是希尔顿管理的，连住四天，让我们又一次体验到国际品牌对这种公寓的功力（虽然我本身也住在由国际公寓管理公司管理的小区）。

I had not expected to stay in a flat in Fiji. This time I was lucky enough to check into one, managed by Hilton, for four days, let us once more experience the contribution of an international brand to this kind of flat, though I am living in a residential community managed by an international flat management company back in China.

对于我们这种纯旅游而不愿意去市场买菜、煮饭的人来说，公寓似乎用到的功能不多！厨房只用了几个杯子/碟子盛小食，冰箱放酒、水，煮了一次打包的剩菜，在餐厅吃吃东西：水果啊、花生啊之类的。露台面海可以BBQ，可惜不会用。倒是到外岛浮潜后一堆堆的盐水衣服可以轻易地用洗衣/干衣设备解决，向南（初判断）的后面小花园露台刚好用作晒衣服。独立的两个洗手间，不会发生争用的场面，浴厕分格更为合理舒服。这样基本的功能我们都用了，只是没有深究它的必要性和是否浪费。近100平方米的大套间，值得我们去学习，简单装饰而功能细致，让你不知不觉感受到“以人为本”的简单哲学，或者你住得更长，就更能体会它的合理、便利和服务！

For us mere tourists unwilling to buy vegetables on the market or to cook, the function of the flat seemed limited! Our use of the kitchen was limited to several cups/ disks to contain snacks, storing wine and water in the refrigerator, boiling of leftovers of a dish taken back from the restauran only once. The dining room was used for eating fruits, peanuts and the like. The balcony facing the sea could be used for barbecue—it was a shame we were unable to use it. To make up for that, the pile of salty clothes after our diving could be easily handled with washers and dyers. The south-facing little garden at the back (according to my initial judgment) was good for sun-drying our clothes. There were two detached bathrooms: unlikely to give rise to a sight of everyone *scrambling for their use. The separation of the shower compartment from the toilet was more reasonable and comfortable. We used all its basic functions, but did not yet go further to study whether it was necessary or wasteful. It is worth our study—the near 100 m^2 suite, simply decorated and finely functional, made you unconsciously feel the simple philosophy of "being human-oriented". Or as you stay longer, you will understand more of its sensibility, convenience and service!*

顺便提一下，这些公寓都是小独栋，一栋4户、6户到8户，我们这个是8户，首层4户，二层4户，其中2户是复式的，连接一部分三层部分，观海感觉更加明显，确是成熟的设计。

By the way, these flats were all small detached units, with 4, 6 or 8 houses each. This flat had 8 houses, with 4 on the first floor and 4 on the second. 2 of them had a suspended floor each. On the third floor, the sea view was more impressive. It was indeed a mature design.

坦白地说，它的复杂在于建筑设计上，错接的户型和管井的合理设计才是值得学习和研究的！

To be frank, its complexity lay in architectural design. What was worthy of study was its staggering house types and reasonable design of tube wells!

这公寓，是对得起希尔顿的品牌的。

This flat merits the brand of Hilton.

斐济希尔顿酒店公寓 *Fiji Beach Resort & Spa Hotel Managed By Hilton*

INTERCONTINENTAL.
FIJI GOLF RESORT & SPA

22/2, 2015 凌晨2:20 [illegible]

转眼已过了三天，到Fiji！，应当说也
让我有些[illegible]，在这个岛国发呆！

初一傍晚上飞机，从香港飞，离香港的
[illegible]，[illegible]，吃了[illegible]
[illegible]，[illegible]
[illegible]第一印象是机场的古朴，机场是
那么的小，[illegible]，而[illegible]也是[illegible]的
只是[illegible]不[illegible]，[illegible]一下！

接机也[illegible]一会才[illegible]上[illegible]"还是
那一句[illegible]！继续努力！一[illegible]
[illegible]

①

Maro Road, Natadola Bay, Viti Levu, Fiji Islands (Private Mail Bag, Nadi Airport, Fiji Islands)
Tel: +679 673 3300 Fax: +679 673 3499 enquiries.fiji@ihg.com fiji.intercontinental.com

INTERCONTINENTAL.
FIJI GOLF RESORT & SPA

[illegible]的酒店，入住[illegible]上午，
[illegible]，[illegible]
让人[illegible]。[illegible]
了一个100多[illegible]
[illegible]一边[illegible]大海的[illegible]！无敌！
[illegible]！一直睡到[illegible]的
下午，[illegible]一个人又去[illegible]了一下[illegible]
对着泳池，对着大海[illegible]
用[illegible] Navo[illegible]
[illegible]晚餐。

大海[illegible]

②

Maro Road, Natadola Bay, Viti Levu, Fiji Islands (Private Mail Bag, Nadi Airport, Fiji Islands)
Tel: +679 673 3300 Fax: +679 673 3499 enquiries.fiji@ihg.com fiji.intercontinental.com

INTERCONTINENTAL.
FIJI GOLF RESORT & SPA

[illegible]，Fiji Beer，
[illegible]
[illegible]
大海[illegible]！

第二天[illegible]
[illegible]，大大的[illegible]
天下了大雨，晚上[illegible]大餐，[illegible]
的Fiji大餐，吃[illegible]，吃
[illegible]
[illegible]！

[illegible]

③

Maro Road, Natadola Bay, Viti Levu, Fiji Islands (Private Mail Bag, Nadi Airport, Fiji Islands)
Tel: +679 673 3300 Fax: +679 673 3499 enquiries.fiji@ihg.com fiji.intercontinental.com

INTERCONTINENTAL.
FIJI GOLF RESORT & SPA

[illegible]！

（小插曲）[illegible]一圈[illegible]，
[illegible]一个[illegible]
[illegible]  [illegible]
[illegible]
[illegible]
[illegible]！

[illegible]
[illegible]！！

[illegible]

④

Maro Road, Natadola Bay, Viti Levu, Fiji Islands (Private Mail Bag, Nadi Airport, Fiji Islands)
Tel: +679 673 3300 Fax: +679 673 3499 enquiries.fiji@ihg.com fiji.intercontinental.com

Three Days in the Twinkling of an Eye

转眼已过了三天

转眼已过了三天，到斐济，应当说也过得挺快的，在这个岛国发呆！初一傍晚上飞机，从香港飞，商务舱，好舒服的享受，看了几部电影，吃了两顿大餐，睡了几段小觉，飞到了位于南太平洋的岛国，应当说第一印象是相当的古朴，机场是那么的小，简陋，而过关也是简简单单的，只是英语不怎么样，勉强应付一下！

In the twinkling of an eye, three days were gone. It is fair to say that the days we stayed in Fiji whiling time away flied quickly! At dusk of the first day of the Spring Festival we boarded the plane, flying from Hong Kong. Traveling business class, I enjoyed myself, watching several films, having two rich meals, napping now and then, until landing on this island country in the South Pacific Ocean. To be fair, it impressed me first as rather shabby and simple: the airport was so small and crude, and it was so simple to pass the customs—my English was barely adequate for the occasion!

接机也折腾了一会才“对上嘴型”，还是那一句，英语真的很重要！继续努力！一个小时的走走停停，才到达位于岛的一端的少有人烟的洲际酒店，入住时间是当地上午，房间还要等，倒是在俱乐部的免费早餐，让人倍感亲切、自在。餐后就顺利入住了一间100多平方米的度假公寓，有厅有房，当然还有一望无际大海的休闲阳台！无敌！懒洋洋地倒时差！一直睡到当地时间的下午，我一个人又去享受了一下免费的下午茶，对着泳池，对着大海（因为我们住在山上），用蹩脚的英语订了海边餐厅Navo（一个小岛的名字）的晚餐。

Talking with the man meeting us the airport, it put me into a lot of trouble struggling to have the right English words come out of my mouth! I have to repeat it here—English is really very important! Continue to make effort! After an hour's driving on and off, we came to the Inter Continental Hotel at one of the island, with few signs of human existence.We checked in in the morning of the local time. We had to wait for the room. However, the free breakfast in the club made us feel friendliness and comfort. After the meal, we smoothly checked into a 100+m^2 holiday flat, with a living room, bedrooms and a recreation balcony looking onto the infinity of ocean! There was nothing like it! Tired from the jet lag, I slept till the afternoon of the local time. Then alone I enjoyed a free afternoon tea. Facing the swimming pool and the sea (we were living in the mountain), with broken English, I ordered supper at the seafront restaurant Navo (named after an islet).

大海鲜餐，当地的蟹做的蔬菜沙拉，大龙虾烧烤，斐济啤酒，小孩点了一个当地的牛扒，美味、满足，还可以去沙滩抓抓螃蟹，戏戏水。黑黑的大海，满天的星星，真是发呆的好地方！

It was a rich seafood meal-vegetable salad with local crab, barbecued big lobster, Fiji beer, a local grilled fillet steak my kid ordered—appetizing and satisfying. You could also catch crabs on the beach and frolic in the seawater. The dark sea and starry sky—a perfect place to kill time!

第二天还是懒洋洋，一样的早餐（不用换餐牌），一样的午餐，大大的午觉（午后天下了大雨），晚上就是自助大餐，当地的斐济大餐，吃撑了。仿佛来这里就是吃、睡、发呆，当然也慢悠悠地看看书！机场买的，还好，有了打发时间的工具！

The next day was equally lazy. It was the same breakfast (you had no need to change your meal card) and lunch. Then we had a big siesta as it rained in the afternoon.In the evening we had a big buffet, a local Fiji buffet. I ate to the utmost. It seemed that we had come to eat, sleep and do nothing. Of course, we also took time reading books bought at the airport—alright, we had something to pass the time with!

满足的一段小假期，明天就要转酒店了，或许会有别样的经历！

It was a satisfactory short holiday. Next day we would change to another hotel where we would have other experiences!

小插曲：

Episodes:

1.下午和小孩子转了一圈酒店，数了很多果子的品种，其中有一种类似四方形的小灯笼似的，不知道名字。

1. In the afternoon, while sauntering around the hotel, my kid and I found many varieties of fruit,among them one like a small square-shaped lantern we could not name.

2.沙滩戏水，发现退潮的沙滩的沙子居然有斜菱图案，以前从没有注意到，或者是习以为常了！

2. While romping on the beach I found that there were slanting diamond-shaped patterns of sand when the tide receded . I had never noticed that before, or had accustomed to it!

时光提醒，知识来自大自然，应当尊重和珍惜！

Time reminds us that knowledge comes from nature and should be respected and cherished!

2015.2.22 凌晨 *2:20* 斐济

February 24, 2015 Fiji

1.飞国外的飞机，可以用手机。于是可以留下从空中俯视斐济各个小岛的照片，非常有趣。

It is allowed to use the mobile phone on a plane flying abroad. So it is possible to have kept air view photos of various islets of Fiji. It is great fun.

2.我们落地的机场非常简陋，有一点点临时搭建的感觉，走在上面有一点点颤抖。
The airport we landed at is quite crude. It looks like a makeshift and shivers a little while we walk on it.

3.“面朝大海，春暖花开”，确是如此。下午的太阳相当耀眼，靠着无边泳池的免费的下午餐令人心旷神怡，正好倒倒时差。

“Facing the vastness of the sea, we see the flowers unfolding in the warmth of spring”, as the line goes. The afternoon sun is rather glaring. Taking a pleasant free afternoon meal near the infinity of the swimming pool, it is a good way to get over the jet lag.

4.我们的房间位于山上，可以一览无遗的沙滩、大海，是个发呆的好地方。

Our house on the hill is a good place for an eyeful of the beach and the sea, as well as for killing time.

5.海边的餐厅，当然是以吃海鲜为主啦，龙虾、大螃蟹，当然还有小孩喜欢的牛排，环境好，蚊子多！

Naturally, the seafront restaurant serves mainly seafood, such as lobster and crab. Of course,it also serves beef steak our child likes. It has a good ambiance but too many mosquitoes!

6.餐后用手机照明走走沙滩，就有不少的收获。

Strolling on the beach after supper we have a big catch, aided by the cellphone flashlight.

7.看到这样的四方形的果实，感叹自己的孤陋寡闻和大自然的无奇不有（经后来查证，这叫作滨玉蕊）。

The sight of the diamond-shaped fruit makes me shamed of my ignorance and wonder at the wonders of the nature (after research, I find it called Barringtonia asiatica (L.) Kurz).

8.适逢中国的春节，酒店的自助餐门前非常热闹，看到了中国的舞狮，相当喜庆！

It is the time of the Chinese Spring Festival. The street at the buffet of the hotel comes to life. The Lion Dance gives a quite festive air!

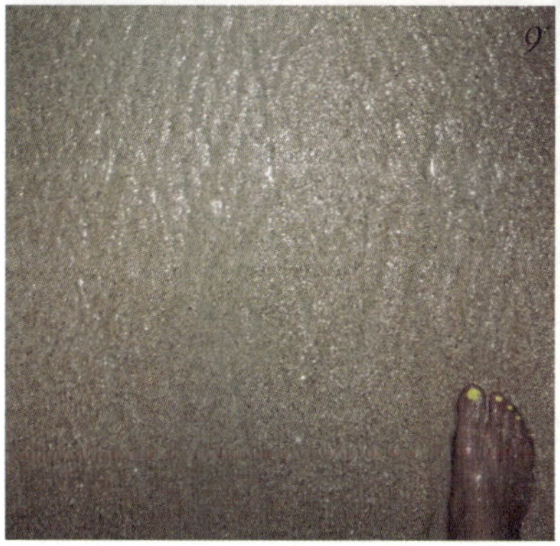

9.以前确实没有留意到，原来退潮后，沙滩上的沙子是有图案的，水与沙的汇合结果。

It turns out that the sand will form certain patterns after the tide recedes——a result of the combination of water and sand,which I had never noticed before.

Fiji Beach Resort & Spa
Managed by Hilton

"不同" [illegible]

[illegible]

Fiji Beach Resort & Spa managed by Hilton, Denarau Island. Nadi, Fiji. Telephone +679 675 6800 Facsimile +679 675 6801

①

Fiji Beach Resort & Spa
Managed by Hilton

世界，[illegible] Stand by me, Don't worry be happy! and some songs Fiji local. [illegible]

[illegible]

Fiji Beach Resort & Spa managed by Hilton, Denarau Island. Nadi, Fiji. Telephone +679 675 6800 Facsimile +679 675 6801

②

Fiji Beach Resort & Spa
Managed by Hilton

"蓝调" [illegible]

今天 take a taxi to Nadi [illegible] Chen's seafood Restaurant [illegible]

Fiji Beach Resort & Spa managed by Hilton, Denarau Island. Nadi, Fiji. Telephone +679 675 6800 Facsimile +679 675 6801

③

Fiji Beach Resort & Spa
Managed by Hilton

[illegible]

[illegible] handcraft shop [illegible] "Made in China"

Fiji Beach Resort & Spa managed by Hilton, Denarau Island. Nadi, Fiji. Telephone +679 675 6800 Facsimile +679 675 6801

④

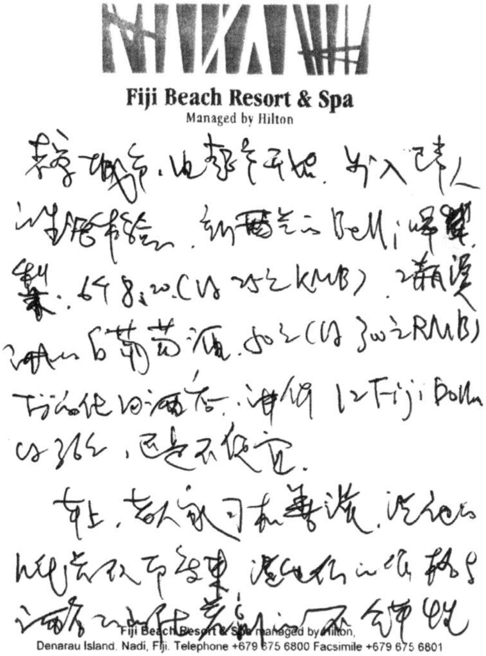
Fiji Beach Resort & Spa
Managed by Hilton
Fiji Beach Resort & Spa managed by Hilton,
Denarau Island, Nadi, Fiji. Telephone +679 675 6800 Facsimile +679 675 6801

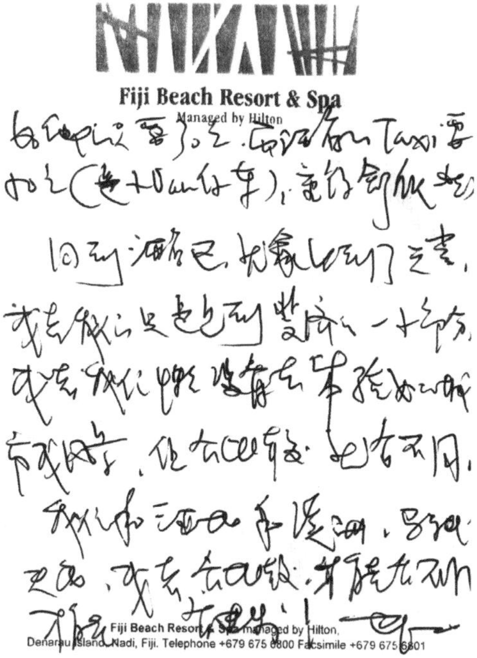
Fiji Beach Resort & Spa
Managed by Hilton
Fiji Beach Resort & Spa managed by Hilton,
Denarau Island, Nadi, Fiji. Telephone +679 675 6800 Facsimile +679 675 6801

Difference

不同

因为有比较，所以有“不同”，或许人就是这样。

Comparison shows “difference”, or it is human to compare things.

转到第二间酒店，希尔顿酒店，在另外一个岛上。岛上黑色沙泥的滩涂，小孩说像踏屎一样，水脏脏的，不像大海，所以第二天（2月23日）就只能参加“一天游”到一个小岛去浮潜。只有少少的小群鱼，还是靠食物引来的，岛上的午餐倒是不错。“景不到，人、物、礼到。”

We were changed to a second hotel, the Hilton Hotel,on another island. The mudflat felt like shit, as my son put it. The water was so dirty that it was not like the sea. Then the next day, February 23, we could only take part in a “day trip” to an islet for snorkeling. There was only a few small schools of fish allured to appear by food. The lunch on the islet turned out to be quite good. The scene was not satisfactory, while the people, foods and hospitality were.

过程：1.5小时的船上，有吃，有喝，有即兴的唱歌：*Stand by Me, Don't Worry Be Happy* 以及一些斐济当地歌谣，非常开心。岛上餐后讲解怎么用椰子叶做手工，开椰子，用椰子做汁、油等等，一方面可以练练英文，还可以了解民风，是不错的经历！

The process: one and half an hour on the boat, eating, drinking and singing off the cuff: Stand by Me, Don't Worry, Be Happy, and some local Fiji songs. It was quite enjoyable. After the meal on the islet, we were given a walk-through of how to make handiwork out of coconut leaves, how to crack open a coconut, how to make juice and oil from coconut, and etc. It offered us a chance to practice English while learning about the local customs—a very rewarding experience!

回程昏昏欲睡，惺惺忪忪中画了一张在海岛潜泳时，第一次看到的好像假的一样的“蓝海星”，真的很蓝，蓝得很不自然，或者在我的认知里面没有这种天然的蓝色，忍不住把每只都翻了个四脚，不是，应当是五脚朝天，很逗，高兴坏了，看到这个！

On the back trip, rather sleepy and drowsy, I drew a picture of Linckia laevigata, which I first saw while snorkeling on the islet, which looked like a fake one—it was really so blue, too blue to be natural, or as far as my knowledge went, there was no such blue in nature. I could not help turning it upside down, with its four legs—no, five legs—upturned. It was great fun. I was really overjoyed to see it!

今天乘出租车去南迪，镇上找中餐馆*Chen's Seafood Restaurant* 吃午饭，很亲切，炒龙虾、炒海参、青菜、鱼头豆腐汤，饭后还从二楼窗口写写画画街道的“风景”。

Today We took a taxi to Nadi Town, looking for the Chinese Chen's Seafood Restaurant. We had our lunch here. It felt quite homey: fried lobster, fried sea cucumber, green vegetable, fish head and Tofu Soup. After the meal, from the window on the second floor I sketched some street views.

落后，并不仅仅是表面街小、路脏、商店橱窗均有防盗网，有比较才知道这是多么“原始”，如中国的六线农村城镇般！

It was poor here, as evidenced not only by the apparently narrow streets, the dirty roads, the theft-proof net on every shop window. Only by comparison did I begin to realize it was as "primitive" as the sixth-tier rural towns in China!

在*Hand Craft Shop*手工艺品商店买了一些特色餐垫作为礼品，还是有特色的，希望不是*"Made in china"*.

We bought some locally characteristic placemat as presents from the Handcraft Shop. We hope they are not "Made in China".

考察城市，由超市开始，步入正常人的生活体验，新西兰的Belli啤梨：6个8.20斐济元（约25元人民币），2瓶澳洲的白葡萄酒50斐济元（约150元人民币），打的士回酒店，计价12斐济元约36元人民币，还是不便宜。

To study a city, start with the supermarket to experience the ordinary people's life. Here, they sold 6 Belli pears for 8.2 Fiji dollars, or about 25 RMB, 2 bottles of white wine for 50 Fiji dollars, or about 150 RMB. The tax fare to my hotel cost me 12 Fiji dollars, or about 36 RMB—not inexpensive, either.

车上，司机老人家善谈，说他几代前从印度来，说他们的价格与酒店的价格差别的不公平性。如他们要30斐济元，而酒店的的士要50斐济元（是小面包车，坐得舒服些）。

The talkative old taxi driver said their ancestors had come from India generations before, and that the fare difference between their taxis and the hotels' taxis was unfair. For example, if they charged 30 Fiji dollars the hotels' taxis would charge 50 as the small vans were more comfortable.

回到酒店区，就像回到了天堂，或许我们只是见到斐济的一小部分，或者我们懒，没有去体验好的城市或风景，但有比较，就有不同。

Back at hotel, it seemed like returning to the paradise. Or what we saw was just a small part of Fiji. Or we were too lazy to experience better cities or scenery. However, wherever there is comparison, there is difference

我们和三亚比，和澳洲、马尔代夫比，或许有比较，才能有不同，才能有进步!

Compare it with Sanya, Australia and Maldives. Or only with comparison will there be difference to be shown, and progress to be made!

*2015.2.24*斐济
February 24, 2015 Fiji

1.去外岛的船上，船员们自弹自唱民歌、经典怀旧歌引起大家的和唱，气氛像空气般的热。

On the way to the island, the sailors are singing folk songs and classic nostalgic songs while playing the guitar, with the tourists singing along—this atmosphere is as hot as the air.

2.在南迪小镇找到相对不错的中餐厅，远处的防盗网林立，近处的电线横飞，与度假胜地格格不入。

We find a relatively good Chinese restaurant in Nadi Town. The thick anti-theft nets in the distance and the crisscross electric lines look quite out of place in the vacation land.

3.海边的房子采用大面积的木饰面还是有问题的，有一股发霉的味道，身体靠上去，小心翼翼地。

It is still a problem to use large areas of wooden veneer for houses near the ocean. The veneer smells moldy and it takes care to lean against it.

4.房间外面有小露台，可以烧烤，往外有草坪，太阳伞下可以看看书，望望海，反而淤泥般的沙滩，感觉脏脏的。

Outside the house there is a small terrace, where a barbecue can be thrown. Further out there is a lawn, where under the sun umbrella you can read books and look out to the sea. The silt-like beach looks rather dirty.

5.很兴奋。第一次吃面包树的果实，炸的，很好吃，很香，真的“很面包”，天赐美食。

Quite excited—it is my first time to eat the fruit of breadfruit. Deep-fried, it tastes palatable and sweet, and is really "very bready"—a gift from heaven.

6.每一次在外旅行，习惯捡一些石头作留念。这一次的石头都是从海中、海滩取的。很庆幸淘到了各种珊瑚。

I have the habit of collecting some pebbles while traveling. This time I have collected them from the beach. Luckily, I have also collected various corals

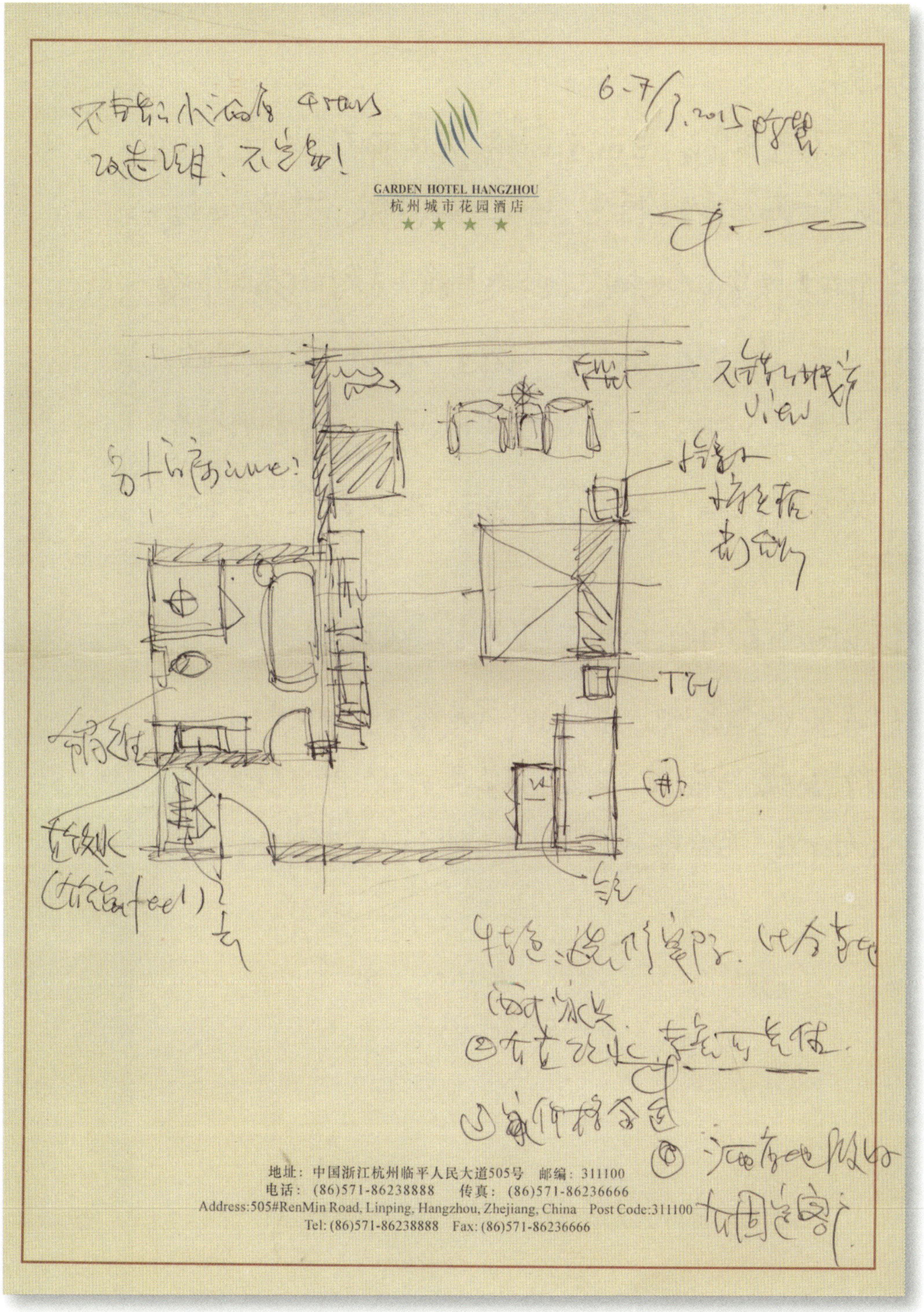

杭州城市花园酒店

17 ★★★★

GARDEN HOTEL HANGZHOU CHINA

Address　: 505#RenMin Road, Linping,Hangzhou, Zhejiang,China
中国浙江杭州临平人民大道505号

Telephone : +(86 571) 8623 8888

Fax　: +(86 571) 8623 6666

饭店感觉的客房

在挑一间好的客房，三两天的客房，会让而归人之享受，你我们假设这个某一酒店在它自己的情况，而且将入住，也许也将靠近海滨有如画般风景的地域，打开房门会让入住客人的种种体验到温暖，吸一口新鲜的空气，舒适，但是洗手台和镜子被安放了浪费的空间，而没有这个必要，如果我们在考虑将这个空间合理利用的话，这个空间也不浪费。

浴室的设计自然也是一个吸引人的地方。

就让我们来看看吧，我们可以看看有这类的特别的酒店，包括设计类、精品类酒店，通过设计，让你的入住也能体验出个性，和家里有温暖的吸引力的话，好吗？！

Flat Guestroom
公寓感觉的客房

这应当是一间改造的房间，三开间变两客房，合并而扩大的房型，似我们做过的某一酒店项目的情况。而且投入有限，也许地段靠近经常有布艺展览的地域，打的是展会的多天居住客的市场，价格实惠。乍一看就像一间公寓，简陋，倒是洗手间相当大，被迫放了浪费的浴缸，着实没有这个必要。如果是我们就会考虑应当怎样让平面丰富的同时，而满足业主的投入要求，这个一点也不容易。

These two guestroom must have been reshaped from three rooms, similar to something we had done for another hotel. With limited budget, and in the neighborhood cloth art regularly took place, the rooms are aimed for guests staying for just a few days, really inexpensive. The first glance saw that it looked like a flat, crude. The bathroom turned out rather big; a wasteful tub was placed there—indeed unnecessary. It is not easy to enrich the design of planes and to meet the input requirements of the owner at the same time.

洗手间配有直饮水，也是一个宜长居的吸引点。

There was direct drinking water in the bathroom, an appealing point favoring long stays.

就让它似公寓吧，看看国外这类四星级类别的酒店，包括设计类、精品类酒店，通过设计，让小小的投入也能做出有个性、有印象、有温暖和吸引力的效果？！

Then design it like a flat. As you see, there are similar four-star hotels overseas, including design hotels and quality hotels. Design also enables small inputs to produce appealing effects, personalized, impressive and homey!

hotel INDIGO

香港英迪格酒店

18 ★★★★

INDIGO HOTEL HONGKONG CHINA

Address : 246 Queen's Road East, Hong Kong 香港皇后大道东246号

Telephone : +(852) 3926 3888

Fax : +(852) 3926 3926

Http : //www.hotelindigo.com/hongkong

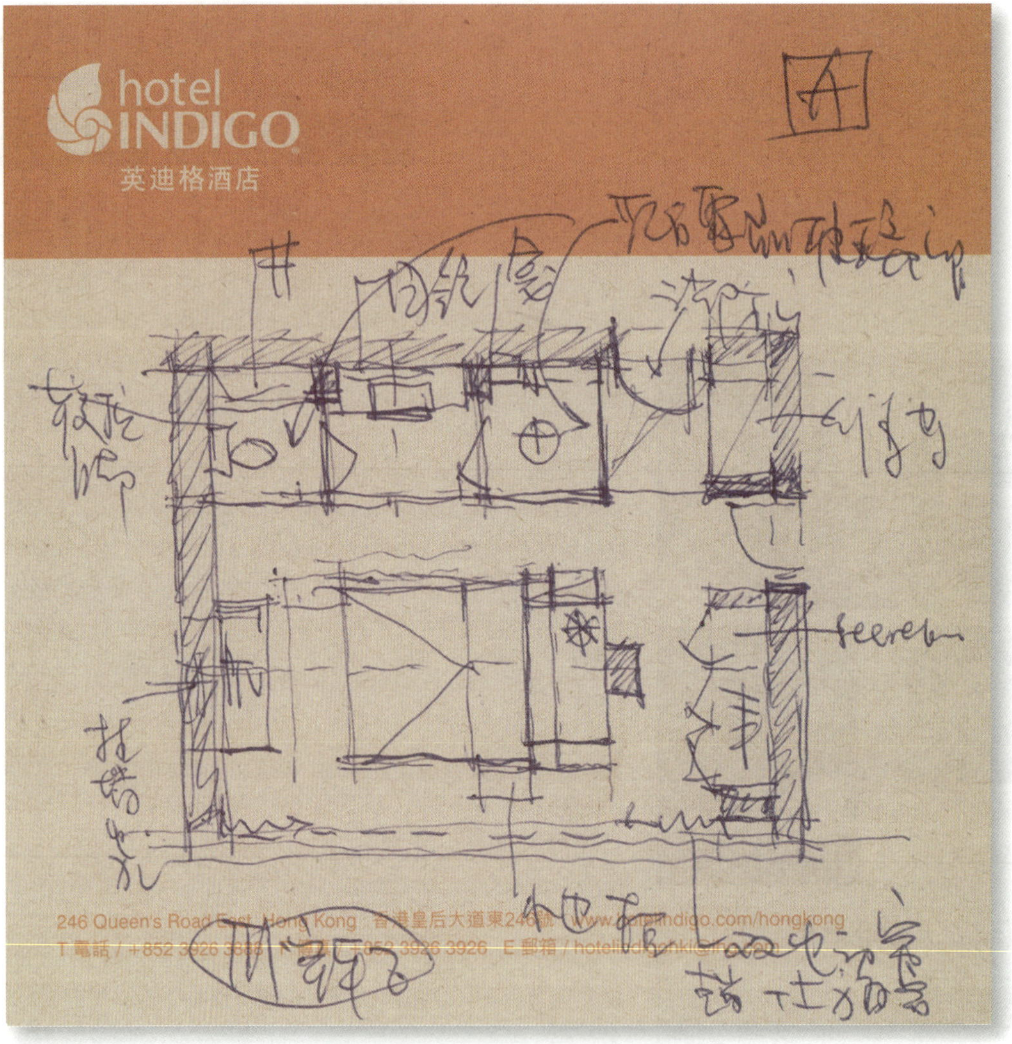

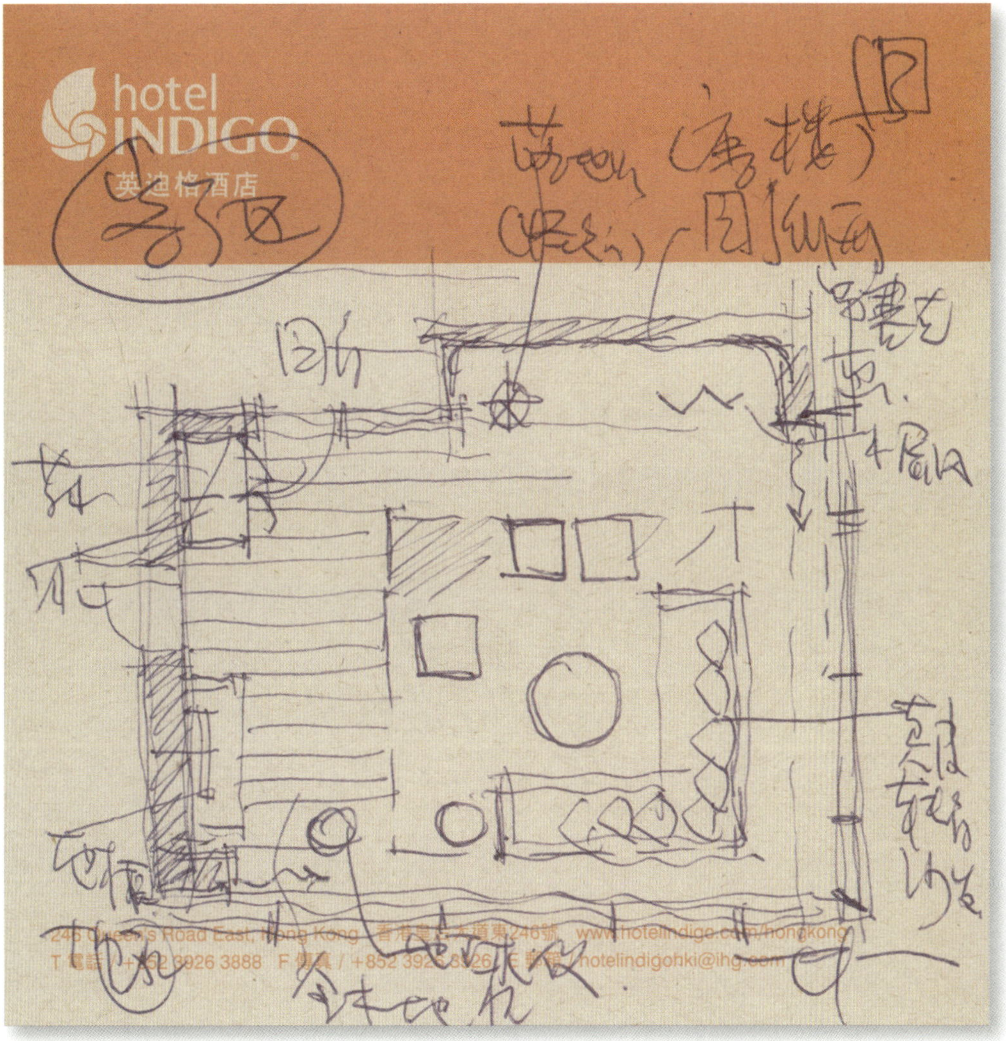
hotel INDIGO
英迪格酒店

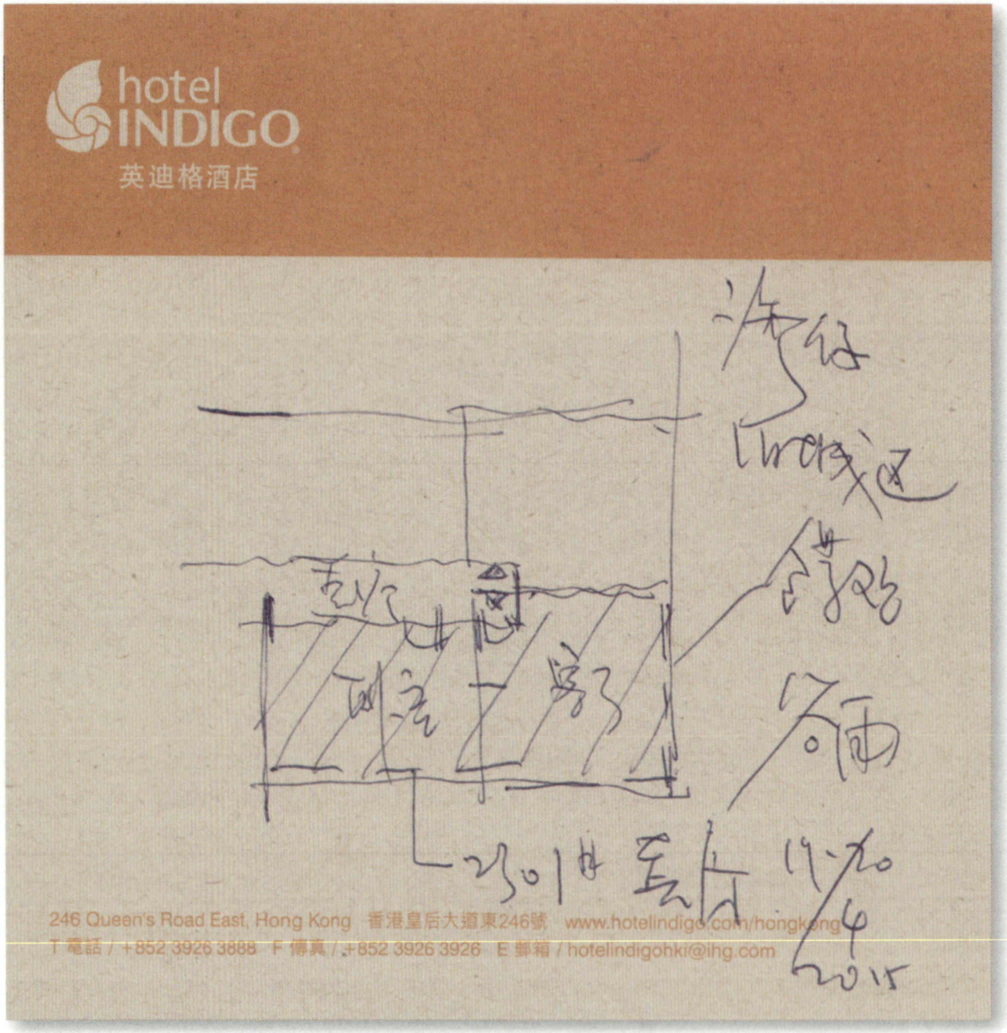
hotel INDIGO
英迪格酒店
246 Queen's Road East, Hong Kong 香港皇后大道東246號 www.hotelindigo.com/hongkong
T 電話 / +852 3926 3888 F 傳真 / +852 3926 3926 E 郵箱 / hotelindigohki@ihg.com

房状大致分开西吧

住了香港湾仔的英皇格酒店本来之外，第一次是住房满，第二次换到了近十天就开业的一间名叫的会店，Tom's 带上十八岁一起。它所以在两间"土豪"地独自立在各种西城的大街里，天啊，转向玻璃之城市"旧梦"，这个西城也是非常好的Indigo的特色，简洁、直接、不烦琐，更是那个全部正面的玻璃会非常有吸引力。

一方一房，相当念情，在香港这么"寸土尺金"的地方，更是令人佩服。

尊了找不到信店，只能用以前的签名开西城来玩一下这个地方，港岛大道的住了那间表面的旧城区的路以酒店（皇后大道东与太原街转角处）

一层只有8间房，客人们不能去接待的地方啊！顶层更有非常好的酒吧和全港唯一一个玻璃泳池，悬入半空，鸟瞰城市，哈哈。

真可惜，下次再来了！

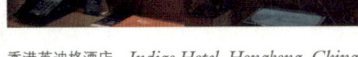

香港英迪格酒店 *Indigo Hotel, Hongkong, China*

The Room is Too Big - Draw It one Part at a Time

房子太大就分开画吧

位于香港湾仔的英迪格酒店声名在外，第一次想住，房满，第二次提前了近十天就预定好一间尽端的套房，可以带上小朋友一起，15岁的谷雨同学“土豪”地独自占据客厅区域的大沙发，无敌转向玻璃的城市“旧景”，主卧区域也是非常有英迪格的特色：简洁，直接，有细节，更是尽用全单面的玻璃窗，非常有吸引力。

The Indigo hotel in Wan Chai, Hong Kong is renowned. The first time when I wanted to stay there, it was full. The second time I booked a suite at the far end nearly ten days in advance. I was allowed to bring along a little friend. 15-year old Gu Yu took the big sofa in the living room like a lord, overlooking the "old sight" of the city outside the glass window. The master bedroom was also quite characteristic of Indigo: simple, direct, detailed and quite appealing, with one-way visibility glass windows.

一厅一房，相当奢侈，在香港这么“寸土尺金”的地方，更是令人佩服。

It was really admirable and luxury to have only one living room and one bedroom in Hong Kong where each inch of land is so valuable.

香港英迪格酒店

Indigo Hotel, Hongkong, China

苦于找不到信纸，只能用记事便签分开区域来画一下这么有趣、港味十足的,位于非凡热闹的旧城区的设计酒店（皇后大道东与太原街转角处）。

As I had difficulty finding any letter paper to write on, I had to draw designer hotel located in the exceptionally busy old urban area (at the corner of Queen's Road and Taiyuan Street), so fascinating and full of Hong Kong features, part by part, on some notepaper.

一层只有8间房，要住的确是要提前订的啊！顶层更有半露台酒吧和全港唯一一个玻璃泳池，潜入水中，鸟瞰城市，哈哈。

There are only eight guestrooms on one floor. You have to book early if you really want to stay there! On the top floor there is a semi-terrace bar and the only glass swimming pool in Hong Kong. Once in the water, you can have an air view of the whole city. Haha!

真可谓，吓死宝宝了！

It really scares me to death!

FOUR POINTS BY SHERATON

Shenzhou Peninsula

神州半岛福朋酒店

19

★★★★★

FOUR POINTS BY SHERATON SHENZHOU PENINSULA HAINAN CHINA

Address : Shenzhou Peninsula Resort District Wanning, Hainan Province China
中国海南省万宁市
神州半岛旅游度假区

Telephone : +(86 898)6253 8838

Fax : +(86 898)6253 8839

Http : //www.fourpoints.com/shenzhoupeninsula

神州半岛福朋酒店

Four Points By Sheraton Shenzhou Peninsula, Hainan,China

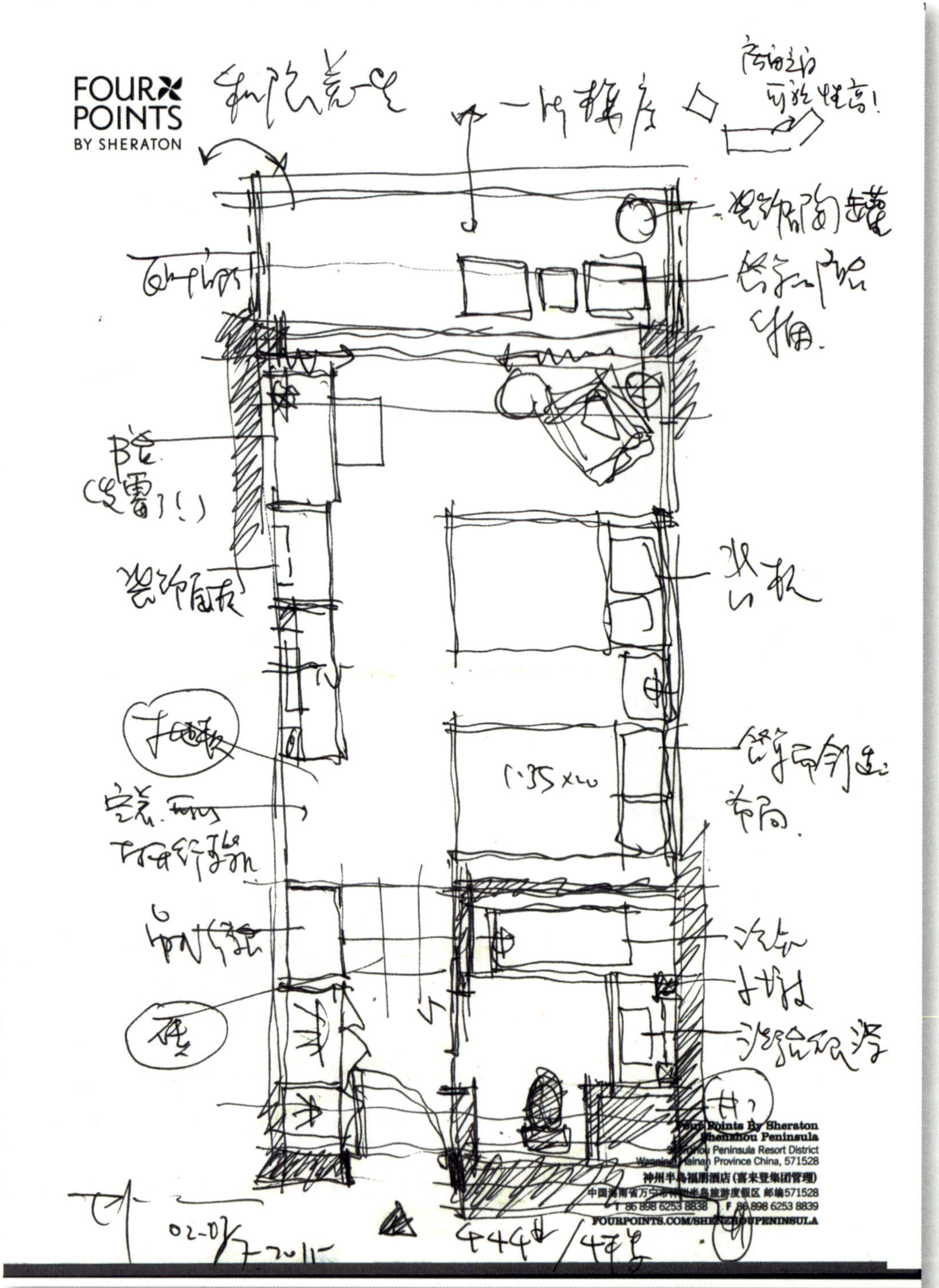

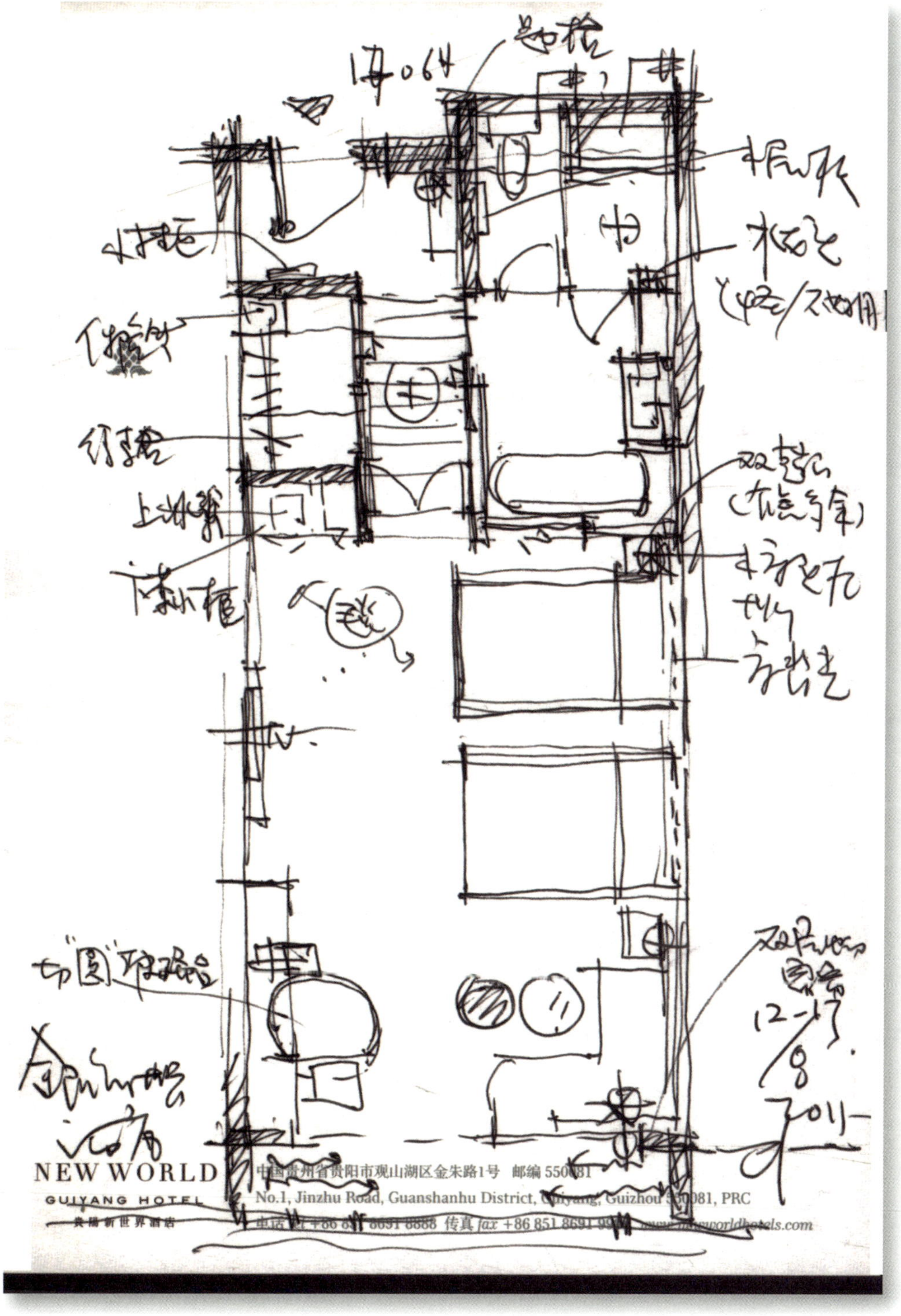

NEW WORLD
GUIYANG HOTEL
贵阳新世界酒店

20 贵阳新世界酒店
★★★★

NEW WORLD GUIYANG HOTEL GUIYANG CHINA

Address : No.1,Jinzhu Road,Guanshanhu District,Guiyang, Guizhou China
中国贵州省贵阳市观山湖区金朱路1号
Telephone : +(86 851) 8691 8888
Fax : +(86 851) 8691 9988
http : // www.newworldhotels.com

Surprise + Stupid（贵阳新世界酒店） 13/8, 2015

是日晚餐，在朋友家门口小馆，使得和张总、张总到"新世界酒店"去坐坐一下。十点到酒店，开始又入住进门！

哇！酒店大堂的玻璃房，倒是很像市内的酒店，如果没有附加到世——原来这里房间的玻璃窗不知什么时候全爆，细小碎玻璃四处，到处都是，这等等，幸亏不是在我们入住时发生的，不然后果不堪设想！（第二天知道天津大爆炸事件，应该有关吧！）

还好，马上换了一间好的房间，入住，试用后，这里房间很大，那边很宽，觉得像个隐藏很多人的空间的感觉。设计、用材、用色、用灯都是很超前的设计，而我到有点像一场梦，感觉出乎意料

NEW WORLD
GUIYANG HOTEL
貴陽新世界酒店
中国贵州省贵阳市观山湖区金朱路1号 邮编 550081
No.1, Jinzhu Road, Guanshanhu District, Guiyang, Guizhou 550081, PRC
电话 *tel* +86 851 8691 8888 传真 *fax* +86 851 8691 9988 *www.newworldhotels.com*

很多、很多、很多的建筑一定也用在这里。到了这样美丽，相信钱是花了不少。但感觉很一般，[illegible]个"四宝村"很好呢！

到贵阳后，更加觉得 stupid：很多的、布局不好。特别是酒店，天花板不能用（这面有酒店[illegible]），要一些有意花边、花边少点（比如挂旗都不如），钱用错了地方。这些更加大的[illegible]酒店、小细节体验像人的地方。（好似太象征了、太酒店一样）

酒店不是没做[illegible]，但我们是人，所以更需要[illegible]，不要太主观[illegible]，除非做到很厉害！

相信我们一定是没有来做过体验过的，不然不会这么[illegible]的！

[illegible]

NEW WORLD
GUIYANG HOTEL
貴陽新世界酒店
中国贵州省贵阳市观山湖区金朱路1号 邮编 550081
No.1, Jinzhu Road, Guanshanhu District, Guiyang, Guizhou 550081, PRC
电话 *tel* +86 851 8691 8888 传真 *fax* +86 851 8691 9988 *www.newworldhotels.com*

贵阳新世界酒店 *New World Guiyang Hotel, Guiyang, China*

The Stupid Surprises

Surprise + Stupid（惊喜得很愚蠢）

到贵阳出差，有几个客户的项目，住得比较远，专门选了“新世界酒店”体验一下，十点到酒店，开好双人房，进门！

On a business trip in Guiyang, I had several clients with outlying projects. I especially chose the New World Hotel for a new experience. At 10 I arrived at the hotel. I opened the guestroom for two and entered!

哗！满地的碎玻璃，住了多年的酒店，从来没有遇到过——原来是淋浴间的玻璃门不知什么时候完全碎了，钢化玻璃四飞，到床头、浴缸、走道等等。好惊险，幸亏不是在我们入住后发生的，不然后果不堪设想！（第二天知道天津大爆炸事件，愿逝者安息！）

Crunch! The floor was strewn with broken glass. For my many years of hotel-staying experience, never before had I seen anything like this—it turned out that God knows when the glass door had broken, sending toughened glass pieces in every direction, to the bed, the tub, the passage and everywhere. What a shock! Luckily it had not had happened when we were there, otherwise the consequences had been unthinkable! (The next day we learned about the Tianjin explosion incident. May the deceased rest in peace!)

还好，迅速换了一间好的房间，入住。试着这里翻翻，那里看看，总觉得像小孩、穿戴大人的衣服鞋帽的感觉，设计、用材、用色、用灯都是“有超前的意识，而控制得一塌糊涂”，感觉发展商是想很多、很多、很多的“好”想法一次用在这里，多了，适得其反。相信钱是花了不少，但感觉很一般，所以是一个“四星标准”呢！

It was no so bad as we were quickly changed into another unaffected room. Once in, I began to file through it here and there. It felt like a child wearing adults' shoes and clothes. The ideas behind the designs, materials, colors, and lamps were far ahead of this age but totally out of control. It felt like that the developer had wanted to apply too many "good" ideas here, only to achieve the opposite. I believe he had been free with his money, yet the effect was mediocre. And it was rated as "four stars".

到使用洗手间，更加觉得“stupid”，很多灯，布局怪。特别是淋浴间，天花花洒不能用（选配与设备维护问题），要一直拿着花洒，花洒水少（比撒尿都不如）。钱用错了地方，应当更加专心地花在设备、水压体验等等的地方（如万达，土豪，但水压、选设备都一流）。

The bathroom felt more stupid: the lamps were numerous, the layout queer, especially the shower compartment—the gondola water faucet would not work (due to problems of fitting and equipment maintenance), so the faucet had to be held all the time; too little water came out of the faucet (worse than when one was urinating). The money should have been spent attentively on other things, such as equipment, water pressure (Wanda, rich as it is, has first-class water pressure and equipment).

（住）客户不是设计师，我们只关心所感所受的东西，不要以太主观的做法去强加市场，除非你真的很厉害！

The guests and customers are no designers. What we care about is what we feel. It is advisable not to impose too subjective ideas on the market, unless you are really formidable!

相信业主一定是没有来住过、体验过的。不然不会是这样的！

I am sure the property owner had never been there to experience it, otherwise it would not have been like that!

INTERCONTINENTAL
KUNMING
昆明洲际®酒店

昆明洲际酒店

21

★★★★★

INTER CONTINENTAL KUNMING CHINA

Address : No.5 Yijing Road,National Tourism Area of Dianch, Kunming,650228 Yunnan Province,P.R.China
中国云南省昆明市滇池国家旅游度假区怡景路5号

Telephone : +(86 871)6318 8888

Fax : +(86 871)6318 6688

Http : //www.intercontinental.com

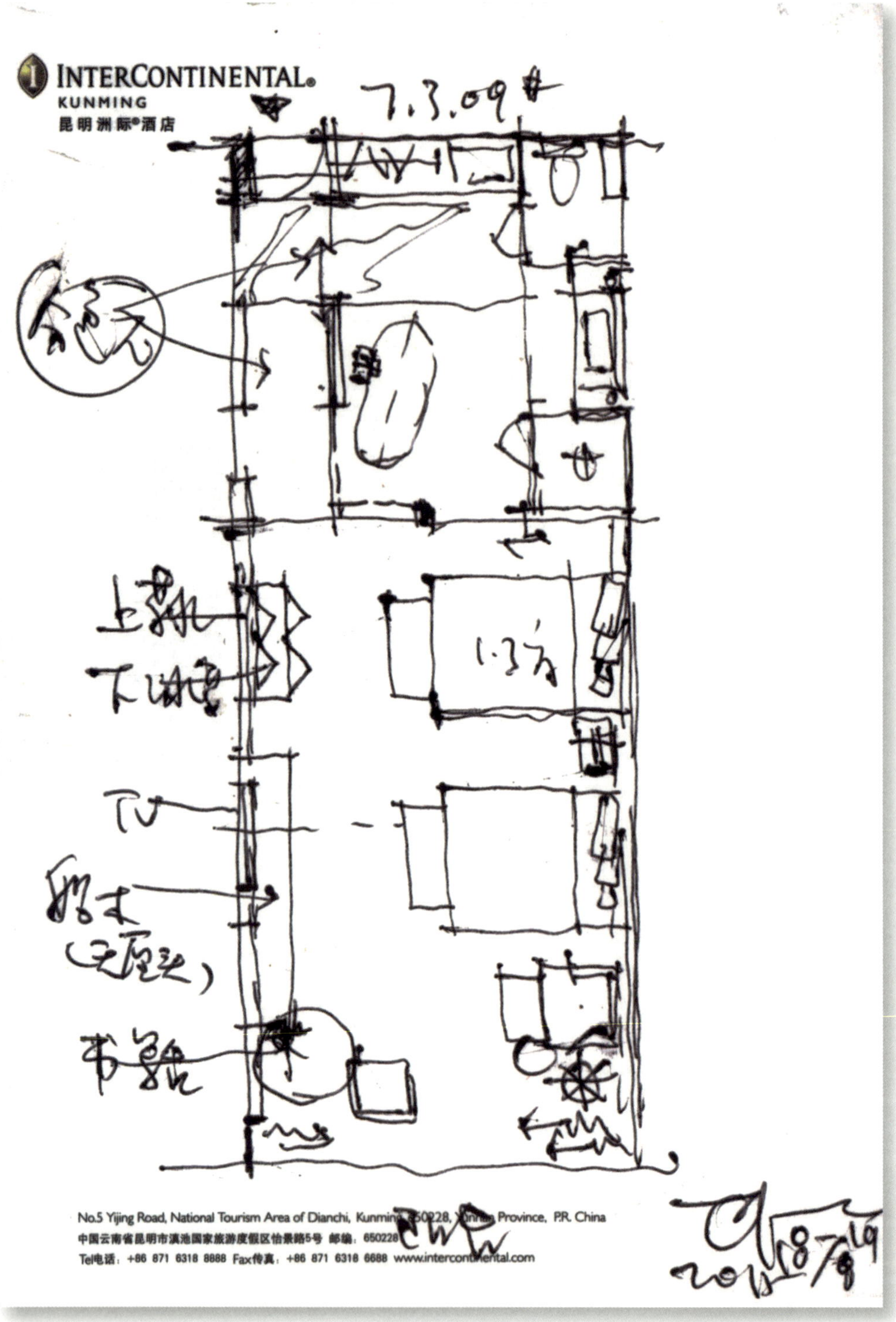

昆明洲际酒店

Inter Continental, Kunming, China

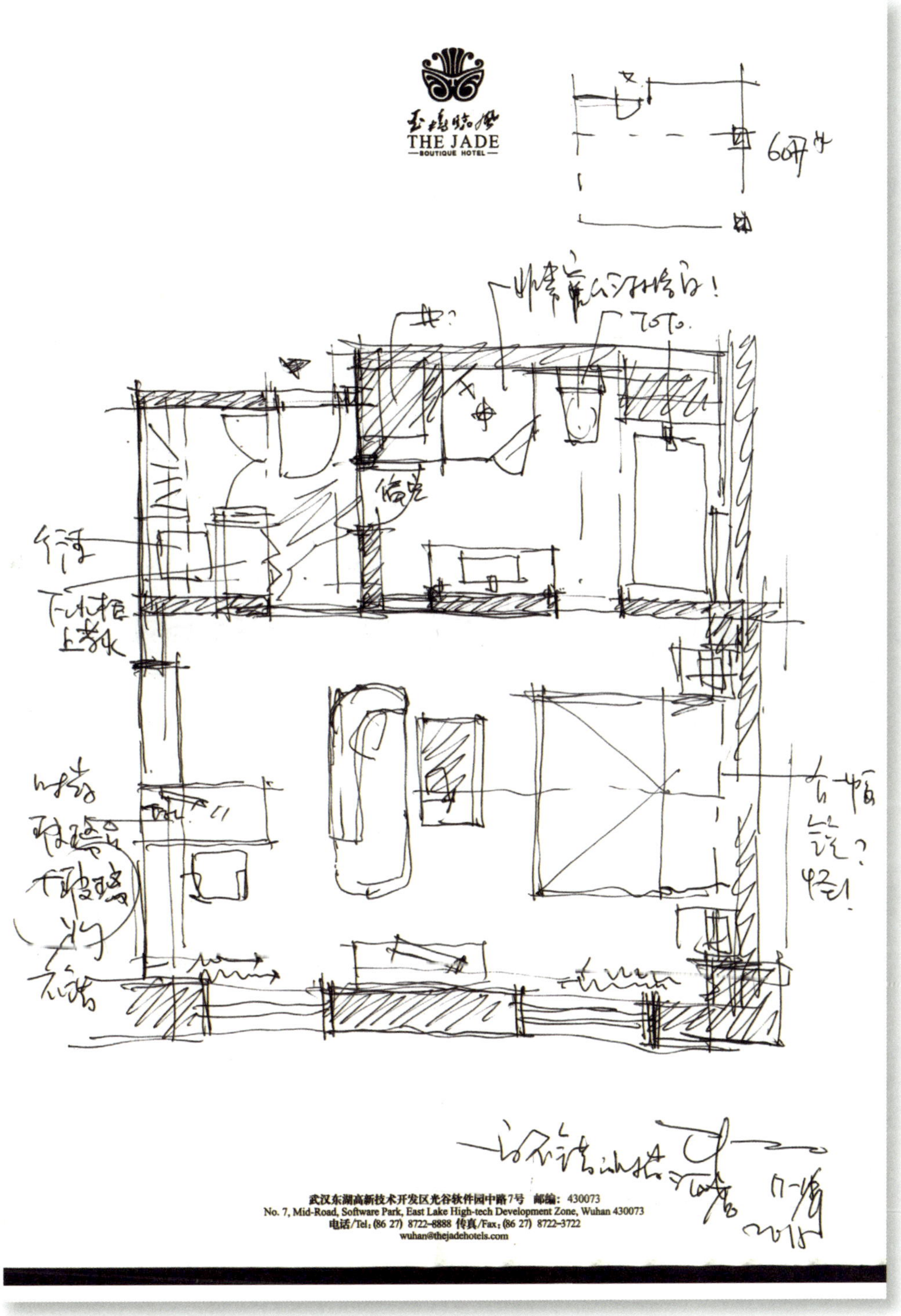

THE JADE
BOUTIQUE HOTEL

武汉玉树临风精品酒店

22 ★★★★★

THE JADE BOUTIQUE HOTEL WUHAN CHINA

Address : No.7,Mid-road,Software Park, East Lake High-tech Development Zone,Wuhan China
武汉东湖高新技术开发区光谷软件园中路7号

Telephone : +(86 27) 8722 8888

Fax : +(86 27) 8722 3722

E-mail : wuhan@thejadehotels.com

发达的精品酒店。

这一次住过的七八个酒店中，印象深刻的是可以归纳为精品、设计型酒店的，但感觉用的"套路"不同。

美国纽约的11 Howard（霍华德11号）材料的力、有格调，深灰色的楼梯扶手的金属铜色，品质细腻；卫浴五金特别，就是精致，设备高级，灯饰精致，不是土豪的金碧辉煌，有种上面的贵族气质；上海的艾本（URBN）酒店注重的是体验感，娱乐性，去强调生态艺术的风格；而璞丽酒店则是在努力塑造对历史的传承和文化的表达，连每一个房间的名字都十分讲究，当然在这么有年份的旧别墅的基础上装修下，处理得很优雅、内敛，为中国人的精神找到的回归情怀。

相反，那家入住位于武汉光谷软件园的封帕风酒店，第一眼非常抓眼球，但下来就会失了深度的感觉。酒店客房的横向布局有一定的建筑优势，一字横成排，直接，算是可以手与当代木桩的排布方式有优化的可能，设计的非常完整，但还是对这类档次的酒店而大部分的保留，使得相对而有人的使用，特别是长住客，使用绝对越零，变成了装饰。这里面有休闲区、会客区、工作区，大大的床，后面有……

横的大镜，成娱乐的）我只享用了几个样子，这个房子之一面积对于我真是浪费，想着精品店设计理念真到了做减法的时候了。

但是可以，增加思考的地方，也算是不错的“好地方”。

Discovery of Quality Hotels
发现精品酒店

这一次住过的70多个酒店中有不少是可以归纳为精品、设计型酒店的，但感觉用的“套路”不同。

Of the over 70 hotels I have stayed in, many can be classified as finely designed, yet in different ways.

美国纽约的11HOWARD（霍华德11号）有张力，有格调，深灰色配搭小小的金铜色，以品质取胜：床好，床品特别，家具精致，设备高级，灯独特设计，不起眼的高投入，骨子里的贵族感；上海的艾本（URBN）酒店注重房间的体验感、娱乐性与现代时尚艺术的混搭；而隐居繁华雅集公馆则注重对历史的传承和文化的表达，连每一间房的名字都十分的考究，当然在这么有年份的旧别墅的基础背景下，处理得优雅、内敛，如中国人的性格般的细思慢嚼。

The 11 Howard in New York, impressive, stylish, with mouse color going with little golden bronze, excelled in quality: the bed was good, complete with special bedding; the furniture was exquisite; devices advanced; lamps uniquely designed; unremarkable high input and inborn noble feeling. URBN Shanghai emphasized how the guestroom felt, and how recreation blend with modern fashion arts. The Bamboo Retreat, in comparison, placed an extra emphasis on continuity of history and expression of culture: each of the rooms were specially named; of course, with such a time-honored villa, everything was designed to show elegance, reservation, to be appreciated with patience like the character of Chinese people.

武汉玉树临风精品酒店
The Jade Boutique Hotel, Wuhan, China

武汉玉树临风精品酒店
The Jade Boutique Hotel, Wuhan, China

相反，深夜入住位于武汉光谷软件园的玉树临风酒店，第一眼非常抓眼球，住下来就会过了激动期。细究这房间，横向布局有一定的建筑优势，一条横线简单，直接紧凑的行李与茶水柜的排布方式有优化的可能。洗手间非常宽敞，但还是对这类档次的酒店配大浴缸有保留，住客极少可放心使用，特别是女住客，使用率绝对趋零，变成了装饰。这里配套非常齐全，写字区、悠闲区、大大的床（居然床背有一幅横向的大镜子，颇娱乐的）。我只享用了几个小时，这么多元的配套对于我真是浪费，想想精品及设计型酒店真到了做减法的时候了。

Contrary was the Jade Boutique Hotel in Wuhan Optical Valley Software Garden. On first sight, it was quite eye-grabbing. Once in, your excitement would be soon gone. A close study showed that the transverse layout gave it some advantage: a transverse line was simple enough; the direct and compact arrangement of luggage and the tea cupboard left room for improvement; the bathroom was quite spacious, however, guests have scruples about the big bath tubs in these luxury hotels and seldom felt safe to use them—especially for women guests, their use was near zero, relegating them to the status of ornaments. The guestroom was complete with a writing area, a leisure area, a very big bed (There turned out to be a large horizontal mirror at the bed back-fun). I spent only several hours there. For me it was a waste of so many components. Thinking it over, I believe it is high time for quality product and design-oriented hotels to do some reductions.

住住可以，增加思考的地方，也算是不错的“发现之旅”。

Those hotels are worth staying as they extra places to think over things. And the trips were fairly good "journeys of discovery".

北京柏悦酒店

PARK HYATT BEIJING

北京柏悦酒店

23 ★★★★★

PARK HYATT BEIJING CHINA

Address : 2 Jianguomenwai Street, Chaoyang District Beijing,China
中国北京朝阳区建国门外大街2号

Telephone : +(86 10)8567 1234

Fax : +(86 10)8567 1000

Http : //www.beijing.park.hyatt.com

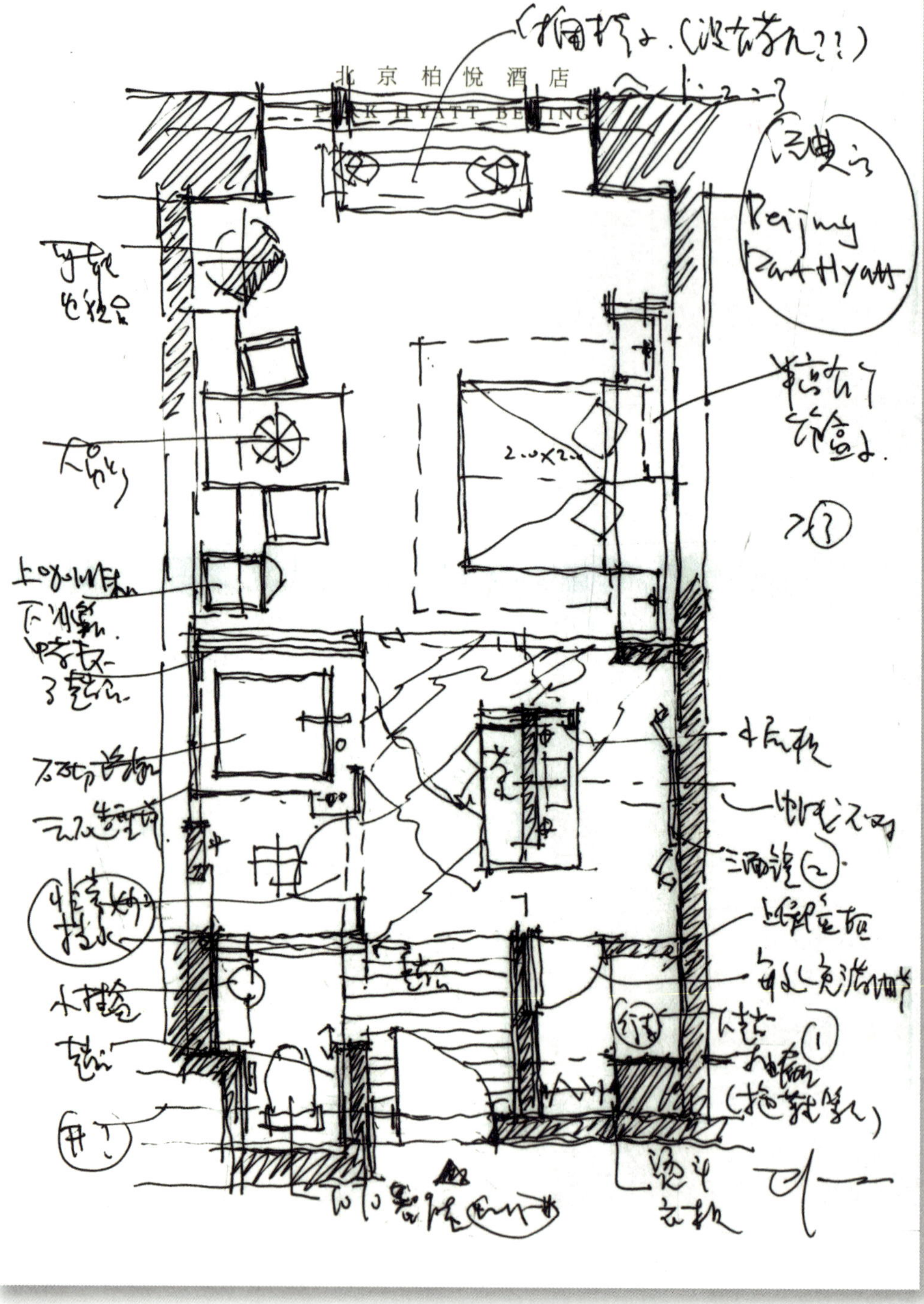

经典之所以是经典，恰恰过了，旧一点而已　19/9, 2015

北京柏悦酒店

PARK HYATT BEIJING

很久没有写一些关于入住酒店的体验了，我感觉新的酒店越开越多，越开越快，而且信息化的普及，设计师们的创意平台化，让"个性""创新"成为难事，倒是反过来愈想念这一些"经典"，它带有那些年代的当年特别风味的东西，还是在传统细节上下功夫，我只好回忆一下这些我们不可忽视的"老酒店"。

人老了，会闻酒，有味道了。酒店久了也会这样，木色变旧了，墙纸变暗了，不过发亮，地毯也会一块儿不平了，但这正是它有的价值，也是它成为经典的原因，应当也会好。

恰恰过了，只是旧一点而已，经典的还是经典！

巴

北京柏悦酒店 *Park Hyatt, Beijing, China*

Classics are Classics, Only a Little Old over Time

经典的还是经典，时间过了，旧一点而已

很久没有写一些关于入住酒店的体验了，或者是新的酒店越开越多，越开越快，而且由于信息的偏平化、设计师的公开平台化，让“个性”、“创新”成为难点。倒是反过来看看这一些“经典”，应当有相当年份的这间当年吸引眼球的柏悦，还是值得动动手、动动口去评价一下，这种住了不少次的“标准房”。

It is a long time since I wrote about my hotel-stay experiences. Maybe what with the increasing number and quickness of new hotels opening, and the mediocrity of information, the openness of the designer platforms, it becomes difficult to be "individual" and "innovative". Look back at these "classics", which are rightfully time-honored. Take Park Hayyt, eye-grabbing back then—this kind of "standard guestroom" I have stayed in for many times is still worth writing about and commenting on.

人大了，有了阅历，有味道了，酒店原来也会这样，木色变黄了。墙纸变柔润了，石头发白了，木地板也有一小点儿不平了，但这正是它存在的价值，也是它成为经典的痕迹，应当也合时。

With age and experience a person gains in taste. So does a hotel. The woodwork is yellowing, the wallpaper softening, the stone whitening, and the wooden floor is getting a little uneven. It is just the worth of its being and the trace of what made it a classic. It is reasonable and in step with time.

时间过了，只是旧一点而已，经典的还是经典！

Outdated as it is, the classic is still the classic, only a little older.

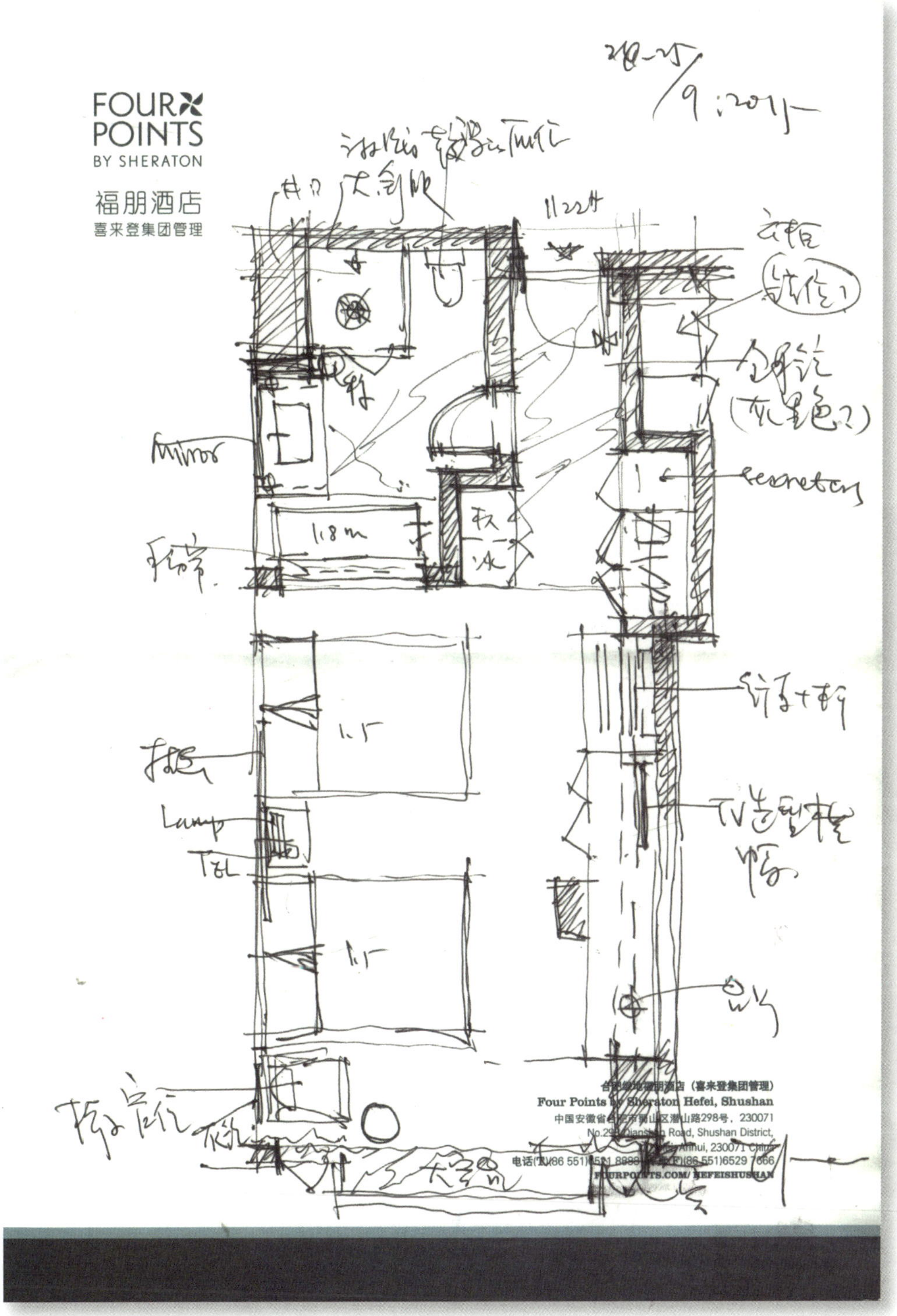

FOUR POINTS
BY SHERATON

合肥绿地福朋酒店

24 ★★★★★

FOUR POINTS BY SHERATON HEFEI SHUSHAN CHINA

Address : 298 Qianshan Road, Shushan District Hefei,AnhuiChina
中国安徽合肥潜山路98号蜀山区
Telephone : +(86 27) 8722 8888
Fax : +(86 27) 8722 3722
http : //www.starwoodhotels.com

合肥绿地福朋酒店
Four Points By Sheraton, Hefei Shushan, China

合肥香格里拉大酒店
Shangri-La hotel
HEFEI

合肥香格里拉酒店
★★★★★

25

SHANGRI-LA HOTEL HEFEI CHINA

Address :No.256 Suixi Road,Luyang District,Hefei,Anhui Province,China
中国安徽省合肥市庐阳区濉溪路256号
Telephone : +(86 551)6550 9888
Fax : +(86 551)6550 9666
Http : //www.shangri-la.com

合肥香格里拉酒店
Shangri-La Hotel, Hefei, China

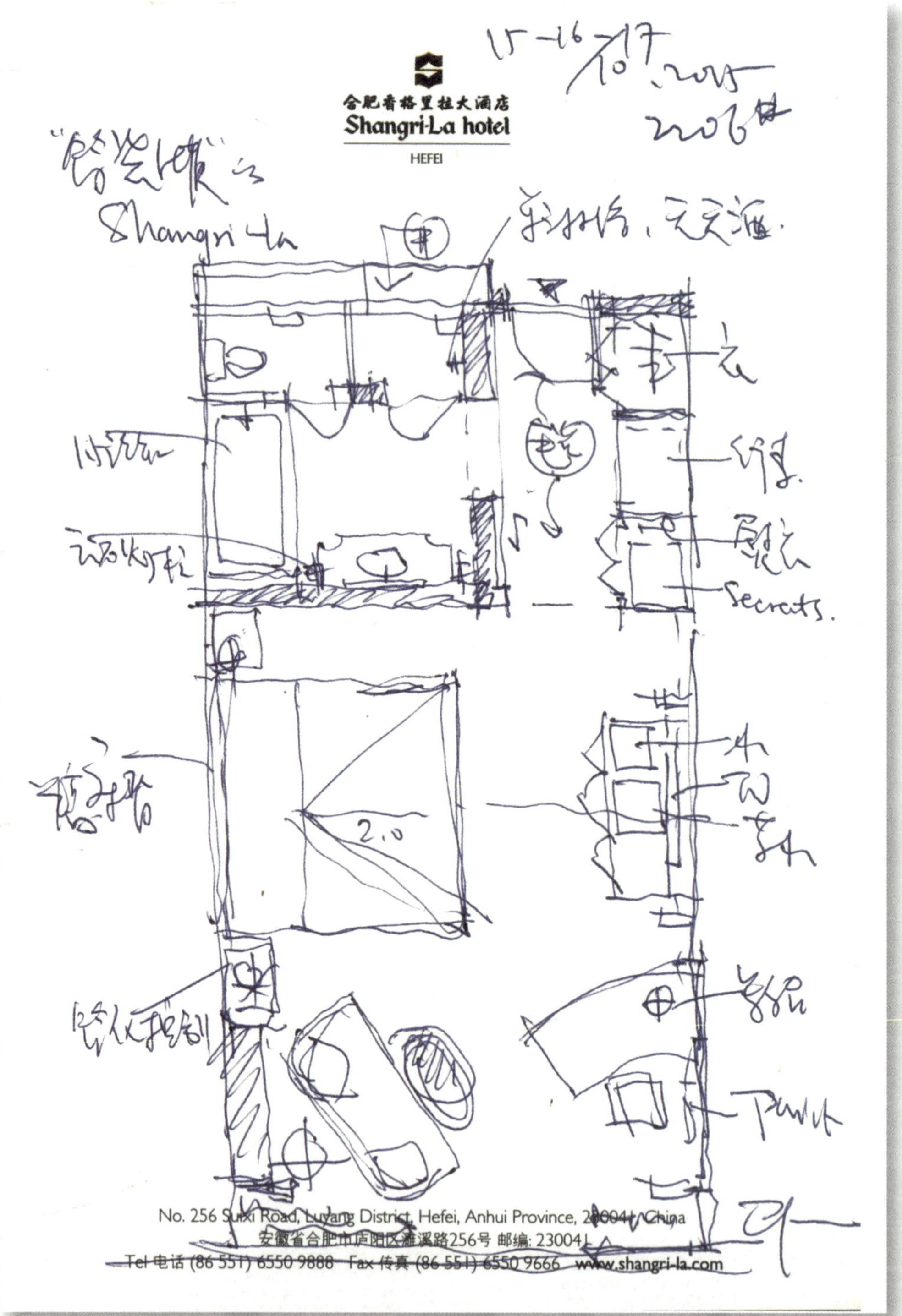

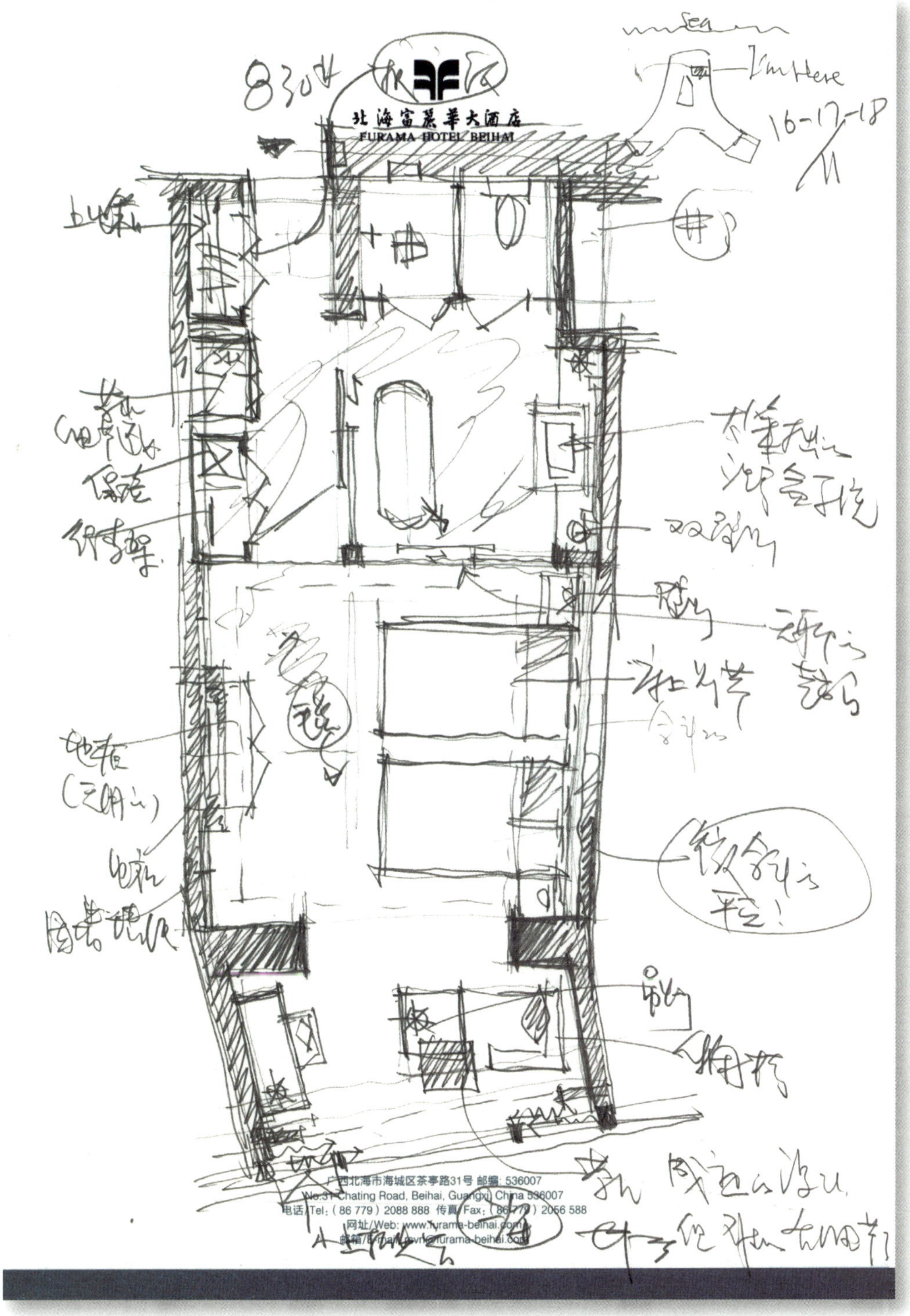

北海富麗華大酒店
FURAMA HOTEL BEIHAI

北海富丽华大酒店

26 ★★★★★

FURAMA HOTEL BEIHAI CHINA

Address : No.31 Chating Road, Beihai,Guangxi,China
中国广西北海市海城区金茶亭路31号

Telephone : +(86 779) 2088888

Fax : +(86 779) 2056588

http ://www.furama-beihai.com

北海富丽华大酒店
Furama Hotel, Beihai, China

URBN

HOTELS & RESORTS

上海艾本酒店

★★★★★

27

URBN HOTELS & RESORTS SHANGHAI CHINA

Address : NO.183 Jiaozhou Road,
(Near West Beijing Road),
Shanghai ,China
200040
中国上海市胶州路
183号（近北京西路）

Telephone : +(86)5153 4600

Fax : +(86)5153 4610

Http : //www.urbnhotels.com

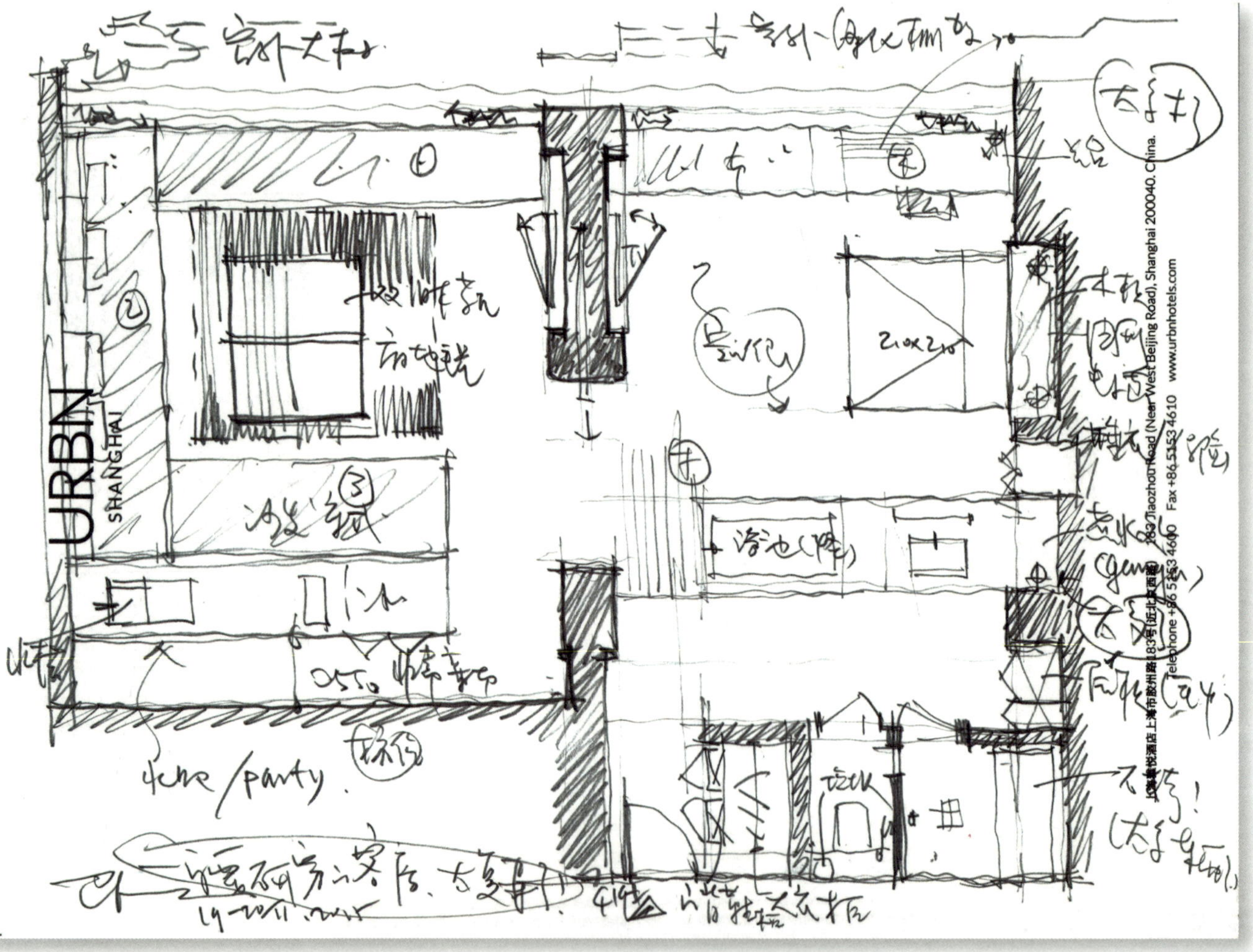

一个房间睡几个人

忘记了怎么会这么聪明地选了这一家在市中心（胶州路）我找了很久才发现入口的酒店，果然是太"大隐于市"，只是一个反光不锈钢的球就是了。URBN，中文名字叫作：艾本，不知道你笨还是我笨。同住着这房，我可是住了这所谓样板的"娱乐套房"

太大了，空位太多了，睡都不知道站在哪里好！沙发太多了，厕所都嫌小，更不要说什么卫浴造型；但是大床2米，显得多余，放在中心位置的不三不四的浴缸，不知如何搞。想象一下：一人泡澡，两个人同时淋浴，一个坐厕所，两个人在床上，两个在沙发也就是在找钟，我唱爬山虎（为事得的），或听鸟，或说书；另一些人则一个在客厅酒吧调酒，"吧型"的沙发，每一边坐我那里是两个，这样算起来，最多都可以容纳十四五个人了，真正的party房，两千多一晚住这么多人还是有点便宜的，不知道是否被允许。也只好在这里说话题，通宵不睡觉就可以了。每一个人都要记得身份证吧，群宿。

像我一个人，真的是不知要在哪里坐好了，可以沙发睡一会，坐起来一会，洗个脸一会，书摊开一会，冲茶喝一会，写稍蹲一会，衣柜里翻一会，猫猫，这样地过一夜了。早上七点才去床上蹲一会。

洗之后，在花园餐厅吃上一顿美美的丰盛早餐。

下午带着一伙人，热闹不消说。

How Many can be Accommodated in One Room?

一个房间能住几个人

忘记了怎会这么聪明选了这一家在市中心（胶州路）而找了很久才发现入口的酒店，果然是太“大隐于市了”，吊着一个反光不锈钢球的就是了。URBN 中文名字难听：艾本，不知是你笨，还是我笨。同事住普通房，我可是住了这个近百平方米的“娱乐套房”。

I forget how I had cleverly chosen this hotel—it was located in the city center (Jiaozhou Road) but took a long time to find its entrance. It turned out to be the one hung with a glistening stainless steel sphere—"a reputed hermit in an urban retreat". The Chinese translation of URBN sounds awkward: Are you stupid or I ? My colleague chose a common guestroom and I checked into this near 100m² "recreational suite".

太大了，空位太多了，脚不知道站在哪里好！沙发太多了，屁股都嫌小，更有这么多的靠垫；倒是大床，2米，显得多余，看着放在中心位置的浴缸，不知所措。想象一下：一个人泡浴，两个人同时淋浴，一个蹲厕所，两个人在床上，两个在窗边矮座，或看书，或嗅爬山虎（有果子的），或听鸟，或望树；另一波人则：一个在客厅酒吧调酒，“U”形的沙发，每一边坐或躺至少两个，这样算起来最起码可以容纳十四五个人了。真正的Party 房，两千多一晚住这么多人还是相当的划算的，不知道是否被允许，也挺有趣的话题，通宵不睡觉就可以了，每一个人都登记身份证吧，群宿。

It was too large. There was too much room-I wondered where to put my feet! The sofas were too numerous—my bottoms were too small for them; so many back cushions. The 2meter bed seemed unwanted. Looking at the out-of-place bathtub at the center, I felt bewildered. Imagine: one bathing in the tub, two showering at the same time, one using the toilet, two in the bed, two sitting before the window or reading books, or smelling at ivies laden with fruits, or listening to birds chirping, or looking at the trees; others: one mixing alcoholic drinks at the bar, at least two sitting or lying on either side of the U-shaped bench. Like this, the room could accommodate at least 14-15 guests. It was a party room in the real sense. It was worthwhile for so many to spend 2000+ yuan for one night there. I wonder if it is allowed. It was a quite intriguing subject. It would do if they wished to stay awake all the night. Each is supposed to register with their own residential ID when a group shares the same room, OK?

上海艾本酒店
Urbn, Shanghai China

像我一个人，真的是不知待在哪里好了，可以沙发睡一会，坐垫靠一会，浴缸躺一会，书椅待一会，淋浴拖一会，马桶蹲一会，衣柜躲一会猫猫，这样就过一夜了。早上七点才去床上蹭一会。

Alone, I really wondered where to stay: I sat on the sofa for a while, lay against the back cushion for a while, lay down in the bathtub for a while, settled down in the study chair for a while, lingered in the shower compartment for a while, sat on the toilet for a while, and hid in the wardrobe for a while. In this manner I spent the night, and it was not until seven in the morning that went to bed to catch forty winks.

洗洗脸，在花园餐厅吃上一顿美美的有机早餐，下次要带上一队人，热闹不浪费。

After washing my face, I had a wonderful organic breakfast in the garden restaurant. Next time, I would come with a team so that the stay would be fun and no waste.

上海艾本酒店
Urbn, Shanghai, China

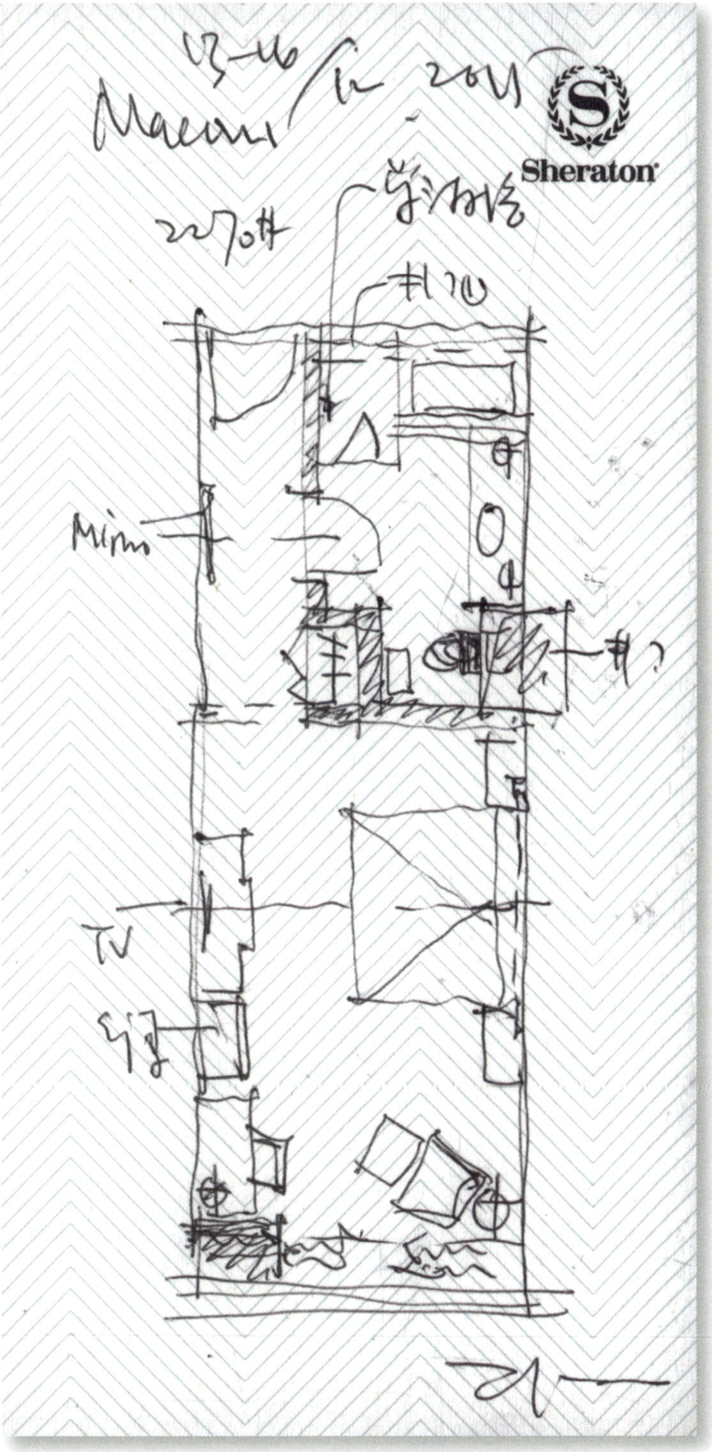

Sheraton

澳门喜来登金沙城中心大酒店

28 ★★★★★

SHERATON MACAO HOTEL COTAL CONTRAL MACAO CHINA

Address : Estrada do Istmo.
s/n, Cotai, Macao, China
中国，澳门，路氹
路氹连贯公路

Telephone : +(853) 2880 2000

Http : //www.sheratongrandmacao.com

Casino（赌场）酒店的客房：吴长一汪雄

衣柜很小，洗手间很大，房间很舒适，与酒店的赌场区的奢华不同，房间非常安静，走廊可以跑步的长，没有自然的采光，让你不分白天黑夜，对，这也是令人无法分辨或是黑夜，赌场的基本特征了——"暗无天日"。

大学同学毕业廿周年聚会，选择了澳门的新葡京酒店，能在餐饮之余，体验一下赌场的客房，房还好，但相比较老，电视节目也丰富多彩，总的说吧，这是一间赌场酒店里面的客房，太舒服了，我都不去赌场"搏杀"了，那我靠什么活啊！

祝健康！

汪之雄

In the Casino Guestroom You Need Only One Bed

Casino(赌场)里面的客房：只求一床睡

澳门喜来登金沙城中心大酒店
Sheraton Macao Hotel Cotal Contral, Macao, China

衣柜很小，洗手间很大，房间很简陋，与酒店赌场区的奢华不同，房间非常的多，走廊可以跑步的长，没有自然的采光，让你不分白天黑夜，对，应当是分不清白天或是黑夜，赌场的基本特征 ——“暗无天日”。

The wardrobe is very small and the bathroom is very large. The room is very crude, without the luxury of the gambling sector of a hotel. There are many rooms and the corridor is long enough to run along. There is no natural lighting, making it impossible to distinguish day from night as it should be—it is typical of a casino.

大学同学毕业25周年聚会，选择了澳门的喜来登酒店，不错的西餐之余，体验一下简简单单的床。床还好，但热水充足，电视节目也丰富多彩，忘记吧，这是一间赌场酒店里面的客房，太舒服了，就不去赌场“搏杀”了，那酒店靠什么活啊！

For the gathering party of our former classmates separated from each other for 25 years, we chose a Sheraton hotel in Macao. After a quite nice western-style dinner, we had a taste of these simple beds. They turned out OK. There was more than enough supply of hot water and a rich choice of TV programs. Forget about it—it is a guestroom of a casino hotel. The bed felt too comfortable. When you are not gambling in the casino, out of what will the hotel make a living?!

天哪，床！

I rely on the bed!

洗洗睡。

I will have a bath before turning in.

the UPPERHOUSE

香港奕居酒店

★★★★★

29

THE UPPER HOUSE HONGKONG CHINA

Address : Pacific Place ,
88 Queenway ,
Hong Kong , China
中国香港
金钟道
88号太古广场

Telephone : +(852) 2918 1838

Http : //www.upperhouse.com

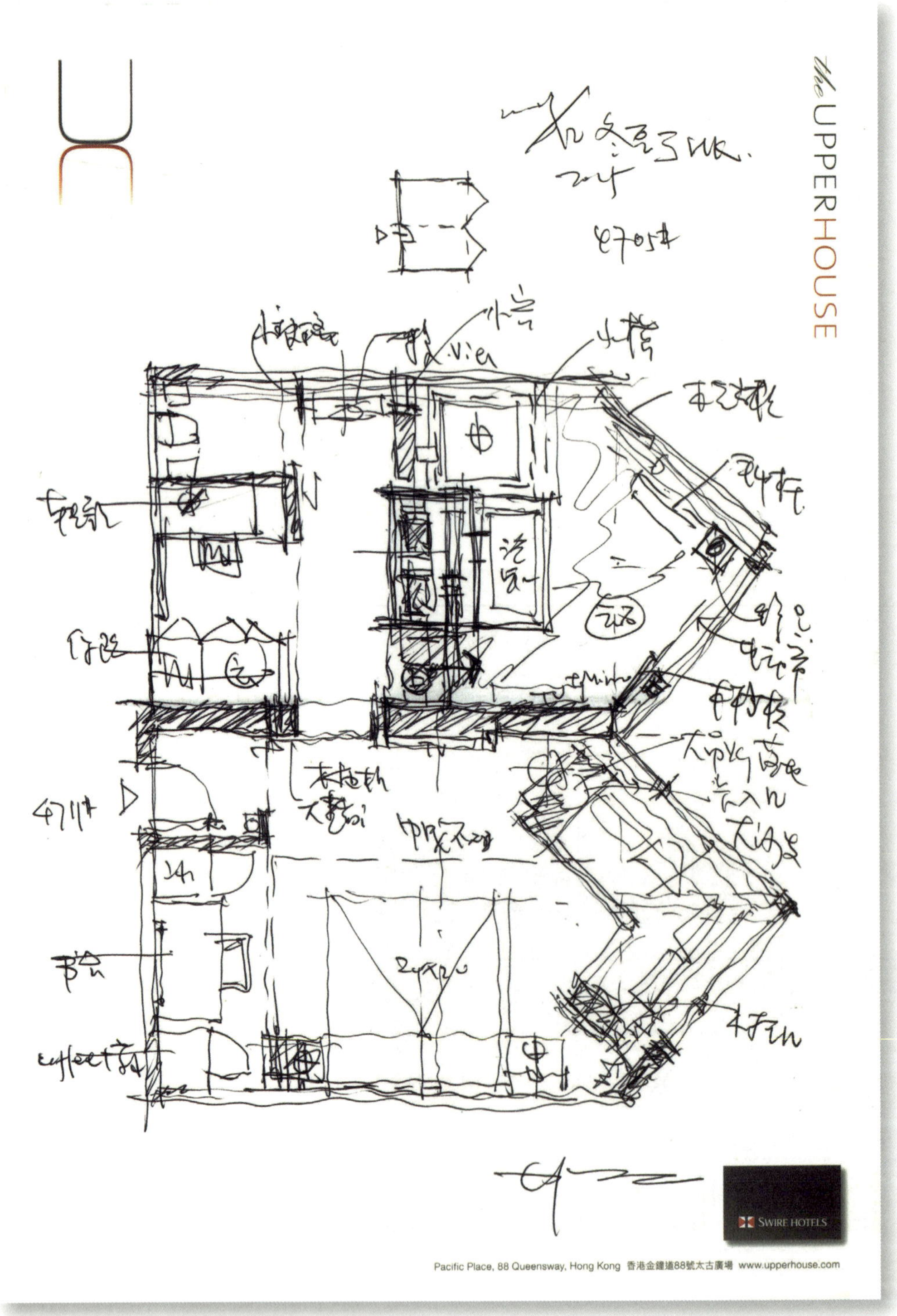

画着一个不完美的平面

作为设计师，有追求极致的完美心理是良好的。

香港 Upper house 可谓其中一个好样，被业内调侃为"完一代"的十余间偶像民宅之完美的设计。再次入住也就随手速写着画一下，对上次去时有不同的房型的思考（上次的见第一本《住胭脂的》P81）。还真不完美的居然让我涂改多次，但还是不美的对的比例。

画了一张如此优美的"研磨"出这个方案，多年前个性的合同所集体情感的……提合的地走让设计的解放了功能重量（我猜的）让厕浴合离，干湿分区，作为物料多样，更重要的是"挤"了不少的面积，与亮处搭上边，我认为美的更好地更合理地设计房的面积更是酒店的客房设计成功的关键！

一边画，一边玩味设计师的好玩，算是佩服了。

一个不完美的平面，让我再一次满足了心愿、手愿！

Drawing a Difficult Plan

画着一个不容易的平面

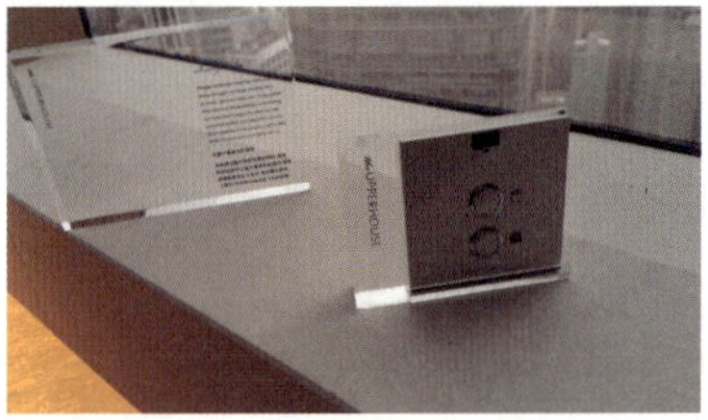

作为设计师，有追求极致的“变态”心理及行为。

Designers are mentally and behaviorally freakish pursuing perfection.

香港奕居酒店可谓其中一个标杆，被业内调侃为“富二代”的小鲜肉傅厚民负责的室内设计。再次入住也就随手想着重温一下与上次有同、有不同的房型的思考（上次的见第一本《住哪？》P81）。还真的不容易，居然让我涂改多次，但还是不怎么对的比例。

The Upper House in Hong Kong, a field rod, had its interior designed by young and handsome Andrew Fu, humorously nicknamed by fellow designers as “rich second generation”. Staying for the second time, I was considering whether the house types had changed since I patronized the hotel the previous time (see Where to Stay 1, P81) . it was not so easy to draw it—the drawing ended up with numerous alterations, and a scale not so correct.

香港奔居酒店

The Upper House, Hongkong, China

想想一张纸慢慢“研磨”出这么有趣、多变而个性的布局所体味的心路历程：提高的地台让洗手台解放了原来的位置（我猜的），让厕浴分离，干湿分区。储物丰富多样，更重要的是“耗”了不少的面积，与“豪华”搭上了边，我认为怎么更好地、更合理地浪费房间的面积，更是酒店的客房设计成败的关键！

Consider it: out of a sheet of paper had grown a course of thoughts embodied in the layout, so fascinating, changeable and individual: the raised platform liberated the wash platform from the former position (it is my guess), separating the toilet from the shower area, the wet area from the dry. Article stored were a rich variety. More important, quite a lot of area was "consumed" bordering on "luxury". I believe that the key to the success of a guestroom design is more about how to better waste the size of the room!

一边画，一边去体味设计师的“好玩”，算是佩服了。

Drawing and tasting the fun of being a designer, I was inspired with admiration for them.

一个不容易的平面，让我再次满足了心瘾、手瘾！

Drawing a difficult plan gave me mental and physical satisfaction again!

香港奕居酒店
The Upper House, Hongkong, China

Holiday Inn

沧州渤海假日酒店

★★★★

30

CANGZHOU BOHAI HOLIDAY INN CANGZHOU CHINA

Address : No.10,Beijing Road,
Yunhe District,
Cangzhou City ,
Hebei Province,China
061001
中国河北省
沧州运河区北京路10号

Telephone : +(86-0317)7566666

Fax : +(86-0317)7590000

Http : //www.urbnhotels.com

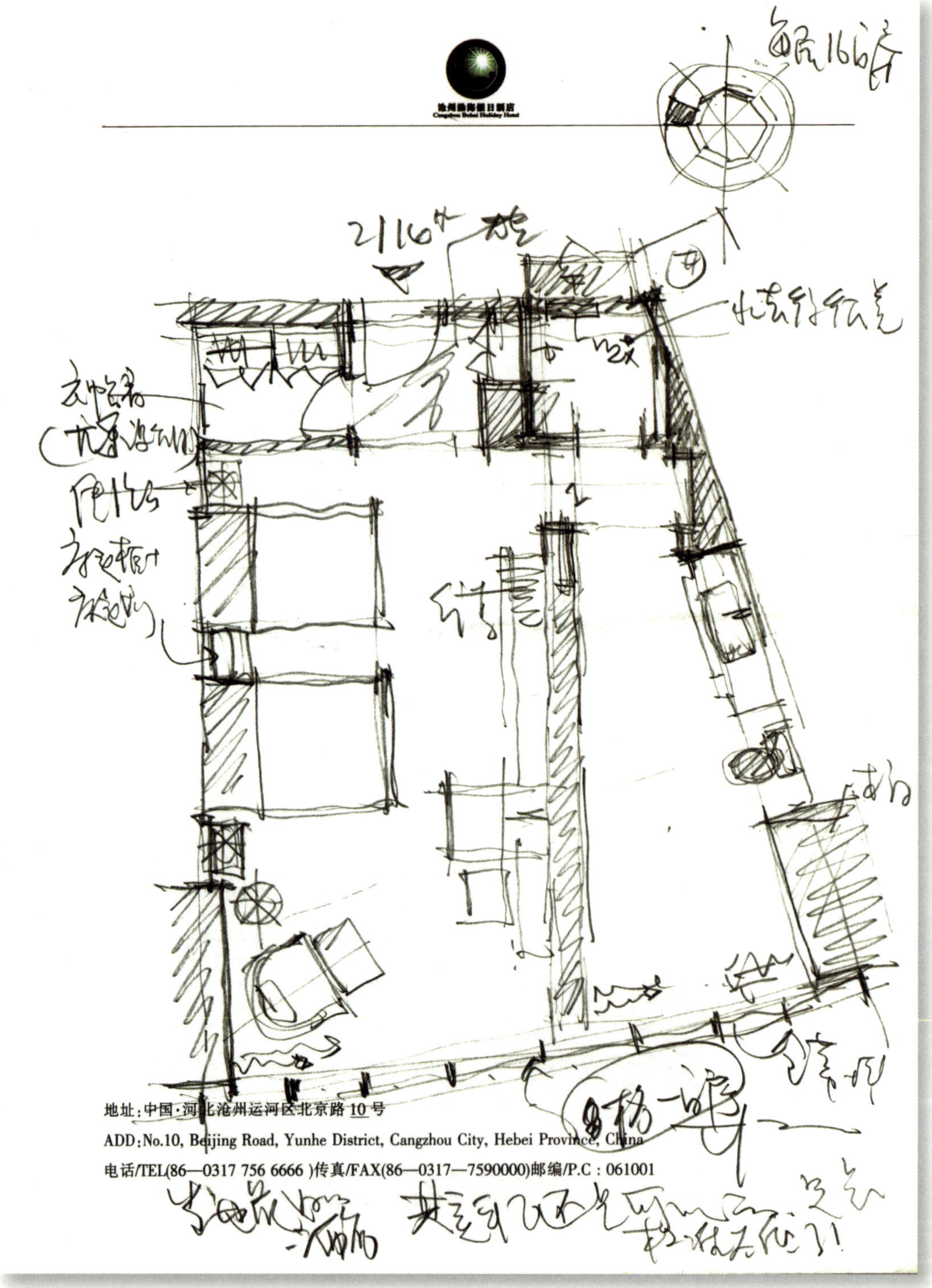

沧州渤海假日酒店
Cangzhou Bohai Holiday Hotel
地址：中国·河北沧州运河区北京路10号
ADD: No.10, Beijing Road, Yunhe District, Cangzhou City, Hebei Province, China
电话/TEL(86—0317 756 6666)传真/FAX(86—0317—7590000)邮编/P.C：061001

难得的"圆美"

做一个圆形的酒店标准层，比较不容易，房间会比较不经济，包括建筑、结构、外墙都是。不过在沧州这个不大不小的城市，也有这么一些特别的需求入住当地唯一一家酒店，一个"圆头圆脑"的酒店，渤海假日酒店。

如果跟我们设计的一样吗？要特别呢还是要打个东西还是一样的设计的东西呢？因为这个是双向的，设计的最后的结果是让"甲方"来接受了。我猜，难以圆美。你说能怎么样呢？

这么大的外侧平面，太少见了，我觉得还是让睡眠区有更大的平面为好。而设计的双方还是要在大的方向上有区别。家具的布置也是一种体验方式的创新。

Rarely Seen
Beautiful Rounded Guestroom
难得的“圆美”

做一个圆形的酒店标准房，以前不少，近年少见，原因首先是不经济，包括建筑、结构、外墙都是。不经意在沧州看一个不大不小的项目，业主热情招呼入住当地最好的一间酒店，一个“圆头圆脑”的酒店：渤海假日酒店。

Rounded standard hotel guestrooms, numerous before, are rarely seen these years. The primary reason is that they are not economical in terms of the building, structure and the outer wall. Visiting a project of average size in Cangzhou the other day I was accommodated by the hospitable property owner in the best local hotel, a "totally rounded" hotel: Bohai Holiday Inn .

如果让我们设计会一样吗？要特色呢，还是要档次，抑或还是一如既往的舒适为先呢？因为这个是双床房，洗手间靠窗的浴缸肯定让经营方“杀掉”了，我猜。难得圆美，你又能怎么样呢？

Should we do the same if we were the designer? What do we want? Character? Class? Or comfort first as usual? As this was a twin room, the bath tub close to the window must have been "eliminated" by the operator, as I guess. It was rare to have such a beautiful rounded hotel. What could you have done otherwise?

这么大的外侧采光，太纠结了，我觉得还是让睡眠区有更大的采光面为妙，而洗手间，双床房还是应与大床房有一些区别，像这样可也是一种体贴与别致。

It was too kinky to have so much offside lighting. I find it better to allow the sleeping area to have more natural lighting. The bathroom in a twin room should differ from that of a double room.

沧州渤海假日酒店
Cangzhou Bohai Holiday Inn, Hebei, China

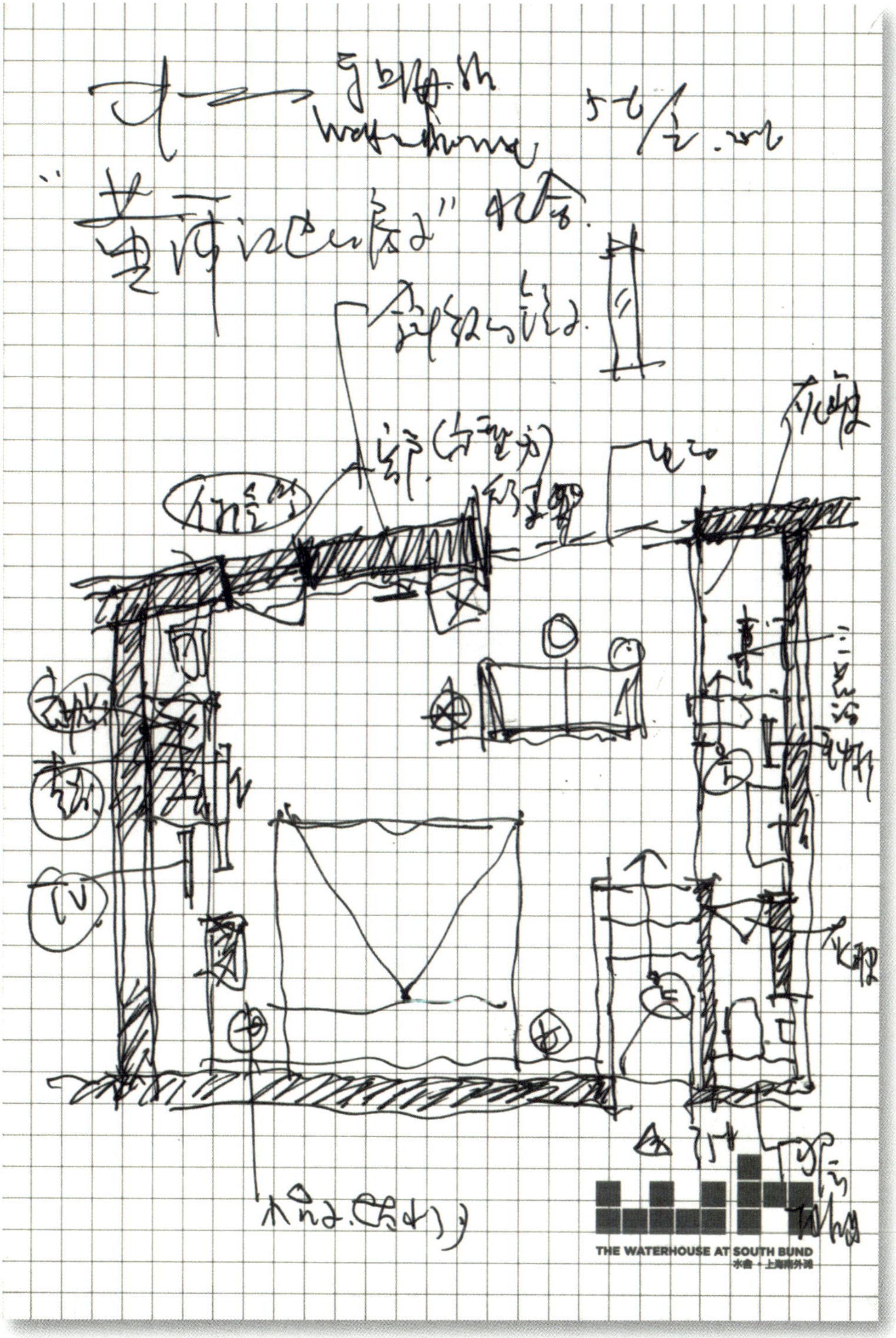

水舍上海南外滩酒店

31

★★★★★

THE WATER HOUSE AT SOUTH BUND SHANGHAI CHINA

Address : Maojiayan Road No.1-3, Zhongshan Road South, Huangpu District Shanghai, 200010,China
上海黄浦区
毛家园路1-3号
200010

Telephone : +(86)21 6080 2988

Fax : +(86)21 6080 2999

Http : //www.waterhouseshanghai.com

E-mail : info@unlistedcollection.com

给了住的了体

吃也、看也、没有住也，对方的同行只是到候去酒店吃个晚饭，而忽略了这种综合服务的客房，是基本的需求——住，除了黑夜。

上一次就是因为提示不够，订不到房间才能去酒店了一次晚餐，也不错；这次特别注意，就住到了。酒店还是很有心思的"设计"了这所让他也获奖无数的"水舍"（上海如恩）也难怪，确定有当下中国酒店业，特别是改造项目所缺少的一面：尊重和个性，当然酒店带给的"生态"也有和住客的充分认同也是这个项目成功的关键。酒店的成功还是在细节中体现，传统材料的巧妙运用。如光、如声，如场所的气味，住过了，体会更加真切。

就我入住的这间房的体会，大多的指标也合，也许每个人的应当推荐入住

Finally I'm Here

终于住上了你

水舍上海南外滩酒店

The Water House At South Bund, Shanghai, China

吃过、看过、没有住过。不少的同行只是习惯去酒店吃吃喝喝，而忽略了让你停留最多的客房，最基本的需求——住，白天与黑夜。

For them, there has been food, sights, but no guestroom. Many colleagues are only used to eating and drinking at hotels, ignoring where they stay the longest—the guestroom, the most basic need, the place to stay in, day and night.

上一次就是因为提前得不够，订不到房间才只能去消费了一次晚餐，也不错；这次特别注意，就住上了。设计师还是很费心思地“设计”了这所让他们获奖无数的“水舍”（上海如恩）。也难怪，确实有当时中国酒店业，特别是改造项目所缺少的一面：尊重和个性。当然设计师坚持的“变态”地方和业主的充分认同也是这个项目成功的关键，深厚的功力还是在细节中体现得淋漓尽致的，如光、如窗、如墙面的处理，住进来了，体会更加真切。

Last time as I had not booked early enough I failed to get one room in the hotel; instead, I only managed to have a supper there—it was nice. This time I paid special attention, so I managed to check in. The designer had put a lot of thought into the designing of The Waterhouse (by Neri & Hu Design and Research Office) which has won them untold awards. No wonder—it indeed had what hotel industry in China, especially those reshaping projects lacked: respect and individual character. Of course, the designer's persistence with its "freak aspects" and the owner's full approval was also the key to the project's success. The superb skill was embodied to the utmost in details, such as the lighting, the windows and the treatment of walls. Once you check in, you'll have a more intimate appreciation.

就我入住的这房间而言，有太多的抬高地面，也许年纪大的应当拒绝入住。

As far as the room I was in, there was too many spots of the floor had been raised. Maybe older people would refuse to stay there.

WESTIN
HOTELS & RESORTS

重庆解放碑威斯汀酒店

32 ★★★★★

THE WESTIN CHONGQING LIBERATION SQUARE CHONGQING CHINA

Address : 222 Xinhua Road ,
Yuzhong District ,
Chongqing, China ,
400010
中国重庆市渝中区
新华路222号，
400010

Telephone : +(86 23) 6380 6666

Fax : +(86 23) 6355 5555

Http : //www.westin.com/chongqing
liberationsquare

重庆解放碑威斯汀酒店
重庆市渝中区新华路222号 邮编: 400010
T 86.23.6380.6666 F 86.23.6355.5555
westin.com/chongqingliberationsquare
THE WESTIN CHONGQING LIBERATION SQUARE
222 Xinhua Road,Yuzhong District,Chongqing,400010,China
T 86.23.6380.6666 F 86.23.6355.5555
westin.com/chongqingliberationsquare
WESTIN
HOTELS & RESORTS

地域风情，也是不错的选择。

心血来潮加去看看前辈梁建国老师（北京集美组公司）设计的重庆中旅地产的会所，中式风格的中航云会，我们一行到访的有日醉的悠信会所上的餐饮部的感叹们溢在可惜，国内的王者从酒店地产修饰入，也是发人深省的，我去感到我们设计师们更是"压力"于太多和当今不同。

重庆是一个有强烈地域文化印记的城市（巴蜀），入住的酒店可谓也是有中式的韵味，当地的"Y楼"连当地也要设计师设计的地吧！

四面布局针对大面积的一层连通的设计，可谓"苦熬思良"地方来善用每一块场地，偏设，选用有"辣椒"风格的也许是一种不错的选择。

结束后，算是接地气的一段序。

Advisable to Position According to Local Features

地域风情，也是不错的定位

心血来潮去看看前辈梁建国老师（北京集美组公司）设计的重庆中航地产的会所：中式风格的中航云会，挑了一间新开的有口碑的定位高大上的解放碑的威斯汀酒店。可惜，国内的五星级酒店比的是投入，也是挺无稽的，或者轮到我们设计的时候更是“屈服”于豪华和与众不同。

On the spur of the moment I went to see the club of Chongqing Avic red estate designed by Mr Liang Jianguo (of Beijing Newsday Design Co). It was a Chinese style club. I chose newly opened reputed and high-rated Westin Hotel at the Liberation Monument. Regrettably, what the domestic five-star hotels compete for is input. It is fabulous, or when it's our turn to design, we are more likely to yield to being luxury and different.

重庆是一个有强烈地域文化印记的城市（巴蜀），入住的客房可谓也是有中式的韵味，入眼的“Y椅”应当是丹麦设计师汉斯的正品吧！

Chongqing is a city of marked regional cultural imprints. The guestroom I stayed in could be described as oozing Chinese features. The eye-catching "Y chair" should be an authentic work by Hans the Danish designer.

平面布局针对大开间做了一些迂回的设计，可谓“苦煞思良”地力求善用每一寸空间，佩服，选用“辣椒”风格也许是一种不错的定位。

In regard to the plane layout, some roundabout design was done to the width of the room. It is admirable that painstaking care had been taken to make good use of every inch of space. And maybe it was desirable to choose the "Chilli" style as a kind of positioning.

简单直白，算是接地气的一间客房。

Simple and with no reservation, it could be counted as a down-to-earth guestroom.

重庆解放碑威斯汀酒店(图片源于网络)

The Westin Chongqing Liberation Square ,China

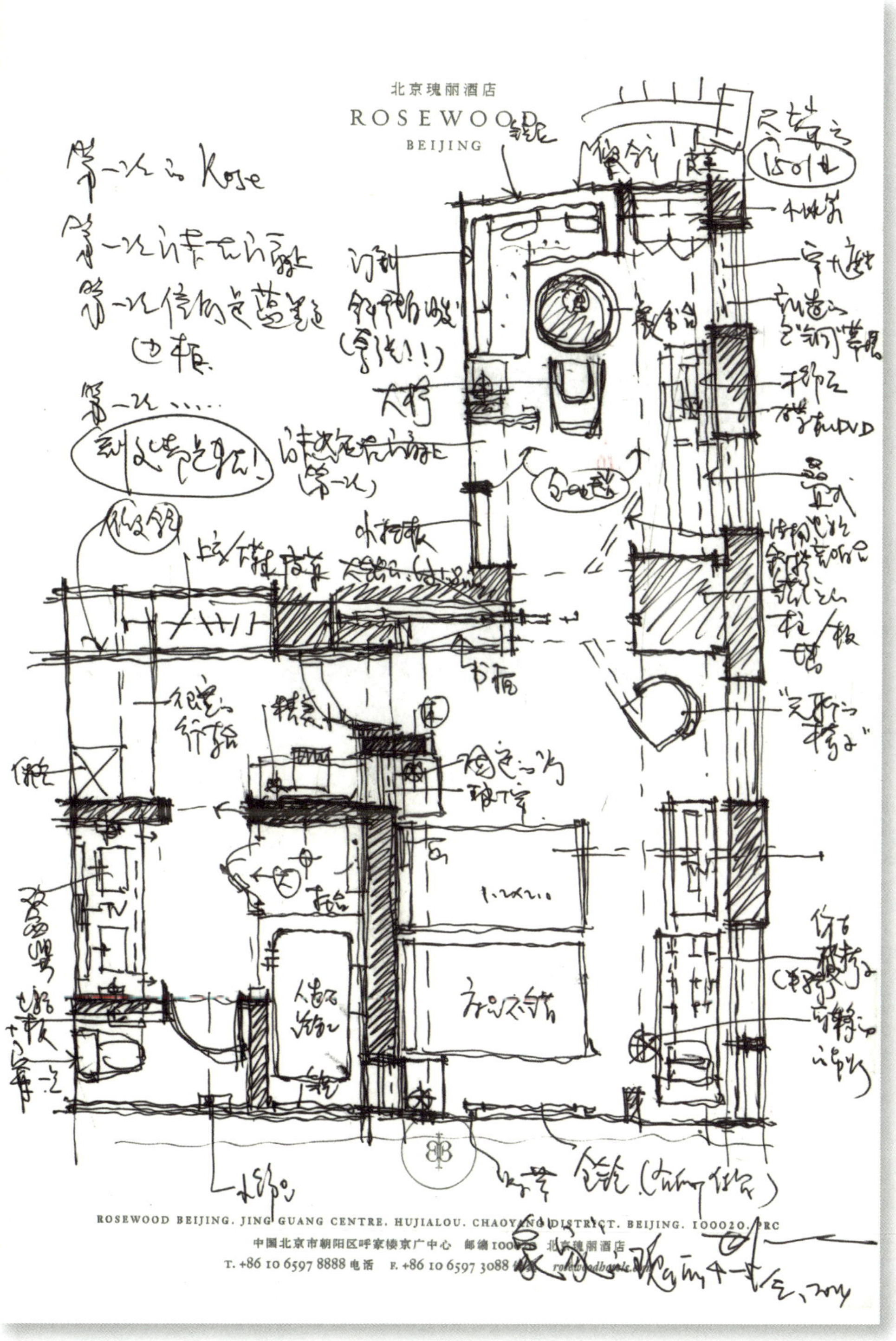

北京瑰麗
ROSEWOOD
BEIJING

北京瑰丽酒店

★★★★★

33

ROSEWOOD BEIJING CHINA

Address : Jingguang Centre, Hujialou, Chaoyang District, Beijing, China
中国北京市朝阳区
呼家楼京广中心，
100020

Telephone : +(86 10)65978888

Fax : +(86 10)65973088

Http : //www.rosewoodhotels.com

第一次。

第一次入住北京后的中国第一个瑰丽酒店 Rosewood。感觉很好，有很多惊喜，很多第一次。

第一次 [illegible] 入住时，我和团队下午5:00到达，可以第二天下午5:00退房。（酒店如何管理啊！）

第一次在"私家车"接去机场，管家陪伴还没有退房。很感谢这种VIP式的体验！

第一次被赞誉：

入房之后就被看作主人（完成接待 [illegible]）很不可思议，这很有意思。

第一次看到全落地式的窗户，卧室面积很大，有近60m²，有小厅、有大卧室、大衣帽间、大浴室，很多很多的细节/设计。作品"很家"，用"家"来形容的"高级酒店"中，可能是第一次。

但是信息化服务，是我要求学习的！

这是香港的信托，第一次在自己国家用"洋葱"设计基本框架，很中国，很轻盈，但能接受吗，中国人？

但是通过体验，在这里我能感受到全球顶尖的酒店品牌，这是我的设计，很不错的第一次！ 叶 5/2014

First Time

第一次

北京瑰丽酒店(图片源于网络) *Rosewood, Beijing, China*

第一次入住改造后的中国第一间瑰丽酒店Rosewood，名字很好听，有所期待，很多第一次。

It was the first time I had been in the first Rosewood transformed in China. Its name sounded nice. There was something to expect. There were many "first times" upcoming.

第一次24小时入住方式，我和同事下午5：30到达，可以第二天的下午5：30退房（酒店如何管理啊！）。

For the first time, I checked into a 24-hr-stay hotel. My colleague and I arrived at 5:30. It meant we could check out at 5:30 next afternoon — how would the hotel manage!

第一次有“私家车”送去机场，写的时候还没有退房，很期待这种VIP式的体验！

For the first time, a private car was available to see us to the airport. When I was writing this, I had not checked out, so I was quite eager for this VIP experience!

第一次继续着：

The first-time experience continued:

入房的门锁在门扇上（无线技术的发展结果），很不习惯，但很有趣。

The door was locked on the door leaf (resulting from the development of wireless technology). Unused to it, we found it interesting.

第一次体验全家居式的布局，平面面积很大，有近60㎡，有小厅，有大卧室、大衣帽间，很多很多的书、杂志、饰品，很“家”，而且是“都可使用的”，这在酒店中亦可谓第一次，倒是信纸没常备，是我要求拿的！

For the first time, I experienced the all-family-stay layout. The room, quite large, nearly 60 m^2, with a small hall, large bedroom, large cloakroom, many books, magazines, ornaments, was quite "homey", and the articles were "all usable". This could be called my first hotel experience. There was no letter paper provided as part of necessities. It was at my request that they fetched me some!

认真看看小信纸，第一次这么有勇气用“深蓝色”将其框住，很中国，很稳重，但接受吗，中国人?

Scrutinizing the letter paper, for the first time I had the courage to frame it in "dark blue". It was quite Chinese and sedate. But will we Chinese people accept it?

倒是这次体验亦真正近距离地接触全球顶级的奢侈酒店，还是旧建筑改造的。

It was this experience that brought me to a really close contact with a top luxury hotel in the world, reshaped from an old building.

很不错的第一次。

It was a fairly good first time experience.

北京瑰丽酒店(图片源于网络)
Rosewood, Beijing, China

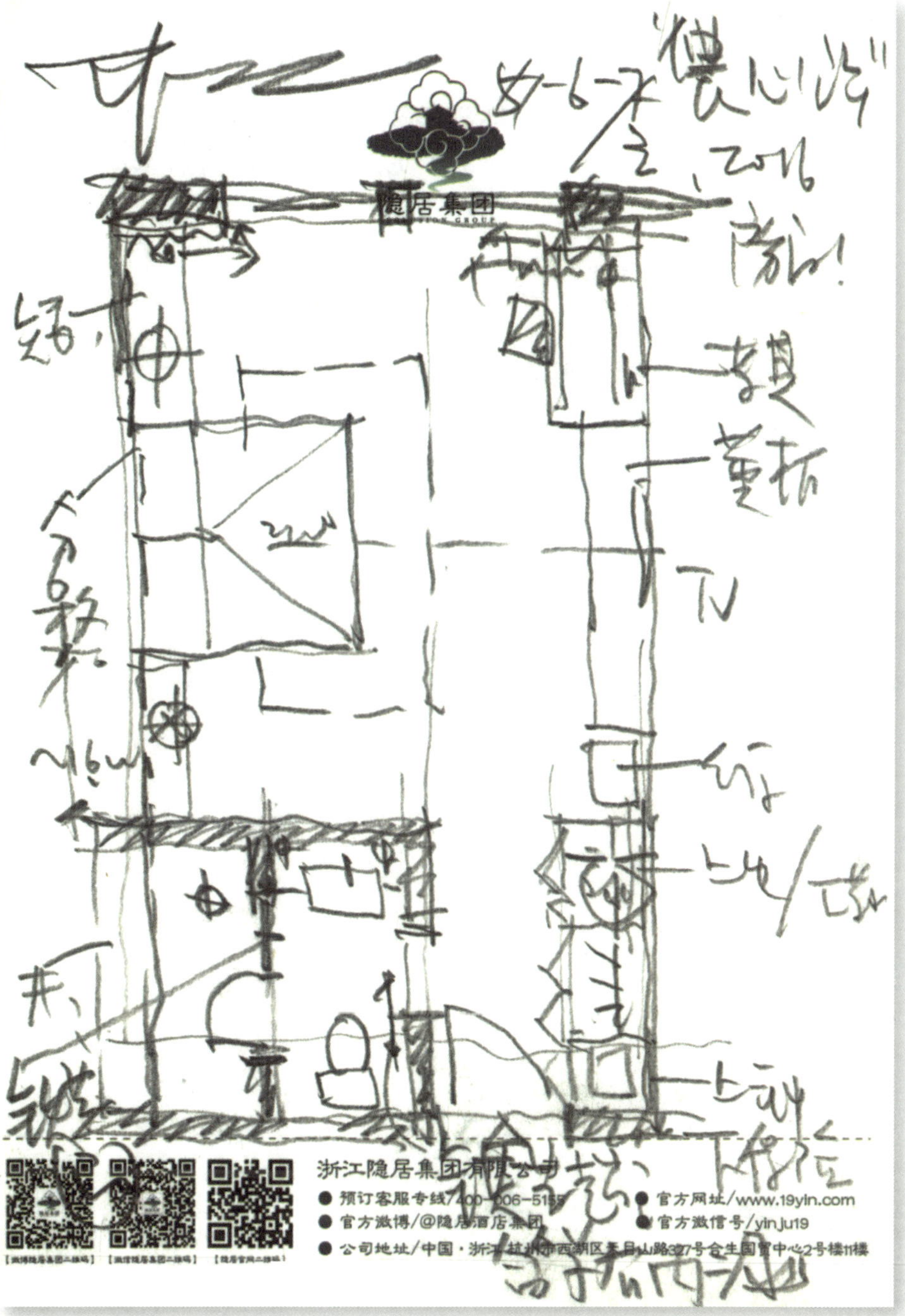

隐居集团
SELUSION GROUP

上海隐居繁华雅集公馆

★★★★★

34

BAMBOO RETREAT SHANGHAI CHINA

Address : No.83-85,Wuxing Road, Xuhui District, Shanghai,China 200030
中国上海
徐汇区吴兴路83-85号

Telephone : +(021)61635500

Http : //www.zkeji.com.cn/hliy4xqz/823180/index

上海隐居繁华雅集公馆

Bamboo Retreat,Shanghai, China

Hakkapark
International Hotel
客天下国际大酒店

客天下国际大酒店

35 ★★★★★

HAKKAPARK INTERNATIONAL HOTEL GUANGDONG CHINA

Address : The Hakka Park , Meijiang District , Meizhou , Guangdong Province, China 514071
中国广东省梅州市梅江区客天下旅游产业园，400010

Telephone : +(86) 753211 8888

Fax : +(86) 753211 6666

Http : //www.ktxhotel.com

E-mail : hakkapark@ktxhotel.com

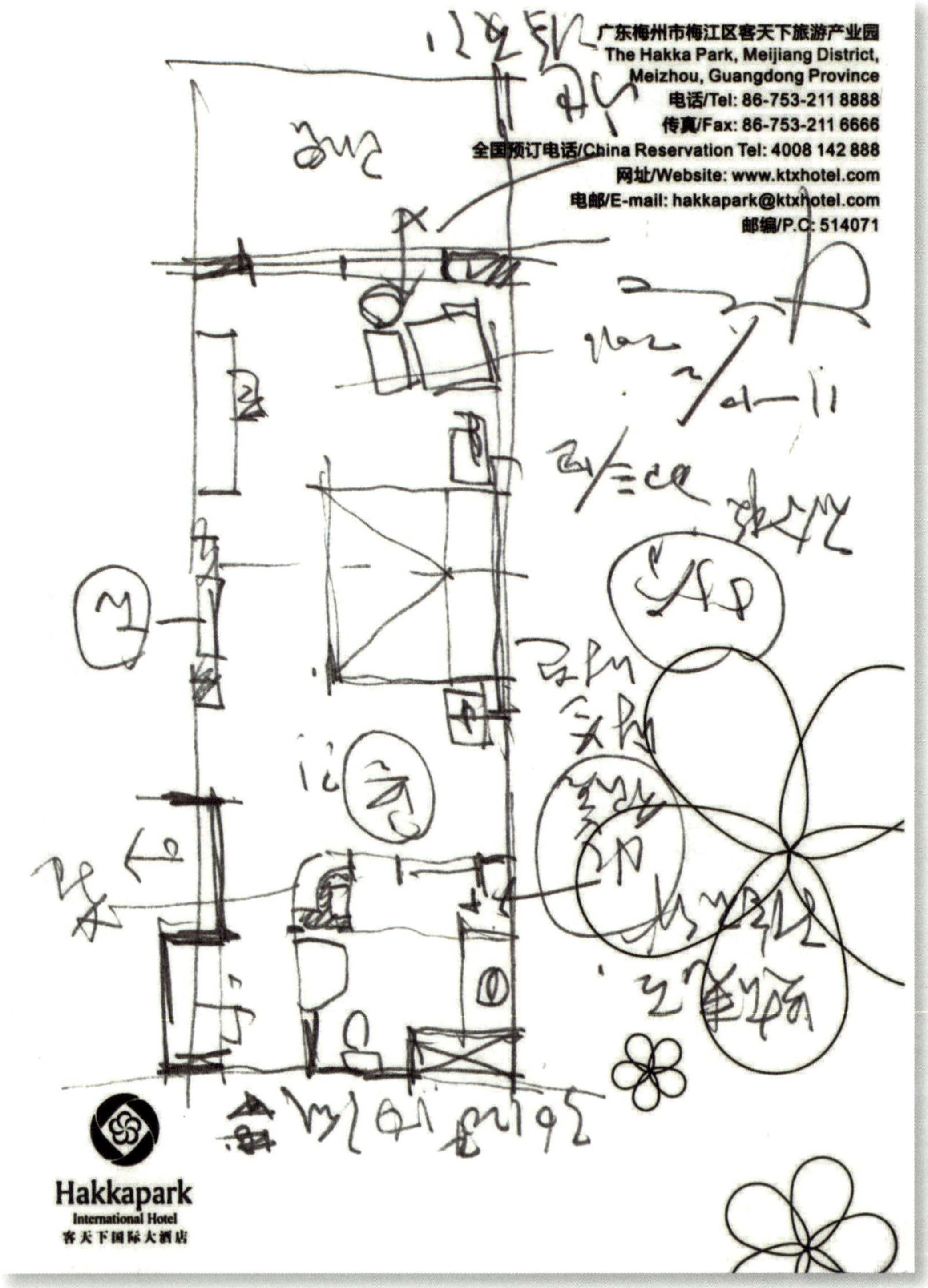

广东梅州市梅江区客天下旅游产业园
The Hakka Park, Meijiang District,
Meizhou, Guangdong Province
电话/Tel: 86-753-211 8888
传真/Fax: 86-753-211 6666
全国预订电话/China Reservation Tel: 4008 142 888
网址/Website: www.ktxhotel.com
电邮/E-mail: hakkapark@ktxhotel.com
邮编/P.C: 514071
Hakkapark
International Hotel
客天下国际大酒店

让林风眠先生作再现的容身。

入住"客天下"酒店的收获，居然是能在房间内看到林风眠先生的画像。以大师的作品装饰的酒店，画作一定用心且必不凡。虽然我努力着但还是没有找到一张原作（可能是太金贵了，没有在酒店的收藏室去找），相信这个业主会有收藏林先生的原作的藏品吧。不过更让我高兴的是他们会懂得用大师的作品（仿品）去装点各个空间呢！

作为国立艺术学院（中国美术学院之前身）的首任院长，梅州人，林风眠先生的彩墨画作我本不会不喜欢。特别是那荷花之趣的，太棒了。但是这个酒店建筑的"客天下"酒店的每个大师空间，没有陈设在公共区域，可能我未见到建筑师专业的打造上了。到室内就算是很好了。所谓设置了这个好的"栖"。

还是把爱投回到林先生的画上吧！

Guestroom Representing Works by Lin Fengmian
让林风眠画作再现的客房

客天下国际大酒店
Hakkapark International Hotel, Guangdong, China

入住“客天下”最大的收获居然是放在房间内的林风眠先生的画集。以大师的作品为主线的酒店，画作的应用无处不在。虽然我努力着，但还是没有找到一张原作（可能是太仓促了，没有去酒店的收藏室看看）。相信这个业主会有些林老先生的原作藏品吧，不然怎么敢这么有底气地用大师的作品（仿品）去装点各个空间呢

It turned out that the best I got from the stay at the Hakkapark International Hotel was the collection of works by Lin Fengmian, placed in the guestroom. For a hotel with the master's works as its main subject, its use was omnipresent. For all my attempts, I failed to find an original work by the master—maybe, pressed for time, I didn't go the collection room to have a look. I believe the owner had collected some of the authentic works by Mr Lin, or how could he have the confidence to use the master's works to decorate each space!

作为国立艺术学院（中国美术学院的前身）的首任院长，梅州人，林风眠先生的彩墨画作我都不是一般地喜欢，特别是荷花主题的，太棒了。倒是这个弧形建筑的“客天下”酒店的客房大而空洞，没有内在的品质可言，可能钱都花到建筑和景观的打造上了，到室内就草草收场了，白白浪费了这么好的“胚”。

Mr Lin, a native of Meizhou, was the first president of the National Academy of Art, predecessor of China Academy of Art. I am more than fond of all his ink and color paintings, especially those of lotuses. They are really too fantastic. However, the guestroom of this arch-shaped Hakkapark International Hotel, spacious yet void, had no inner quality to recommend itself. Perhaps all the money had been spent on the building and landscape, leaving nothing for the interior. So good a "fetus" had been thus wasted.

还是把爱放回到林先生的画上吧！

It is better to give love back to the works by Mr. Lin!

客天下国际大酒店
Hakkapark International Hotel,Guangdong, China

Sheraton
Shanghai
HONGQIAO HOTEL
虹桥喜来登上海太平洋大饭店

虹桥喜来登上海太平洋大饭店

36 ★★★★★

SHERATON SHANGHAI HONGQIAO HOTEL SHANGHAI CHINA

Address	: 5 Zunyi Nan Road, Shanghai, China, 200336 中国上海 遵义南路5号， 200336
Telephone	: +(86 21) 6275 8888
Fax	: +(86 21) 6275 5420
Http	: //www.sheraton.com/hongqiao

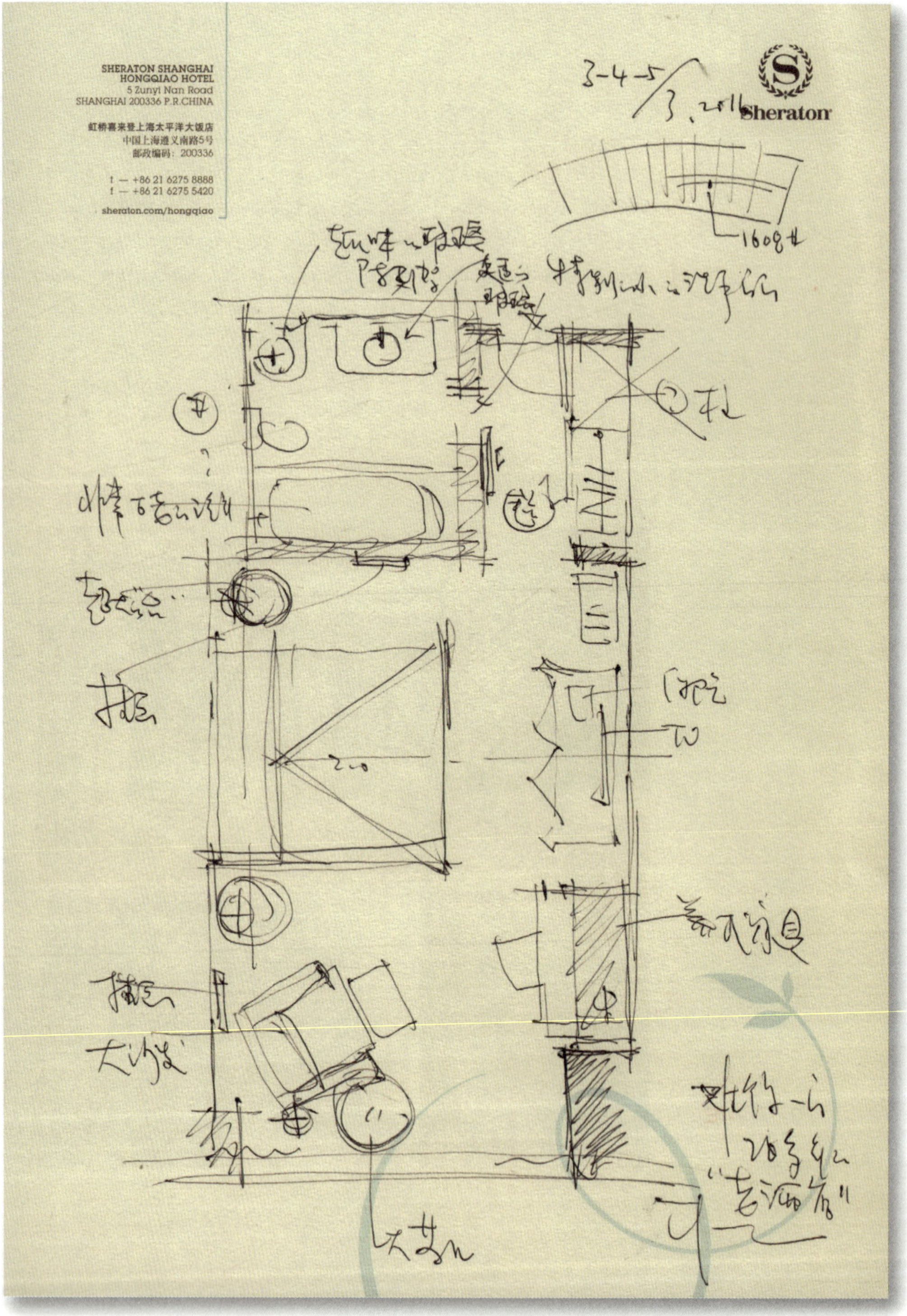

感受老酒店的细节

参加客户龙湖地产集团的合作商大会，"让员工过上体面的生活"这句话触动到我的心里，也激励着我去把公司做得更好，不断地好，就像我们的客户一样。

会议安排的住宿是老酒店——由锦江集团酒店管理，据说当年某些名人住在这里。房间本身不是很精致，大格局是非常"老旧"的，洗手间相当的宽大，传统的中式走道，一侧是洗手间，另一侧是衣柜。

房间尺度合适，简洁且实用，但是洗手间的设计颇有心思，它采用的是玻璃台盆洗手台（或是当时的爆款），精巧，特别之处在转角处设有的挂毛巾的钩架，为这个传统的空间添色不少，细节的用心才让老酒店的价值彰显。

有所限制是事实，但用心才是营造一个让人舒心的细微之处，感受到随手所得的方便也是难能可贵的。

老酒店，就感受细节吧。

Look at the Details of an "Old Hotel"

让我们看看"老酒店"的细节

虹桥喜来登上海太平洋大饭店
Sheraton Shanghai Hongqiao Hotel,China

参加客户旭辉地产集团的合作商大会，"让员工过上体面的生活"这句话铭刻在我的心里，也激励着我去把公司做得更好，不断的好，就像我们的客户一样。

"Let our employees live a decent life", the words I heard at the meeting of cooperators held by my client CIFI Group, have been imprinted in my mind, inspiring me to have my firm do better and better, just like our client.

会议在有20多年历史的老酒店——虹桥喜来登酒店举行，顺理成章地住在这里，房间本就不抱惊喜，大格局是非常"老旧"的，洗手间相当的紧凑，传统的走道，一侧是洗手间，另一侧是衣柜。

The meeting was held at Sheraton Hongqiao, an old hotel of a history of over 20 years. It was only too natural to stay there. I had expected no surprise in the first place. The overall layout was quite "outdated"; the bathroom was rather compact; a traditional passage was flanked with the bathroom on one side and a wardrobe on the other.

房间尺度合适，够用且实用，倒是洗手间的设计颇有心思，经典的玻璃洗手台（或是当时的爆款），精巧、特别之处在转角处设有的玻璃陈列架，为这个传统的空间添色不少，细节与用心才是老酒店的价值沉淀，有局限是事实，但用心机去营造一系列宜人的贴心细微之处，感受到随手而得的方便也是难能可贵的。

The guestroom was of a moderate size, adequate for use and practical. Noteworthy was the design of the bathroom, showing great care and deliberation—a classic glass wash platform (maybe popular back then), and a glass stand installed around the corner, delicate and special, adding to the taste of this traditional space. Detail and consideration is what runs in old hotels. Limitations are facts. However, it is invaluable to take painstaking care to create a series of close-to-heart spaces, making guests feel convenience at hand.

老酒店，就感受细节吧。

Feel the details of old hotels!

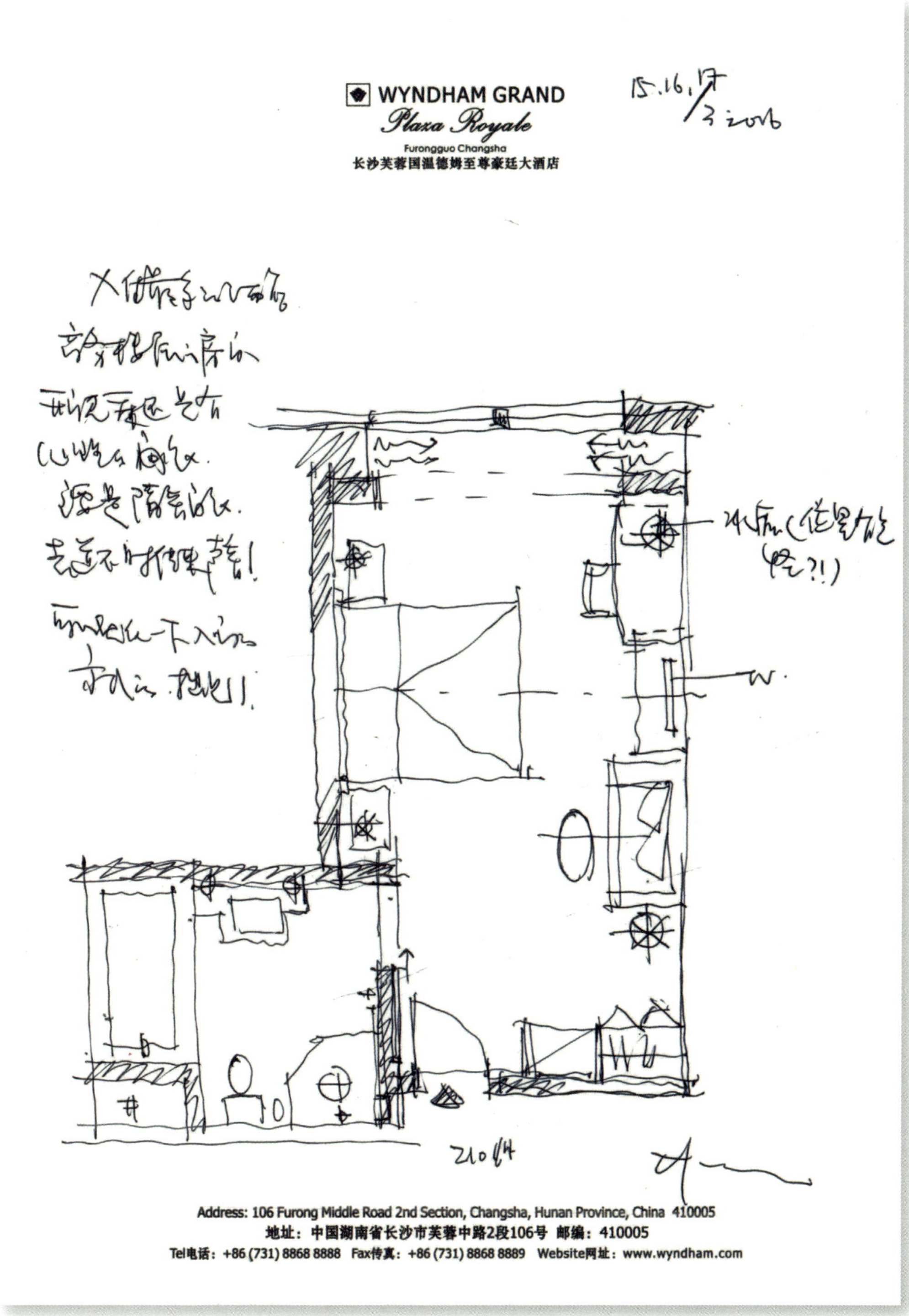

WYNDHAM GRAND
Plaza Royale
Furongguo Changsha
长沙芙蓉国温德姆至尊豪廷大酒店

长沙芙蓉国温德姆至尊豪廷大酒店

37 ★★★★★

WYNDHAM GRAND PLAZA ROYALE FURONGGUO CHANGSHA CHINA

Address : 106 Furong Middle Road 2nd Section, Changsha, Hunan, China 410005
中国湖南省长沙市芙蓉中路2段106号，410005

Telephone : +(86 731)8868 8888

Fax : +(86 731)8868 8889

Http : //www.wyndham.com

WandaVista
Hohhot
呼和浩特万达文华酒店

呼和浩特万达文华酒店

38 ★★★★★★

WANDA VISTA HOHHOT NEIMENGGU CHINA

Address : 26 Xinhua East Street,
Hohhot,
Inner Mongolia ,
P.R.CHINA
010020
中国内蒙古呼和浩特市
新华东街26号，
010020

Telephone : +86 (0)471 518 8888

Fax : +86 (0)471 516 3333

Http : ///www.wandahotels.com

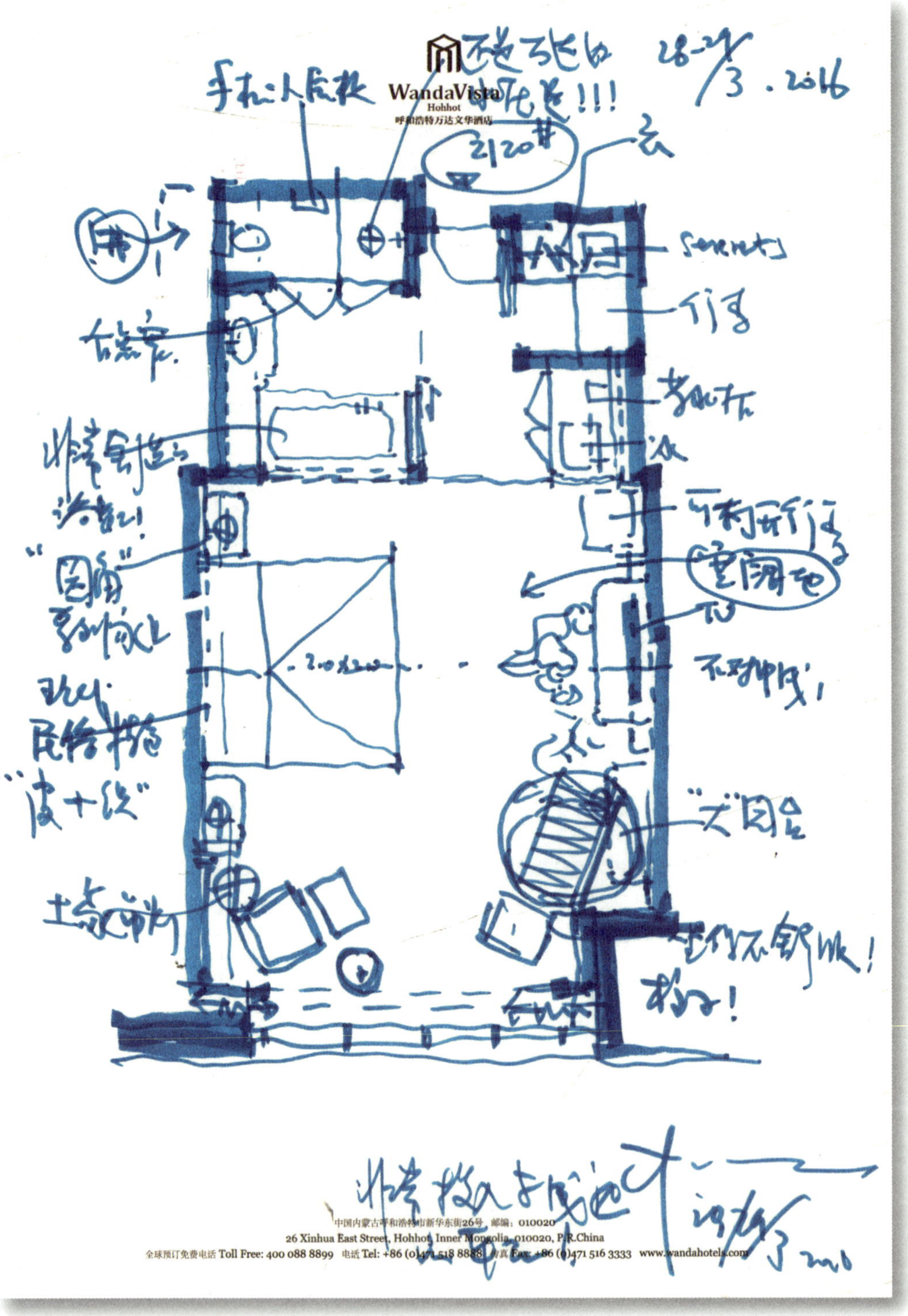

WandaVista
Hohhot
呼和浩特万达文华酒店

还是万达

去呼和浩特出差，住了一夜！

选了一家刚刚开业的万达文华酒店，原本想着这是比较低调的城市（楼价不到一万）不会投入太大，谁知万达文华级别的白领吧！

半夜check in入住，还是被他们惊喜了，就是一些大堂格局、家具、装饰性、大小尺寸，客房，一些布艺在一个城市的调子，包括选用的石材，不封边型都是"大手笔"，当然咱们最关注还是"入乡随俗"，从"宫"这个在呼市人民包括少数民族的性格气质的之事：前台挂毯到位，客房绿色调，缝纫；蒙古包，加号称号在哈萨，手毛的图案件系以大甚至，细腻的文化与传统融到行住好地融合在一起，游客们用享以世外"宫"的好舒子上半辈子！

我在以后哪儿一批世，每一家酒店的好坏，我"是

中国内蒙古呼和浩特市新华东街26号　邮编：010020
26 Xinhua East Street, Hohhot, Inner Mongolia, 010020, P.R.China
全球预订免费电话 Toll Free: 400 088 8899　电话 Tel: +86 (0)471 518 8888　传真 Fax: +86 (0)471 516 3333　www.wandahotels.com

[illegible]金秋之相聚、浩荡沙漠、天际花海、水流急、水面大，令人眼[illegible]。当代建筑[illegible]之房子设计，（达茂旗）近四年之[illegible]，[illegible]。

让我再次感叹这个年轻的城市[illegible]，[illegible]开阔之[illegible]。[illegible]再一个惊人的成就，更加还可追赶，更别说超越之酒店[illegible]！

还是不错，好样的！

[illegible]

中国内蒙古呼和浩特市新华东街26号　邮编：010020
26 Xinhua East Street, Hohhot, Inner Mongolia, 010020, P.R.China
全球预订免费电话 Toll Free: 400 088 8899　电话 Tel: +86 (0)471 518 8888　传真 Fax: +86 (0)471 516 3333　www.wandahotels.com

呼和浩特万达文华酒店(图片源于网络)
Wanda Vista Hohhot, China

Wanda is Wanda
还是万达

去呼和浩特出差，简单的一夜！

A simple business night stay in Huhehaote!

选了一家刚刚开了几个月的万达文华酒店，原来想着这类消费比较低的省会城市（楼价不到1万），不会投太大就能继承万达文华系列的血脉了吧！

I chose this Wanda Vista Hotel opened just several months before. I had thought, in this provincial capital with a relatively low consumption level, it was not quite likely to invest too much to carry on the lineage of Wanda Wenhua.

半夜check in入住，还是挺有惊喜的，首先是大堂格局豪华，装饰性的大水晶灯、高空间，一切都是在一线城市的调子，包括选用的石材，石材造型都是“大手笔”。当然吸引我的还是“入乡随俗”引入“马”这个在呼市人民包括少数民族的性格象征物的元素，主幅挂毯到顶，草原绿色调，清新，蒙古包加羊群，马在吃草。马毛的组合，休息区大背景，细腻的变化与马毛粗犷很好地组合在一起，祥云的图案引入也为“马”的犷野系上精彩！

Checking in at midnight, I was still surprised. First, the hall was luxury —the ornate quartz lamps and the space height, all was as stylish as in the first-tier cities, including the stone used; the shapes of stone were all works of skilled craftsmen. Of course, what attracted me was the introduction of the "horse", the element symbolic of the character of Mongolian people. On the tapestry hung from the ceiling was the grassland, green, fresh, yurts, hordes of sheep and horses grazing. The subtlety of the change of the lounge area as a whole went well with the roughness of the horsehair. The pattern of auspicious clouds also added to the splendor of the unrestrained "horse"!

我在住《住哪?2》提过，看一家酒店的好坏，“水”是非常重要的标准，深夜洗澡时，天花花洒，水流急，水量大，舒服。当然，“土豪”级的房间装修，（商务层）近4米的房间高度，霸气。让我再次有了画画平面的冲动，虽然还是那样的平面，但看到一个更加成熟，更加不可追赶，更别说超越的酒店霸主。

As I mentioned in Where to Stay? 2, to judge a hotel, water is a very important criterion. Showering in the depth of night, I found the gondola water faucet spouting massive jets of water—it felt quite comfortable. Of course, the room, extremely ornate, nearly 4m high for the business room, was imposing. Again I felt compelled to draw its plans—though they were the same plans, I saw a more mature lord of hotels, not to be emulated, let alone to be overtaken.

还是万达，好样的！

Wanda is Wanda—Good for you!

呼和浩特万达文华酒店(图片源于网络)
Wanda Vista Hohhot, China

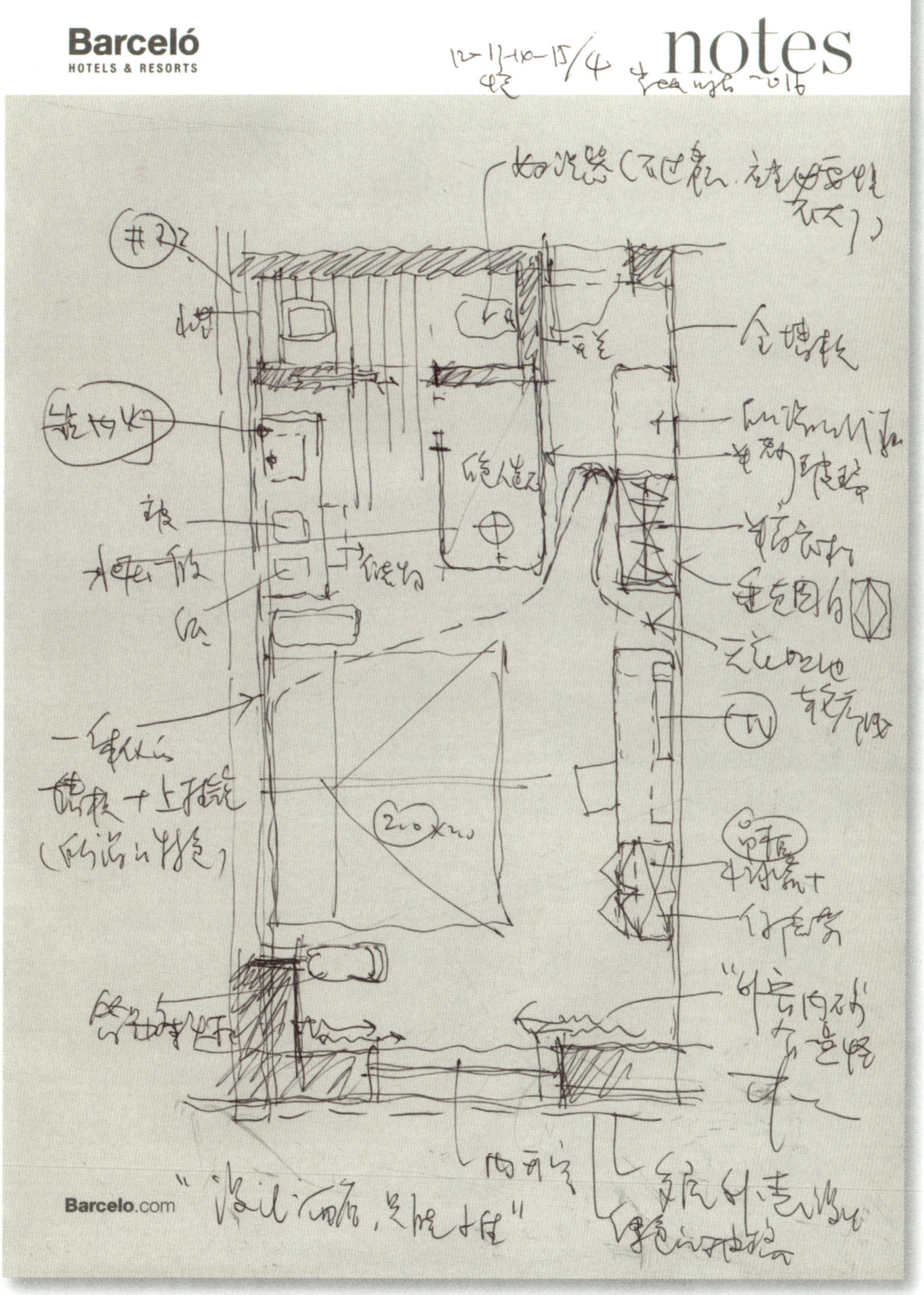

Barceló
HOTELS & RESORTS
巴塞罗酒店
★★★★

39

BARCELÓ HOTELS & RESORTS MILAN ITALY

Address : Via Stephenson, 55, milan, Italy
Telephone : +(39 02) 9390 0552
Fax : +(86 20) 8513 3999
Http : //www.bercelo.com

设计型酒店之夜

入住设计型酒店总是令人兴奋的，这回还是住了华丽的这个有标杆意义的巴塞罗(Barceló)酒店，大堂公共区域充斥个性张扬，甚至喜欢到想什么认知。例如整个用了绿松石作主题，结合大堂的法语文字，意境不错，好在其中反馈给我们惊喜的，“设计”之夜在这里还是让人愉悦的。

房间不太大，却充满趣味，开放式空间包含不同的功能，地面的圆角家具，墙面流动的造型与灯光融为一体，充满自由与动感的曲线贯穿出强烈的个性，半开放式的浴室让浴缸成为焦点，浅木色与绿色的组合让居住感轻松而可谓一气呵成。

可能是倒时差的缘故，每一天能体验一点点不同的细节，除了一个性感的按摩器外。

夜很深，刚好让每一天遇到新的我们难以入睡。

这个以设计为标志的酒店可谓细节十足，收纳、灯光等等，投入甚大，值得一住。

设计之夜要不讨人厌才行

Appropriateness of the Design Hotel
设计型酒店的度

米兰巴塞罗酒店
Barceló Hotels & Resorts, Milan, Italy

入住设计型酒店总是令人兴奋的，况且还是位于米兰的这间有标杆意义的巴塞罗（Barcelo）酒店。大堂公共区域就个性张杨，喜不喜欢就看你的认知了，倒是餐厅用白桦树作主题，结合大量的清镜反射，意境不错，行走其中挺有戏剧性的，“设计”的度在这里还是让人愉悦的。

It it always exciting to check into a design hotel, let alone this benchmark Barcelo Hotel in Milan. The public area of the hall was individually showy —whether you like it, it all depends on your perception. Walking through the dining hall, with silver birch as the subject, combined with a lot of clear mirror reflection felt artistic and dramatic. The appropriateness of design here can be rated "pleasant".

房间不太大，却充满趣味，开放的空间有着不同的功能，纯白的圆角家具，墙面流动造型与灯光融为一体，天花的自由走动的曲线颇有章法与个性，半砂全玻璃的淋浴间让淋浴的过程成为焦点，浅木色与绿色的组合让居住空间氛围颇轻松，可谓一气呵成。

The room, not so big, was full of fun. The open spaces had various functions. Pure white smooth-angle furniture, the blend of the fluid shape of the wall surface and the lighting, the free curvy lines of the hung ceiling, both constrained and characteristic, the all-glass shower compartment that made the showering process become the focus and seems a little erotic, and the combination of gray wood and green which makes for a comfortable stay —it was a coherent whole.

米兰巴塞罗酒店
Barceló Hotels & Resorts, Milan, Italy

可能是住了四天的缘故，每一天能体验一点点不同的细节，除了一个怪怪的妇洗器外。

Maybe it was because I stayed for four days, I appreciated a little bits of different details every day, except a queer vaginal cleaning device for women.

床很大，刚好让每一天逛到腿软的我瘫倒就睡。

The big bed was just good for me —I would fall asleep once touching it as sauntering around would wear me out every day.

这个以设计自豪的酒店可谓细节十足，收纳、灯光等等，投入也充分，值得一住。

This hotel, proud of its own design, with sufficient input, which pays more than enough attention to details such as intake, lighting and etc, is stay worthy.

设计的度要不讨人厌才行。

Offensiveness is what can not be accepted in a design.

哥本哈根万豪酒店
Copenhagen Marriott Hotel, Denmark

MARRIOTT

哥本哈根万豪酒店

40 ★★★★★

COPENHAGEN MARRIOTT HOTEL COPENHAGEN

Address : Kalvebod Brygge 5, Copenhagen, 1560, Denwmark
Telephone : +(0045) 8833 9900
Http : //www.marriott.com

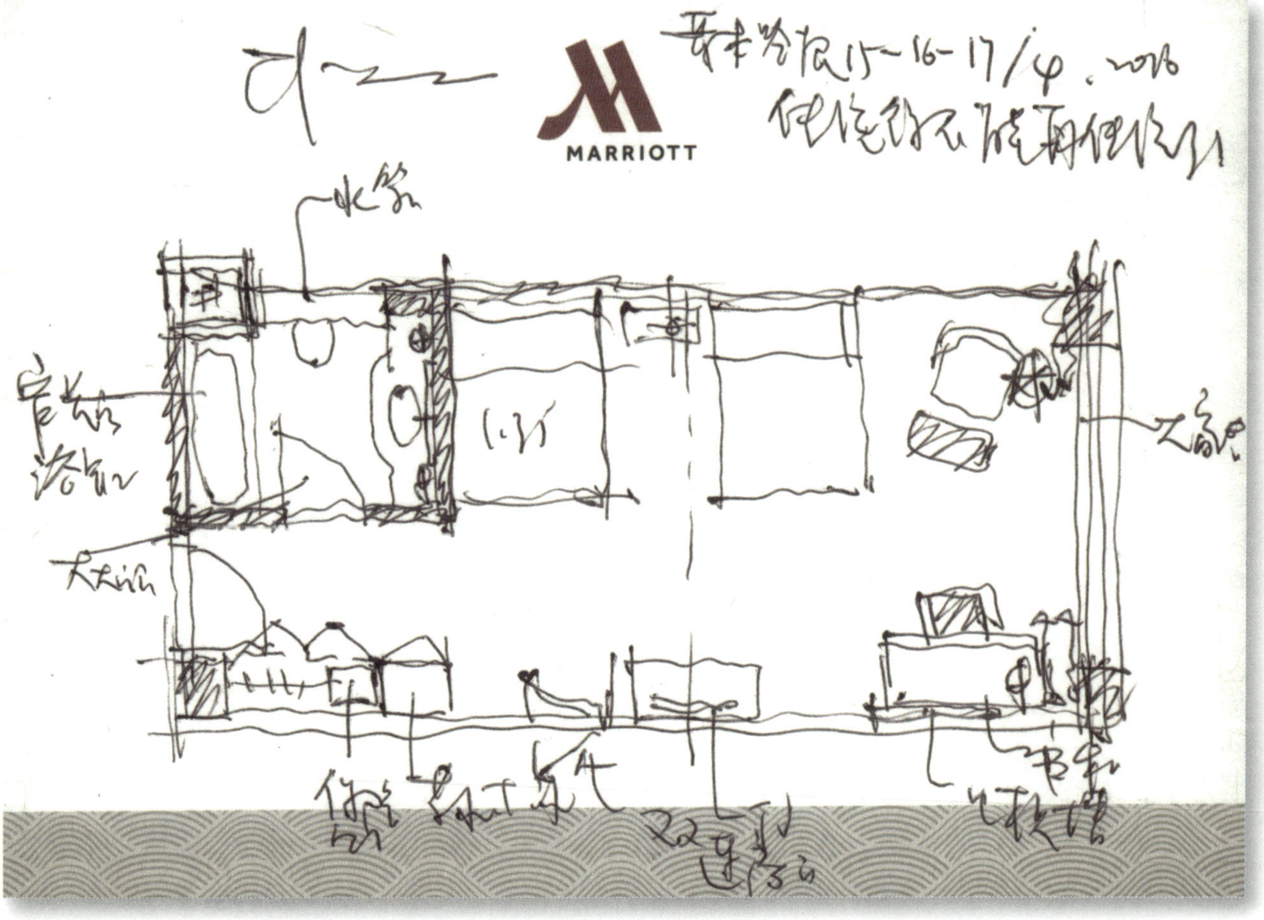

Hilton
HOTELS & RESORTS

赫尔辛基希尔顿酒店

★★★★

41

HILTON HELSINKI KALASTAJATORPPA FINLAND

Address : Kalastajatorpantie 1, Helsinki 00330, Finland

Telephone : +00 358 9 45811

Http : //www.hilton.com

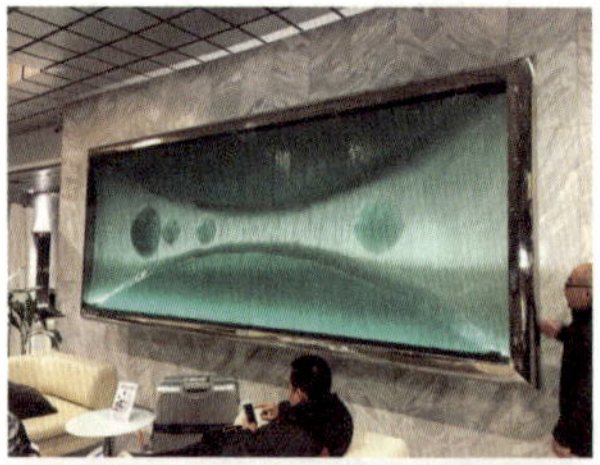

赫尔辛基希尔顿酒店

Hilton Helsinki, Kalastajatorppa

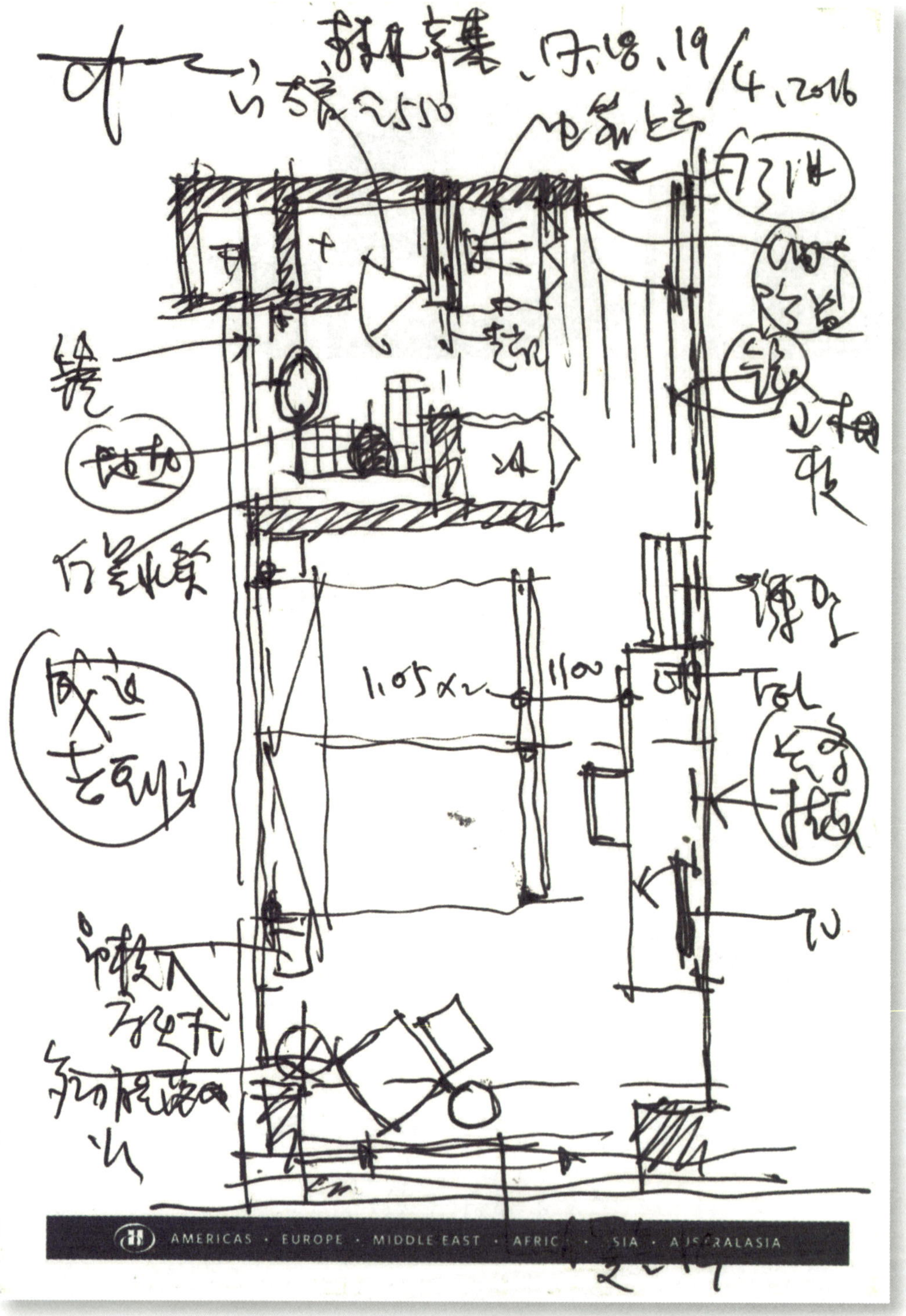

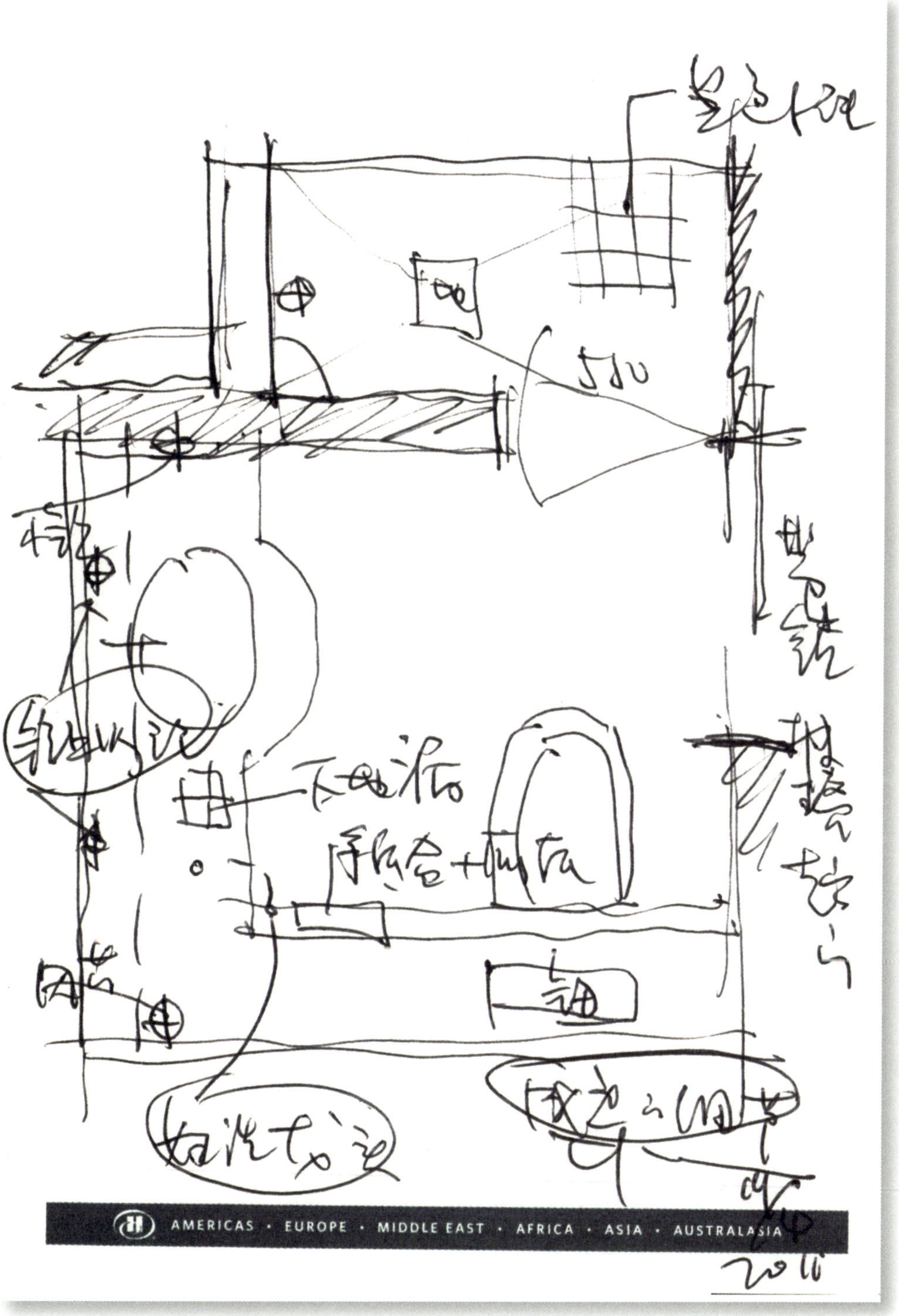

赫尔辛基希尔顿酒店
Hilton Helsinki, Kalastajatorppa

上海礼遇・二更呆住酒店

42 ★★★★★

LIVO HOTEL ER GENG SHANGHAI CHINA

Address : 3F, No.3 ,
JianGuo middle Rood ,
Shanghai Area, 200025 ,
China
中国上海市
黄浦区建国中路
3号三层
20025
Telephone : +(86) 18210631192
Http : //www.livoshanghai.com
E-mail : //res@livohotel.com

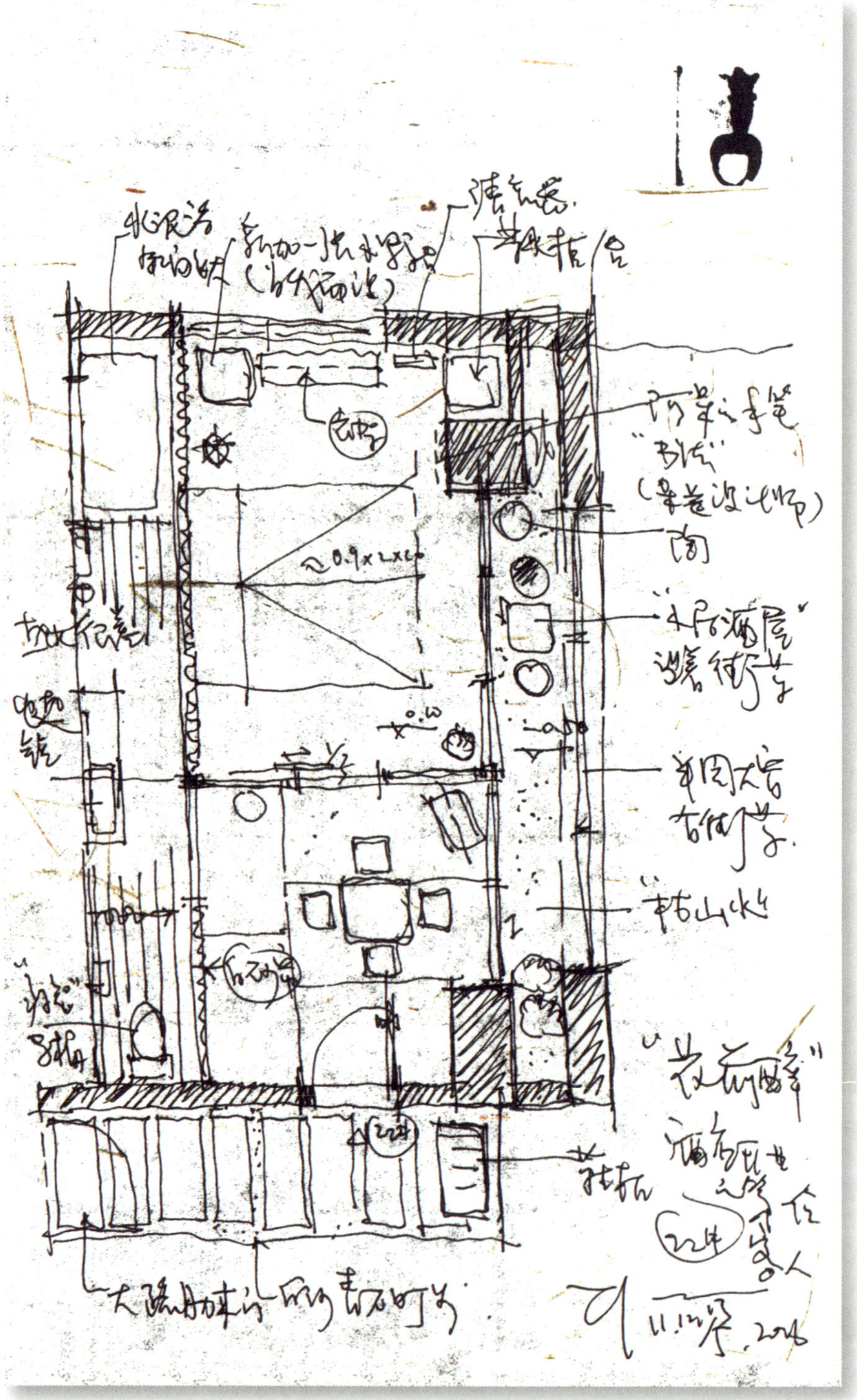

"遇"

从没有想过会成为酒店开业的第一个客人，可谓不可多得的"机遇"，真称是一种机遇！

因为要往杭州去上海，在ctrip上找了住处，从古街到市区，从西湖到外滩开业，非常向往……"机遇·一天·半佳"酒店。在他们当年都会留意的心，准备在那个不太，一定不重复，一定揣摩，当然一定很好奇心！结果成了开业后，他们酒店的第一位。

地址就在建国路上，绕了一圈又一圈，西子"卷帘"特地在门前等候。从后门到了三楼，经过"会所"进入大堂，打开电梯（因为下雨了，大厦只能从后面入住的客）在房，他才进入到了房间——古色古香的房间，因为不完善，惊人也和小西子聊了这家酒店是落地于上海的"机遇"酒店，这里面的房间似乎我们在推进的老宅会落地的"2016"的情况，我来这是一个古旧的梦，古梦的日子，老宅的缘由。

我们没有……三十年，相信能让自己的项目落地始终是一件好事情，相信这个"机遇"团队也会慢慢建造"民宿"之梦。

①

对于我们之建筑而言是非常好的"批评"标准之判别之处，一是旧楼改造，尚限多，二是手工艺，或是中国传统吧，不讲究，再谈到之际，更是"不堪"，不是说中国是我们自己，我们之祖先，更加我们传统的控制(掌控)能力。欧洲市中心之源头与我们最大区别是它建筑会有机地随着历史而来相继出现，一种情怀是不足以支撑之操作的，更加不用说"气韵"之技巧与分寸了。水泥不行，设计上要选择要多些的心理感受，才情熟之材质也得强调了——一点点"本末倒置"，Hermes之讲究用的，太精美了，不知我要是什么。

"遇"到是一种缘分，它就是很从容之人之安宁因素，认真之也走了一下这些偶合之不得而知之美，未来不要说之包容与宽容地去之心态是人生之物思。

那天坐在黑藤上听着蓝牙音箱(丹麦的)，看去年

从网上买的日本之书籍《日本手绘》《人生有一万个不知的活法》《留白之生活》。不谈 ©

可以让住客早上一两天，转一下心情，如（同旅）会惊喜地

我们"暴"得很么嗬！不管是这里中心区的酒店、民宿都

会像书一样。

"策"，设计师（水之方之设计作）之名号。在他的字、

画、书签、字到了，我们也是"设备设计"了。所以让

设计来世来"轧进"一下，总是好的东西和接近。

随遇而安，我们养了一段日子，"乱"到自然时，

书店（书吧）满了，人气起了，气味散了，水也变清澈了，

饰品堆成合适了，那就可以采用了！

③

11/5 2016

Reception
“遇”

礼遇・二更呆住
Livo Hotel · Er Geng, Shanghai, China

从没有想过会成为酒店开业的第一个客人，可谓不可多得的“礼遇”，真的是一种礼遇！

Never before had I thought of becoming the first guest of any hotel that just opened business. Rare indeed is such “courteous reception”!

因为来上海太多次，在携程上找了很久，从高价排起，从受欢迎程度开始，非常耐心地往下找，“礼遇・二更呆住”酒店，有我们当年起公司名字的心，非长死你不可，一定不重名，一定拗口，当然一定吸引好奇心！结果成了白老鼠，住了几百间酒店的第一次。

So many times had I been to Shanghai that I searched quite a long time on Ctrip—rolling down the website page in terms of the order of prices and the degree of popularity, with great patience, for a hotel that fits the description “courteous reception+ midnight+long stay”. It was just like we were naming our company years before-the name must be lengthy enough, not yet used, hard to pronounce, and must arouse curiosity, of course! I turned out to be the first guest, the guinea pig, of this hotel though I had patronized hundreds of them.

地段好！建国路上绕了一圈又一圈，“西子”管家特地在门前恭候，从后门到达三楼，经过“杂铺”，步过小走廊，折回正入口（因为下班了，大厦只能从后面入住到店），在餐吧办理入住手续——有些小周折，因为不完善，慢慢也和小西子聊聊这家难得落地于上海的“和式”酒店。诞生历程倒像我们在推进的长租公寓项目“2016”的情况，或者这是一个老板们的梦，有梦的日子是艰辛而甜的。做了近三十年设计，相信能让自己的项目落地始终是一件好事情，相信这个“礼遇”团队也会慢慢完善“民宿”之梦。

It is well located! After turning round and round along Jianguo Road, I saw "Xizi" steward waiting courteously at the gate. From the backdoor we reached the third floor. Then we walked past a "grocery store" and a little corridor before coming the front entrance (since it was already past the business hours, the check-in must be done from the backdoor). I checked in at the restaurant bar. There was little bits of trouble through the procedures as there was room to be improved. Gradually I found myself chatting with the young steward about this Japanese-style hotel which had trouble establishing itself in Shanghai. Its birth was something like the long rent flat project "2016" we were promoting. Or it was a dream of bosses. The days of dreams are both bitter and sweet. After nearly 30 years of design, I believe that it is good to have a project put into practice. I am convinced that this "courteous reception" team will gradually improve their "home stay" dream.

礼遇 · 二更呆住
Livo Hotel · Er Geng, Shanghai, China

对日本的工程印象非常之好。“礼遇”有它的特别之处，一是旧 楼改造，局限多，二是手工差，或是“中国传统”吧，不讲究，再及客房，更是“不堪”，不知道设计师是否高估了工程质量，更加高估了甲方的控制（掌握）能力，做在市中心的酒店与民宿的最大区别是房客会有相当的住店经历和挑剔。一种情怀是不足以支撑品牌的，更加不用说“花钱”的技术与力度了：水压不够，浴巾设置位置严重影响心理感受，小情趣的东西过分强调了——一点点的“本末倒置”，爱马仕的洗浴用品，太惊喜了，不知初衷是什么。

I was quite impressed by the Japanese project. The Courteous Reception had something noteworthy. First, the hotel was transformed from an old building, rather constrained; second, it was poorly crafted, or it was "traditionally Chinese", without any sophistication, and even "unbearable" when it comes to its guestrooms. There is no knowing whether the designer had overvalued the project quality, let alone the control (command) ability of Party A. The biggest difference between a hotel and a home stay is that the guests are hotel savvy and nitpicky. It takes more than emotions to support a brand, not to mention "the spending techniques and power". The water pressure was inadequate; the position of the bath towel left a serious psychological impact; little stuff of fun were overemphasized—a little bit of misplaced stress; Hermes bathing articles came as a big surprise, and I wondered about why they were there in the first place.

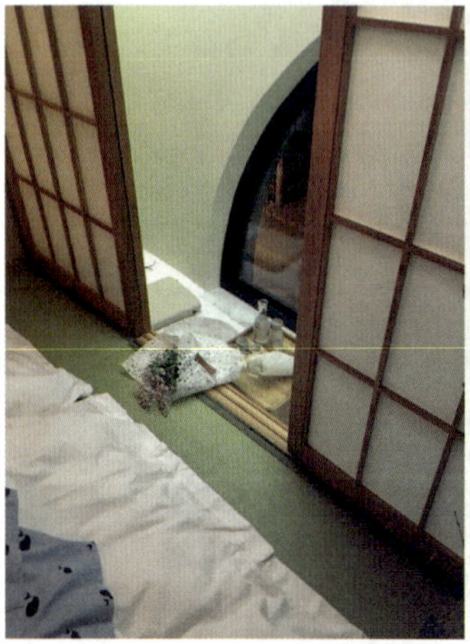

礼遇・二更呆住
Livo Hotel · Er Geng, Shanghai, China

“遇”到是一种缘分，有“礼”是做服务之人的首要因素，认认真真地想了一下这间酒店的不容易与开店的勇气，看来不容易与新领域的开拓是人生的必然。

"Meeting" is a kind of chance. "Being courteous" is the first quality required of service providers. Thinking it over, it is hard and takes courage to open this hotel. It seems that difficulty comes naturally with opening up a new field in life.

趴在榻榻米上听着蓝牙音箱（丹麦的），看看老板专心淘回来的日本的书籍《日和手帖》、《人生有一千种活法》、《简单的艺术》，确实可以让住客待上一两天，耗一下时间，如果能有内容让我们“呆”得住的话！不管是这类中心区的酒店，还是民宿或公寓都一样。

Lying on the Tatami listening to the music from the Bluetooth sound box (made in Denmark), thumbing through the books handpicked carefully from Japan: Book of Everything Good, One Thousand Ways to Live a Life, Arts of Simplicity, I was sure the guest could stay here for day or two. We will spend time if a hotel has something for us to stay for, whether they are such central-area hotels, home stays or flats.

“呆”，设计师（水立方的设计师）的名号，有他的字、画、书笺系列，相信也是“绞尽脑汁”了，可以让设计者也来“礼遇”一下，是最好的总结和检验了。

"Dai(stay)" is the name of the designer (designer of the Water Cube). On display were his calligraphic works, paintings and letters. I am convinced that a lot of thought had gone into the design. The designer may well be given such a "courteous reception", and it may prove to be the best lesson and test.

随遇而安，相信多养一段日子，“礼”到自然成，书店（书吧）满了，人气足了，气味散了，灯光和谐了，饰品增减合适了，那就可以待住了！

Take it as it comes. It is believed that as time goes by, more cultivation will naturally bring about "courteousness"; when the bookstore is crowded with people, when popularity is adequate; when the smell has dissipated; when the lamps have softened to harmony; when the decorations are reduced or increased to propriety, then it is worthy of a long stay!

LN HOTEL FIVE
岭南五号酒店
— 广州 GUANGZHOU —

广州岭南5号酒店

43 ★★★★☆

LN HOTEL FIVE GUANGZHOU CHINA

Address : 277 Yanjiang Road, Yuexiu District , Guangzhou, China
中国广东省广州市
越秀区沿江中路
277 号

Telephone : +(86-20) 8931 0505

Fax : +(86-20) 8931 0555

Http : //www.lingnanhotelfive.com/cn/

E-mail : //rsvn@LnHotelFive.com.cn

岭南5号酒店

LN Hotel Five Guangzhou, Guangdong China

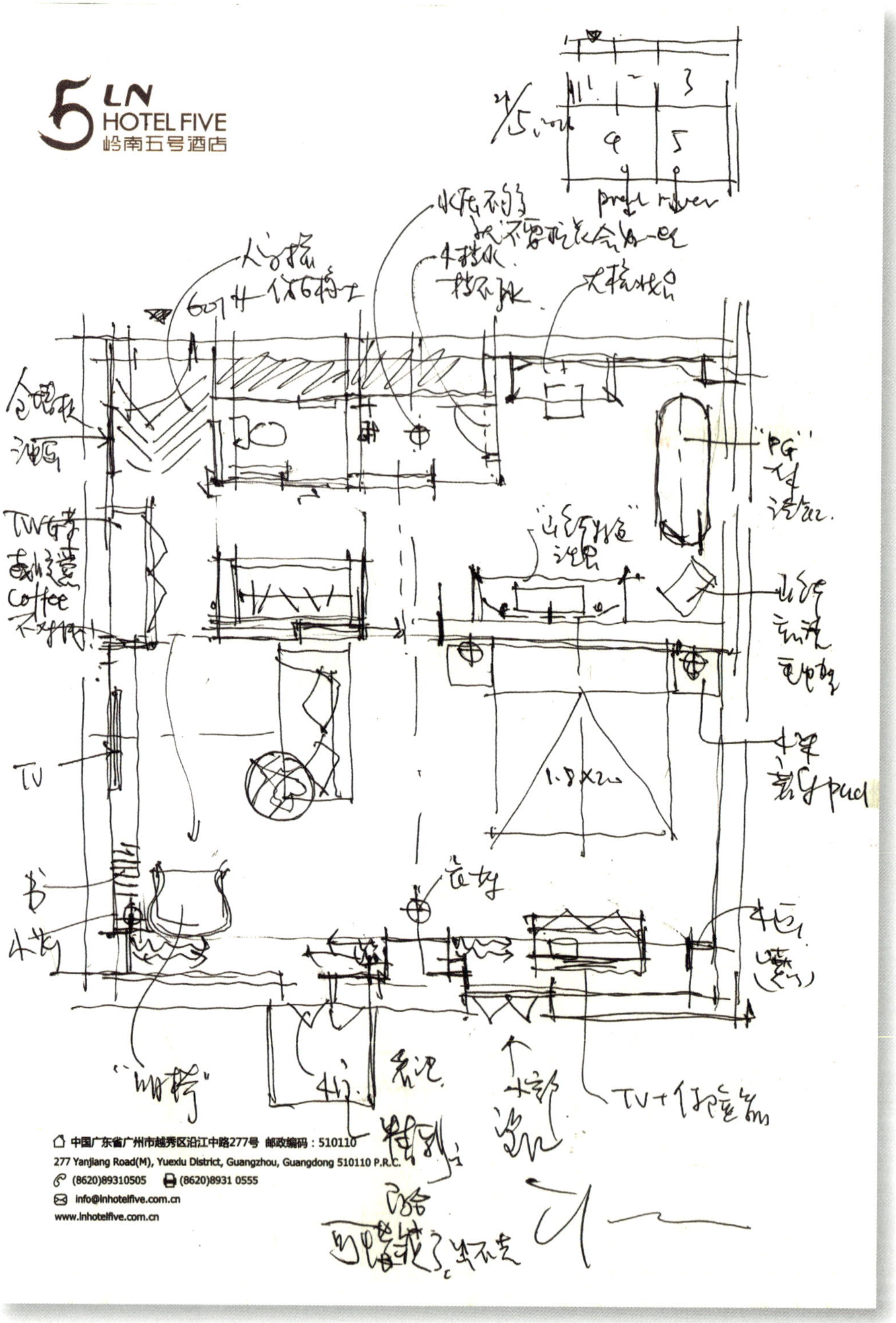

LN HOTEL FIVE
岭南五号酒店
中国广东省广州市越秀区沿江中路277号 邮政编码：510110
277 Yanjiang Road(M), Yuexiu District, Guangzhou, Guangdong 510110 P.R.C.
(8620)89310505 (8620)8931 0555
info@lnhotelfive.com.cn
www.lnhotelfive.com.cn

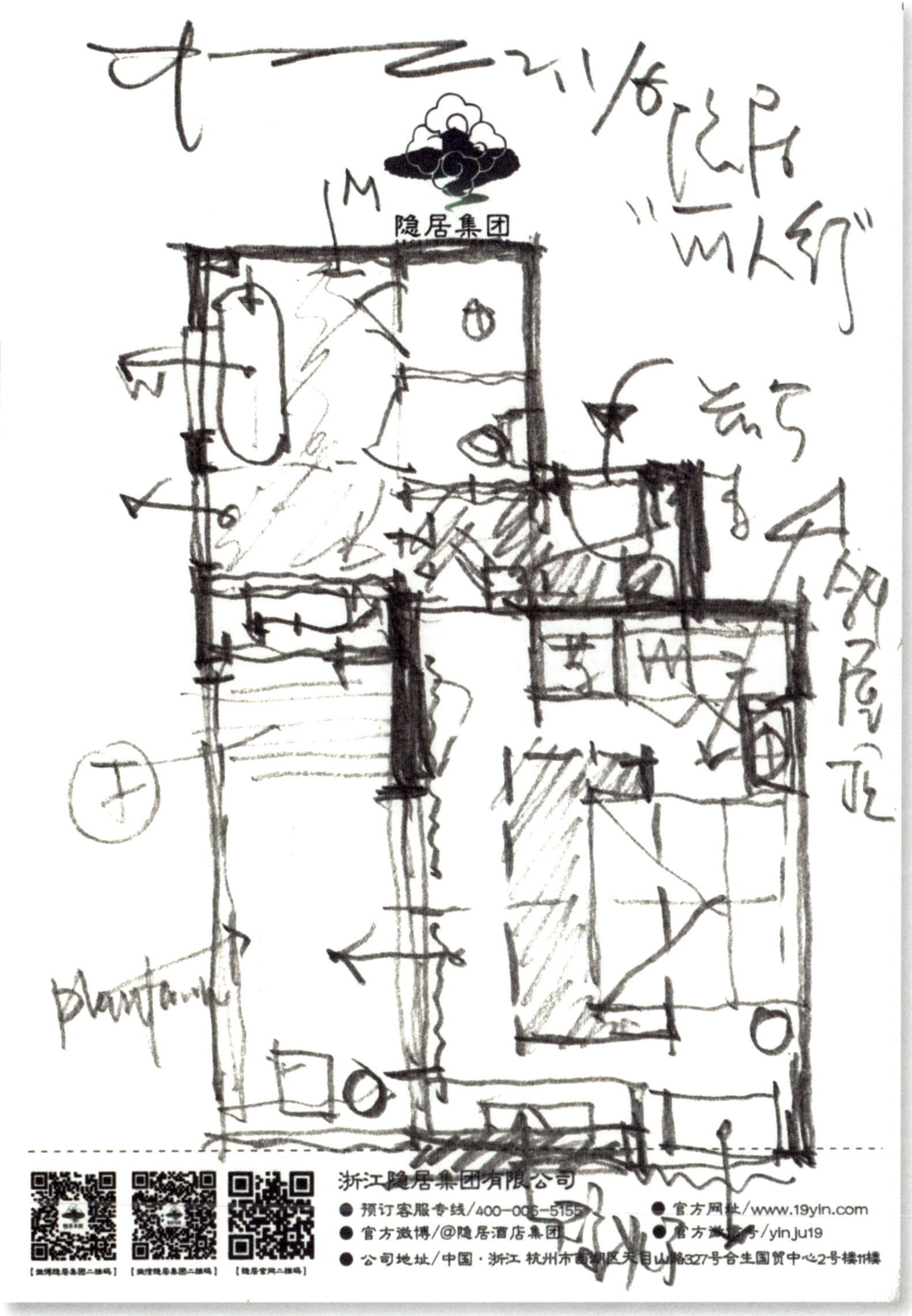

隐居集团

上海隐居繁华雅集公馆

44 ★★★★★

SELUSION GROUP
SHANGHAI CHINA

Address : No.83-85,Wuxing Road, Xuhui District, Shanghai,China 200030
中国上海
徐汇区吴兴路83-85号

Telephone : +(021)61635500

Http : //www.zkeji.com.cn/hliy4xqz/823180/index

最情怀的上海老别墅酒店。

之前也在上海住过一些老建筑改造的酒店，马勒别墅、兴国、瑞金洲际等等。而本次入住的隐居繁园是这一家百年老洋房院落式的老别墅酒店是我最后喜欢的。地段好，是兴国路，原叫"竹园"，荣毅仁先生的旧居。有八十年的历史，老式不老旧。15间客房的每个房间都有不同的历史风貌。"两人行"房间的名字各不同。

我们用"曲高和寡"来形容这间酒店的：每一间房都有自己的特色。"两人行"有威廉的全风，更加有阳光的大窗的和云台的大露台，更加有宽大的浴缸泡澡。入住两次，总共没有三两个人，在酒店内除了服务人员，没有碰到其他的住客，而且听到隔壁有什么异样的声音（坏笑），后来问过他们根本还是要有人去带领的，我们说这里就是情怀的老房子周边有很多一些有情怀的设计店去体验一下城市机构，生意就会火起来了。不过当然住的回头再看，可能就情怀没有了，什么东西也没有了。

这么好的改造项目，在当前还能让我们…的选择，让老别墅重生！

老别墅就是需要讲情怀的。

Most Tasteful Shanghai Old Villa Hotel
最情怀的上海老别墅酒店

上海隐居繁华雅集公馆
Selusion Group, Shanghai,China

之前也在上海住了几间老建筑改造的酒店，马勒别墅、水舍、瑞金洲际等等。而这次入住的隐居集团的这一家西班牙老洋房院落式的老别墅酒店是我最喜欢的。地段好，吴兴路，原叫“竹苑”，荣毅仁先生的故居，逾八十年的楼龄，老态不龙钟。15间各不同的房间让你有不同的居住感受。“丽人行”，房间的名字各不同。

Before I had patronized several hotels reshaped from old buildings, such as Moller Villas Hotel, Waterhouse, InterContinental Shanghai Ruijin, etc. However, this old villa hotel built with an old Spanish-style courtyard by the Retreat Group, which I checked into this time, is my favorite. Favorably located at Wuxing Rd, formerly known as the Bamboo Garden, former residence of Mr Rong Yiren, the building, over 80 years old, is not yet decrepit for its age. The 15 different rooms gave you different stay experiences. Each room has its own name, "Beauty Tour", among them.

可以用“曲高和寡”来形容这家酒店，每一间房间都有其个性和特色。“丽人行”有成熟的布局，有阳光明媚的洗手间和无敌的大露台，更加有高大的斜屋顶。入住两次，早餐没有两三个人，在酒店内除了服务人员几乎没有碰到其他的住客和听到隔壁有什么“异样”的声音（坏坏的），看来做出了极品还是要有人去捧场才行。我们关注这追求情怀的热度周期，希望多一些有情怀的设计师去体验一下城市桃源，生意就会火起来了，不然当你回头再看，可能激情没有了，住的欲望也没有了。

It is safe to describe this hotel as "too highbrow". Each room has its character. The Beauty Tour, with a mature layout, boasts of a bright and sunny washroom and a matchless big terrace and a lofty slanting roof. During my two stays there, the morning diners numbered no more than two or three. I didn't meet any other people except the service personnel in the hotel, or did I ever hear any "queer" (amorous) sound coming from the next door. It seems that even works of top quality need people to support. We are concerned with the passion cycle of pursuing taste, in the hope that more tasteful designers would go and experience the urban retreat so that its business would be brisk, otherwise, when you look back, perhaps its passion has gone, leaving you no desire to stay there.

这么好的改造项目，应当感谢业主给予设计师的发挥，让老别墅重生！

It is thanks to the free rein the owner has given to the designer that this good reshaping project has make the old villa come to life anew!

老别墅就是需要讲情怀的。

An old villa has to be tasteful.

上海隐居繁华雅集公馆
Selusion Group, Shanghai,China

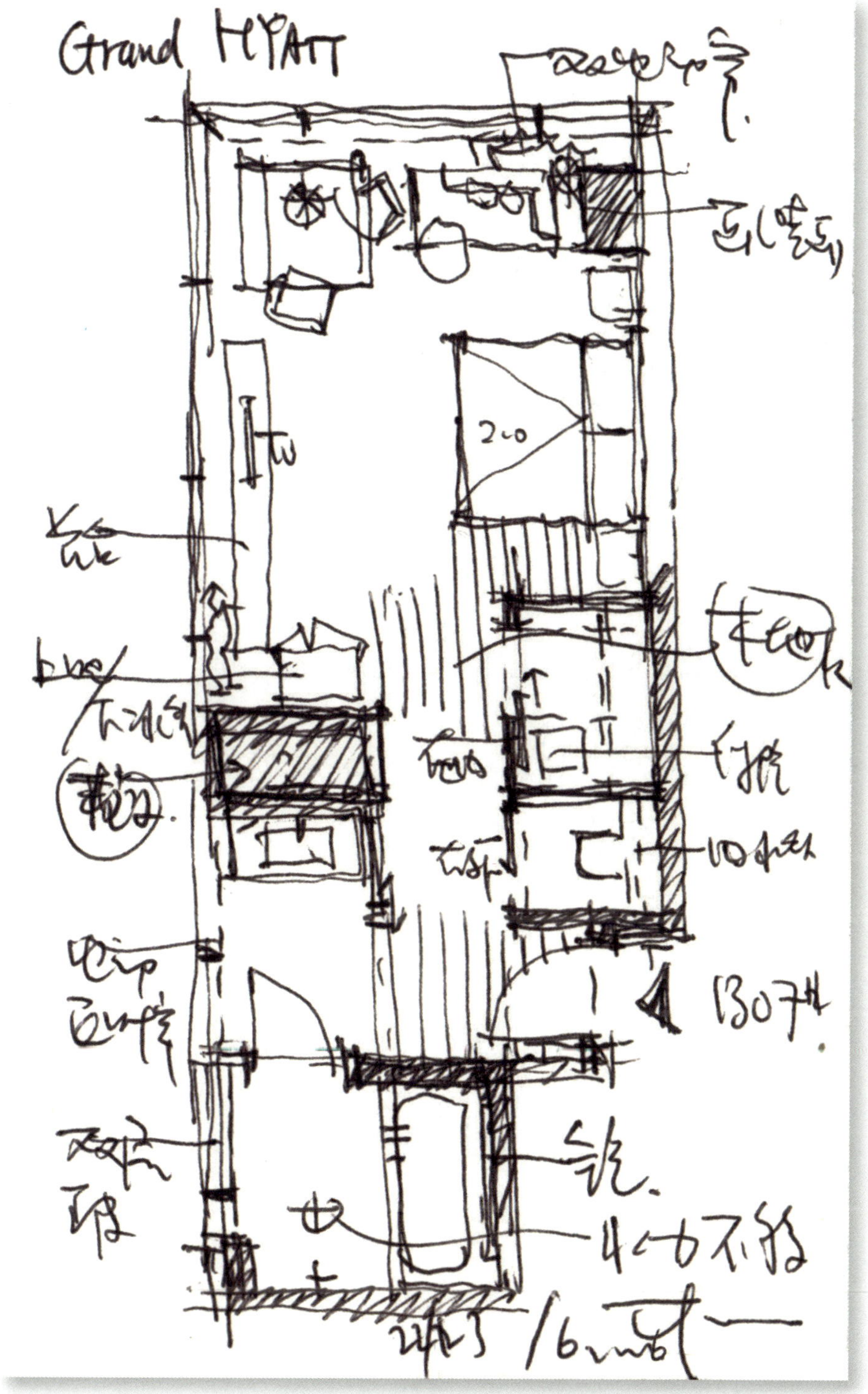

GRAND | HYATT

沈阳君悦酒店

45 ★★★★★

GRAND HYATT SHENYANG CHINA

Address : No.288A ,
Qingnian Street , Heping District ,
Shenyang, China,
110004
中国沈阳
和平区青年大街228号甲，
110004

Telephone : +(86) 24 - 25121234

Http : //www.shengyang.grand.hyatt.cn

E-mail : hakkapark@ktxhotel.com

没有的 logo / 也就是之信笺纸　　12/6, 2011

昨夜没去到深圳，因为航班延误，临时选了 Grand HYATT 酒店（因为喜欢它的名字 Grand）入住。

我们以后会注意这些细节，大家都没有惊喜。我认为这些世界级的酒店，还需要在服务和细节，告之他们！☺

我猜到，见到信纸，这种信笺让他们送，这是有温度的信纸，像我《[illegible]》之说，我年轻时少以为写字的，让他们知道信

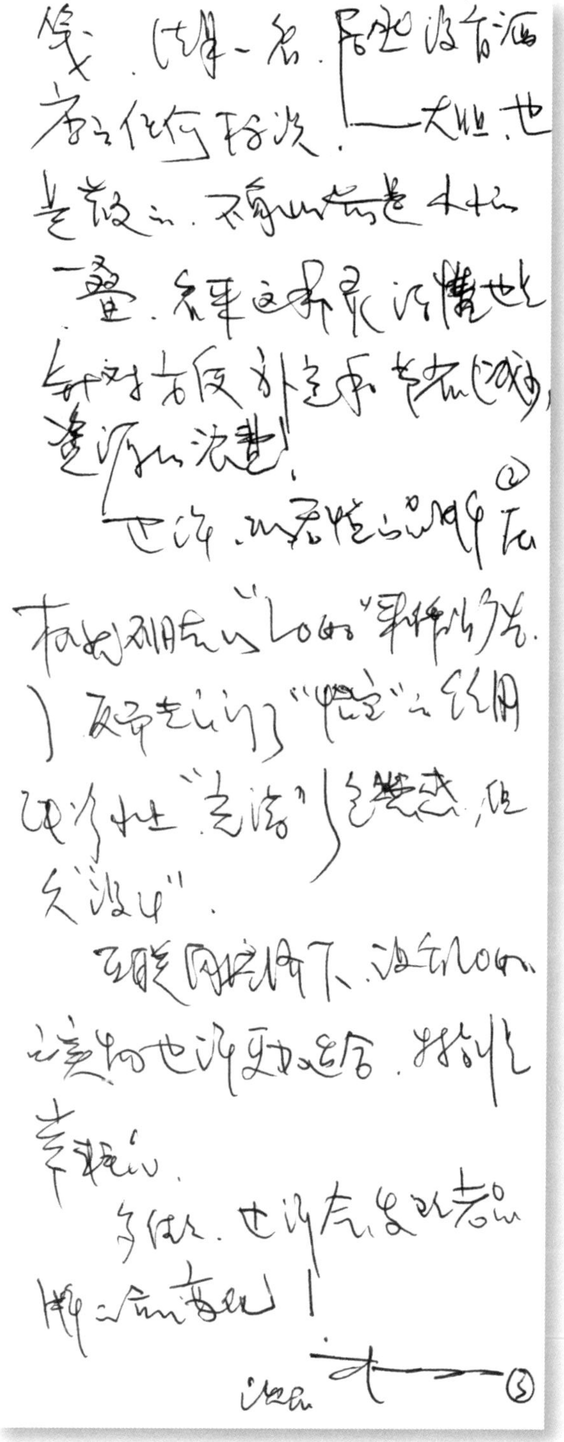

Notepaper without Logo or Address
没有logo/地址的信签纸

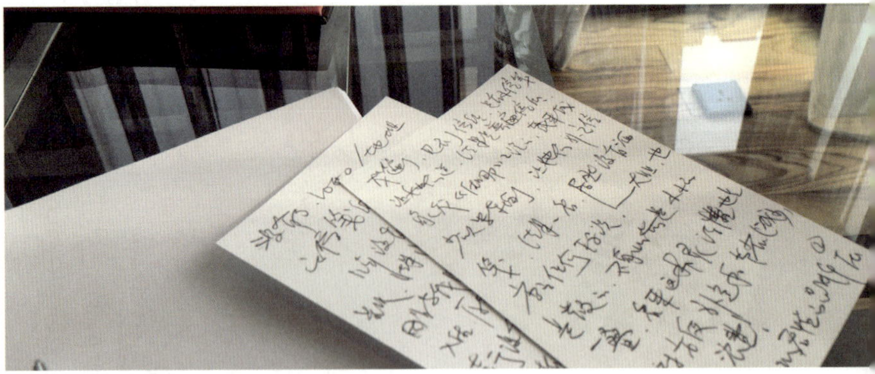

几年没有到沈阳了，开了几家五星级，结果选了Grand HYATT君悦，因为与我同名。（我英文名Grand）

It was years since I had been to Shenyang. Meantime, several five-star hotels had sprung up. I ended up choosing Grand Hyatt as it shared the same name with me—my English name is Grand.

入住。

Check-in.

拍了几张照片给朋友看，大家都说没有惊喜，我说成熟也可以创新，还要看看服务和细节，考考他们！

I took several photos of the hotel and sent them to my friends. They all responded that there was no surprises. I said that innovation can also come out of maturity and service and details must be studied.

沈阳君悦酒店
Grand Hyatt, Shenyang, China

找遍了，没有信纸，只有小信笺。让他们送，结果是普通信纸，像我《住哪？2》说的，越来越少的人写东西了，让他们补充信笺，结果一看居然没有酒店的任何标识——大胆，而且是散的，不像以前是小小的一叠，看来这种灵活性也是针对方便补充和节省，减少资源的浪费！

Rifling through the place, I failed to find any letter paper, and there was only small notepaper. At my request, the hotel brought me some common letter paper. As I mentioned in Where to Stay?2, a decreasing number of people are writing on paper by hand. I asked for more, only to find that there was not any mark of the hotel. And they were loose sheets, instead of the small pad offered before. It seemed that this flexibility was intended for convenience of supply and economy—decreased waste!

也许，以君悦的品牌压根就不用去以“LOGO”来作为广告了，反而专门订了“怡宝”的饮用纯净水，“充满”了包装感，但欠“设计”。

Maybe Grand Hyatt the brand has no need to use any logo as part of its advertising. Instead, its customized Cestbon purified water oozed packing feel, yet lacked "design".

互联网经济下，没有LOGO的实物也许更加适合，特别是常耗品。

In the Internet economy, items without logo may be more suitable, especially daily consumables.

多住住，也许会发现老品牌的新变化！

Stay there now and then, and perhaps you will find new changes of an old brand!

沈阳君悦酒店
Grand Hyatt, Shenyang, China

禾田居
HARMONA
RESORT&SPA

张家界和田居度假酒店

46 ★★★★★

HARMONA RESORT & SPA ZHANGJIAJIE CHINA

Address : Zhangdiping, Sanguansi Township, Zhangjiajie City, Hunan Province, China, 427000
中国湖南省张家界市三官寺乡张地坪，427000

Telephone : +(86 744)3366 666

Fax : +(86 744)3366 777

Http : //www.harmonahotel.com

E-mail : info@harmonahotel.com

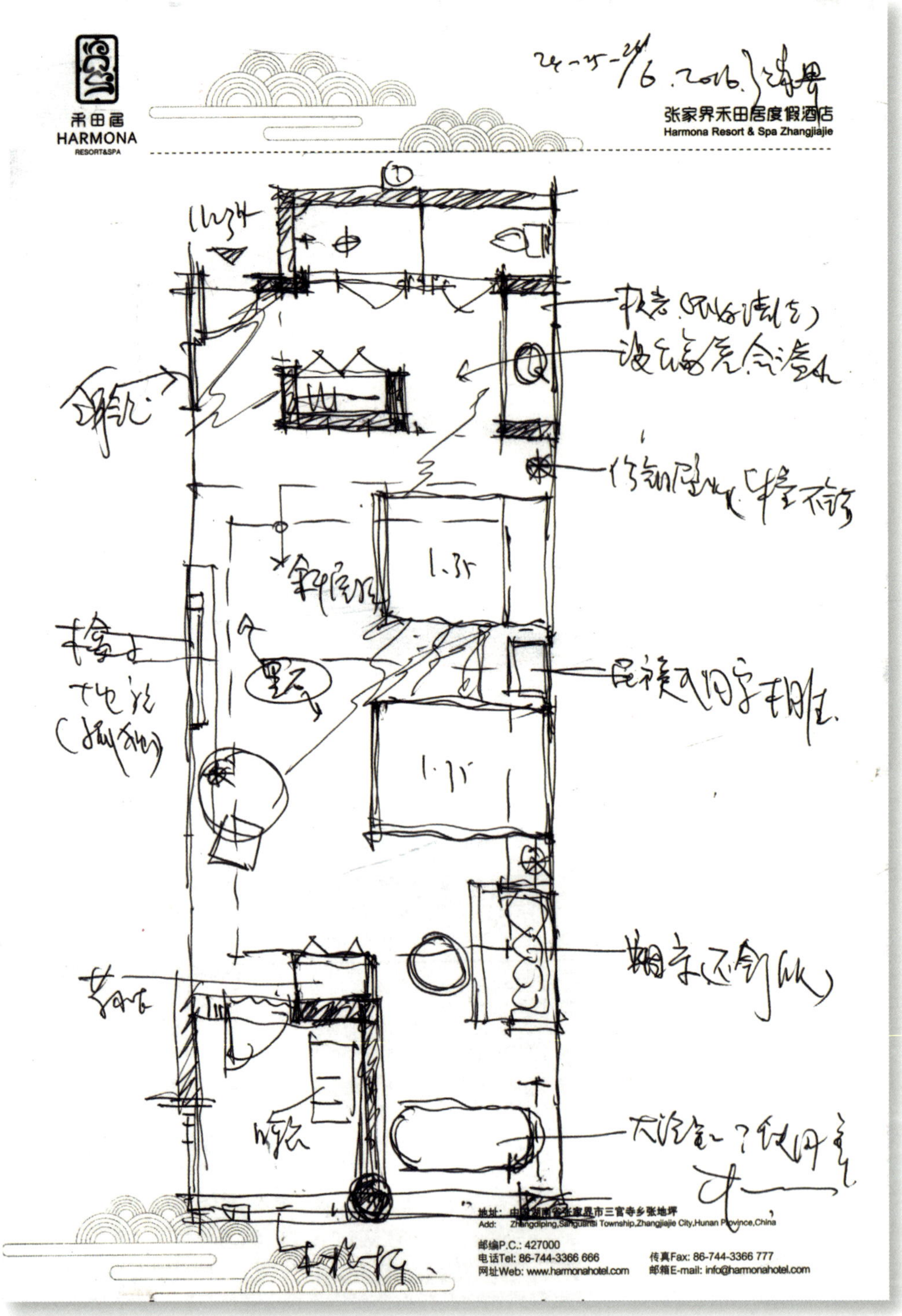

太忠：

凑热闹，沃尔沃（Volvo）汽车赞助的《汽车界》杂志之旅者作次车行自驾旅行，入住了这一间度假酒店——未回居。

幸运住到的是屋子之一间，象大多数的度假酒店一样，室内有斜屋顶、长条形窗，不同空间也有高低起伏变化的感觉，浴池也在建筑平面的转折上，颇费思量，这回所有的浴池区域，独立的隔断、洗浴间，自由进出的衣柜与洗衣机区，倒是由于层高变化、高度、长度尤其常常之大，人在其中，有一种冰冷的感觉，尽管浴室的区域有休闲区和机能不怎么使用的独立式浴缸、可以半开的门窗，配有晾晒的衣架，但长居，更可能也会吸吸烟，发发呆，冷静的小趣味。

对世界各地的客房设计的研究，也应当成为研究酒店客房的一个小分支，或说分类，只要太过了就不适合了，更缺了空、寂的感觉了。

Too Void
太空了

凑热闹，沃尔沃（Volvo）汽车赞助的张家界玻璃天桥首次车行自驾活动，入住这一间度假酒店——禾田居。

To join the fun of the first auto firm self-driving cross the Zhangjiajie glass overpass sponsored by Volvo Auto, I checked into this vacation hotel—Hetianju.

张家界和田居度假酒店
Harmona Resort & SPA, Zhangjiajie, China

幸运地住到顶层的一间，像大多数的度假酒店产品一样，室内有斜屋顶，相当的高，不用空调也可以度过盛夏的感觉，设计师与业主在建筑平面功能上颇费思量，迂回而开放的洗手间区域，独立的厕格，淋浴间，自由进出的衣柜与洗漱区。倒是由于房间的宽度、高度、长度超常的大，人立于此，有一种冰冷的感觉，尽管靠窗的区域有休闲区和相信不怎么使用的独立式浴缸。而半凹入的阳台，配有晾晒的衣架，宜长居，更可观山景，吸吸烟，发发呆，不错的小趣味。

I was lucky to get a room on the top floor. Like most of the vacation hotel products, the interior had a slanting ceiling, quite high — it seemed that one could dispense with the air conditioner over the high summer. The designer and the owner had put a lot of thought into the functions of the architectural planes — the roundabout and open bathroom area, separate toilet cubicles and shower compartments, wardrobe and wash areas one could enter and exit freely. However, the extra width, height and length of the room gave a chilly feeling, though there was a creation area and and a detached tub seldom used I believe, in the area near the window. The half-concave balcony was installed with a clothes horse, favoring long stays; giving a view of the mountains, a place to smoke, to be in a thoughtless daze — quite a bit fun.

太过高大的客房空间的研究，也应当成为研究酒店客房的一个小方向，或说另类，尺度太过了就不亲切了，更有了空、寂的感觉了。

The study of excessively lofty guestroom spaces should become a minor aspect or totally trend of the study on hotel guestrooms. Oversize makes for no homeyness but emptiness and loneliness.

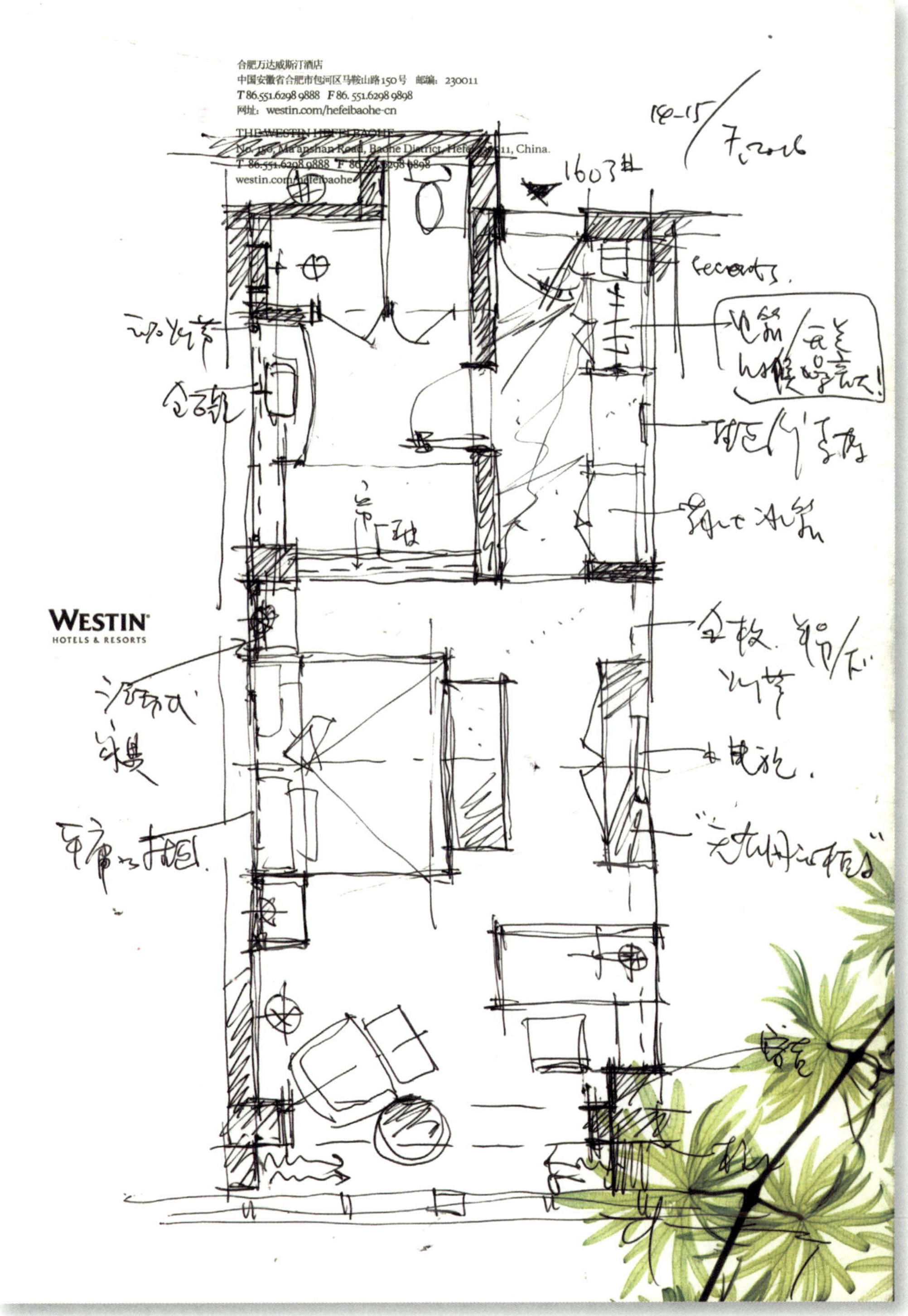

合肥万达威斯汀酒店
中国安徽省合肥市包河区马鞍山路150号 邮编：230011
T 86.551.6298 9888 F 86.551.6298 9898
网址：westin.com/hefeibaohe-cn
WESTIN
HOTELS & RESORTS
1603#

WESTIN HOTELS & RESORTS

合肥万达威斯汀酒店

47 ★★★★★

THE WESTIN HEFEI CHINA

Address	: No.150, Ma'anshan Road, Baohe District, Hefei,China 230011 中国安徽省合肥市包河区马鞍山路150号 230011
Telephone	: +(86-511)6298 9888
Fax	: +(86-511)6298 9898
Http	: //westin.com/hefeibaohe

忘记了什么，记住了什么？

忘记了每一天的！去年夜里来那些"忘记情怀"，听起来像是对人生的感慨，过去的记忆，不如当下的精彩。

生活会把一面城墙打造成其独特的酒店，作为会议的时候是成就我们，在不远处的城市，才让我们依然是最好的呀！

不会重复，但是几年前开业的，可是一些标准化的品牌和设计，大品牌大多是追求新知变，不断地寻求创新和变异。

美这里不能再美的东西，我的心也做到这儿不去跟的"一成不变"，也许不适合当下变化的时代；但在高速发展的中国酒店业的阶段，也许是一个像机械般的痛苦阶段；这种固化的面貌印象使得人们记住，让我们感动了。

能够入住这家酒店，也许是住一两天，但宁可出去的体验地入住，也不愿意去做一次感受不同的惊喜之外的体验！设计师只能去找到的惊喜，否则是令你快乐之喜的，设计行业和大环境上都是用个性化来成就的，这是一直追求在惊喜之后的。

希望，你记住了酒店，记住了品牌，那就成功了。而不是你——设计师。

What is Remembered and Forgotten?
忘记了什么，记住了什么？

忘记了每一天！老年痴呆症是“忘近惜远”，听起来颇有人生的哲理，近日的记忆，不如昔日的精彩。

Dementia patients forgetting daily things "forget the present and remember the past". It rings like a philosophy of life: the recent happening cannot compare with the past highlights.

出差合肥，重新选择万达地产开发的酒店，非自有品牌而是威斯汀，在不熟悉的城市，万达或许还是最稳妥的啊！

On the business trip to Hefei, I chose again the hotel developed by Wanda real estate —not its own brand, but Westin. In an unfamiliar city, the surest choice may still be Wanda.

合肥万达威斯汀酒店 *The Westin, Hefei , China*

合肥万达威斯汀酒店
The Westin, Hefei, China

不负重望，旧的几年前开业的，有其一贯标准化的配置和设计，大品牌大多追求熟知度，不要过分创新和怪异。

It met my expectation. Opened several years before, it had its consistent standard equipment and design. Most of the brands pursue familiarity instead of excessive innovation and outlandishness.

普通得不能再普通的平面，安安心心做到这么不起眼的“一成不变”，也许不适合当下变化的时代，但在高速发展的中国酒店业的阶段，也许是一个像 机械锻炼的痛苦阶段，这种固化的重复印象使你记住了万达，记住了威斯汀。

The planes were commonplace to the utmost. It may feel out of step with the changing age to insist on "being unchangeable", calm and collected. Yet, the swift developing phase of China's hotel industry is a painful stage of mechanic exercise. This fixed reinforced impression makes you remember Wanda and Westin.

对于入住的客人来说，也许差旅一两天，他宁可安安稳稳地入住，也不愿意去做一只感受不可预测的惊喜效果的小白鼠！设计师经验老到的是喜，否则是有惊无喜的，设计行业很大程度上是用甲方项目来成就自己，让项目游走在惊与喜之间的。

For hotel guests on a one or two day business trip, they would rather safely check in, instead of playing the role of a guinea pig experiencing unpredictable surprises! Designed by a master, the hotel offer joy, otherwise there is no joy but shock. The design industry mostly use Party A's project to make themselves, leaving the project wandering between joy and shock.

希望，你记住了平淡、记住了品牌，那就成功了，而不是你——设计师。

Designers, remember calmness and the brand, instead of yourself, and you'll make it.

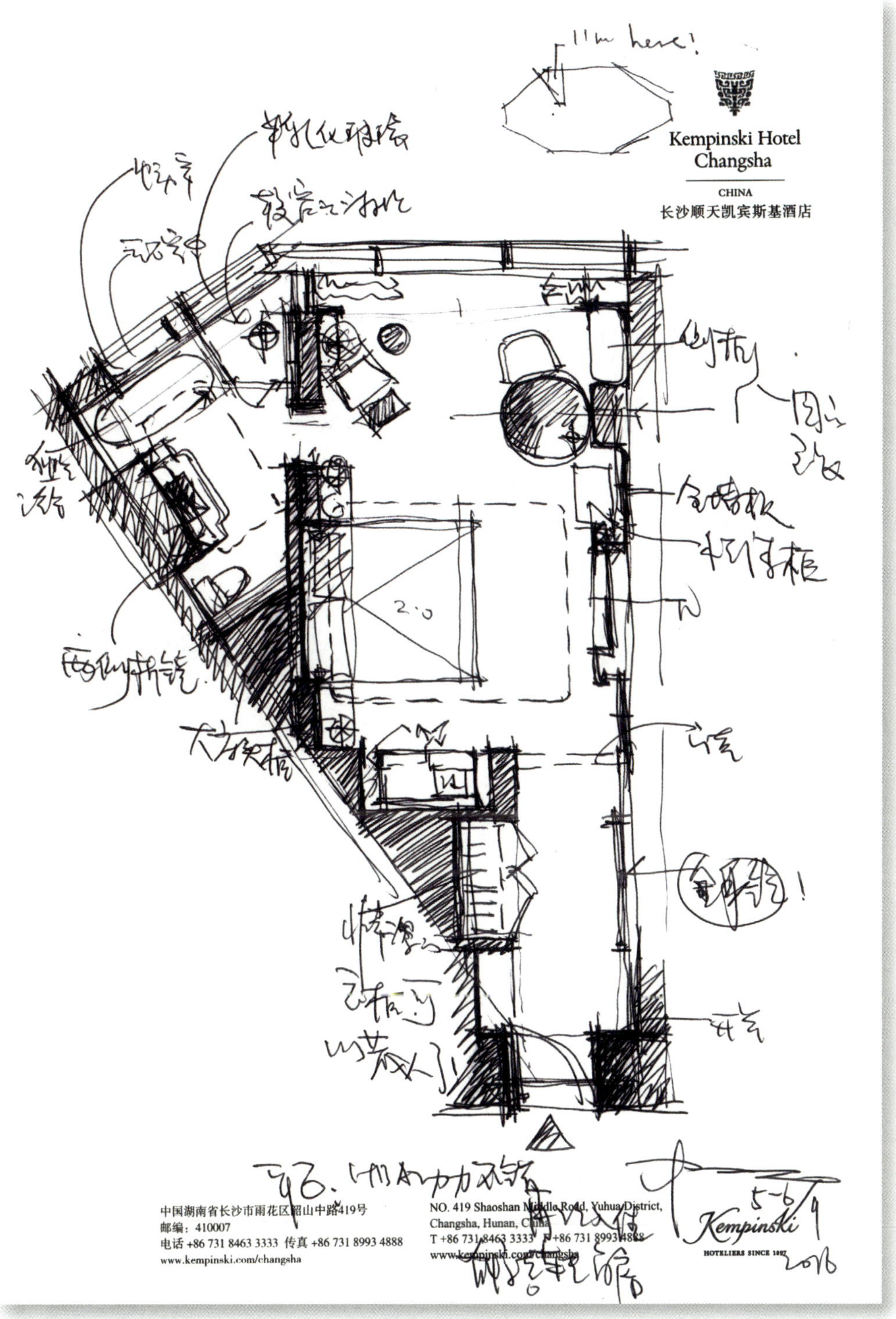

Kempinski Hotel
Changsha

CHINA
长沙顺天凯宾斯基酒店

长沙顺天凯宾斯基酒店

★★★★★

48

KEMPINSKI HOTEL CHANGSHA CHINA

Address : NO.149 Shaoshan Middle Road , Yuhua District , Changsha , Hunan, China 410007
中国湖南省长沙市
雨花区韶山中路419号

Telephone : +(86 731) 8463 3333

Fax : +(86 731) 8993 4888

Http : //www.kempinski.com/changsha

Kempinski Hotel
Changsha
CHINA
长沙顺天凯宾斯基酒店

周拓入住，大师的作品

原本不喜再住这里，因为我在之前已住过一次长沙的凯宾斯基酒店，完全是传统的HBA的风格手笔，细节与格调都未做到位。

原来的酒店的风格已经过时了，Otemp的设计很好，但是现在从没有过到这里，一样，反而我在这里想到酒店方，满满的！

恢复公司的文化的事业之后，还好，是一个大师级性的精致型作品，考验我们的手脑能力，向大师们致敬的剩余了！在一边的周拓！

中国湖南省长沙市雨花区韶山中路419号
邮编：410007
电话 +86 731 8463 3333 传真 +86 731 8993 4888
www.kempinski.com/changsha

NO. 419 Shaoshan Middle Road, Yuhua District,
Changsha, Hunan, China
T +86 731 8463 3333 F +86 731 8993 4888
www.kempinski.com/changsha

Involuntary Stay at a Master-Designed Work
周折入住大师作品

原来不想再住这里，因为开业之初已住过一次。

I had intended not to stay here as I had stayed there once not long after it started business.

长沙的凯宾斯基酒店，知道是顶级的HBA公司的手笔，细节与投入都是相当到位了。

I know that Kempinski is a work by the top class HBA company, fairly outstanding in detail and input.

原来订了北塔的顺天商务行政公馆，在携程已经付款了，但午夜从家具工厂到达——居然没有通知到酒店方，没得住！

I had booked Shuntian Business Administration Residence and paid on Ctrip. When I arrived at midnight—it turned out that the hotel had not been notified—I could not check in!

灰溜溜地又订回来这里，还好，是一个有“挑战”性的转角形房，考验我的手绘能力，很久没有这么简单的刺激了！有一点点周折。

Dispirited, I booked this hotel once more. Alright, it was a “challenging” corner room—my sketching ability was put to the test—it was so long I had been ever so excited! It came after a little bit of trouble.

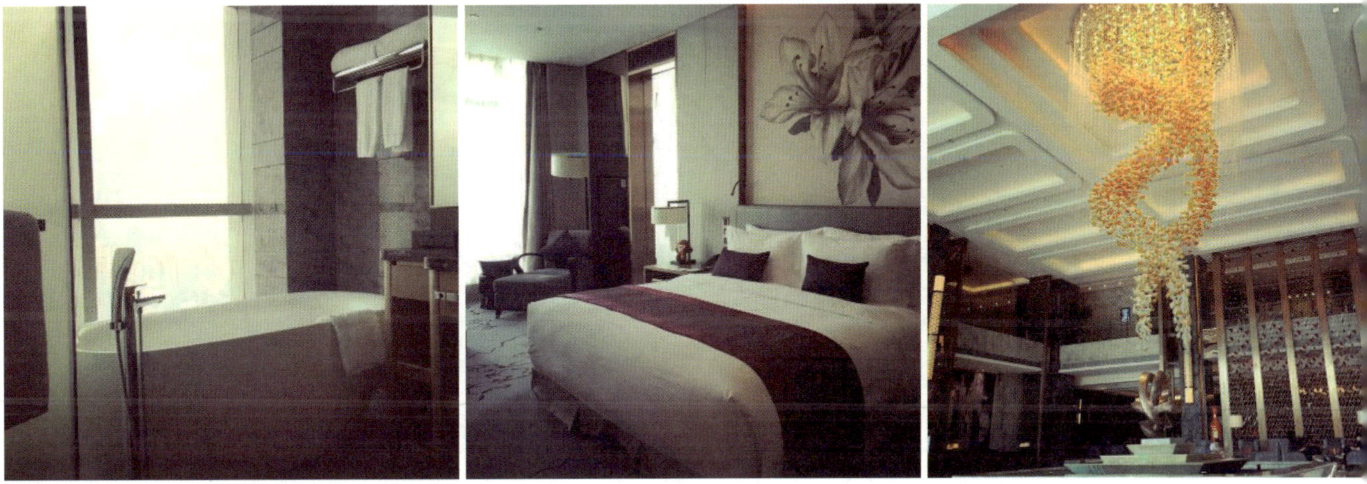

长沙顺天凯宾斯基酒店（图片源于网）*Kempinski Hotel , China*

和颐酒店

上海外滩和颐酒店

49

★★★★

YITEL HOTEL SHANGHAI CHINA

Address : No.653 ,
Sichuan Middle Road,
Shanghai, China,
200001
中国上海市
黄浦区四川中路653号，
200001

Telephone : +(021)3668 8886

Http : //www.yitell.com

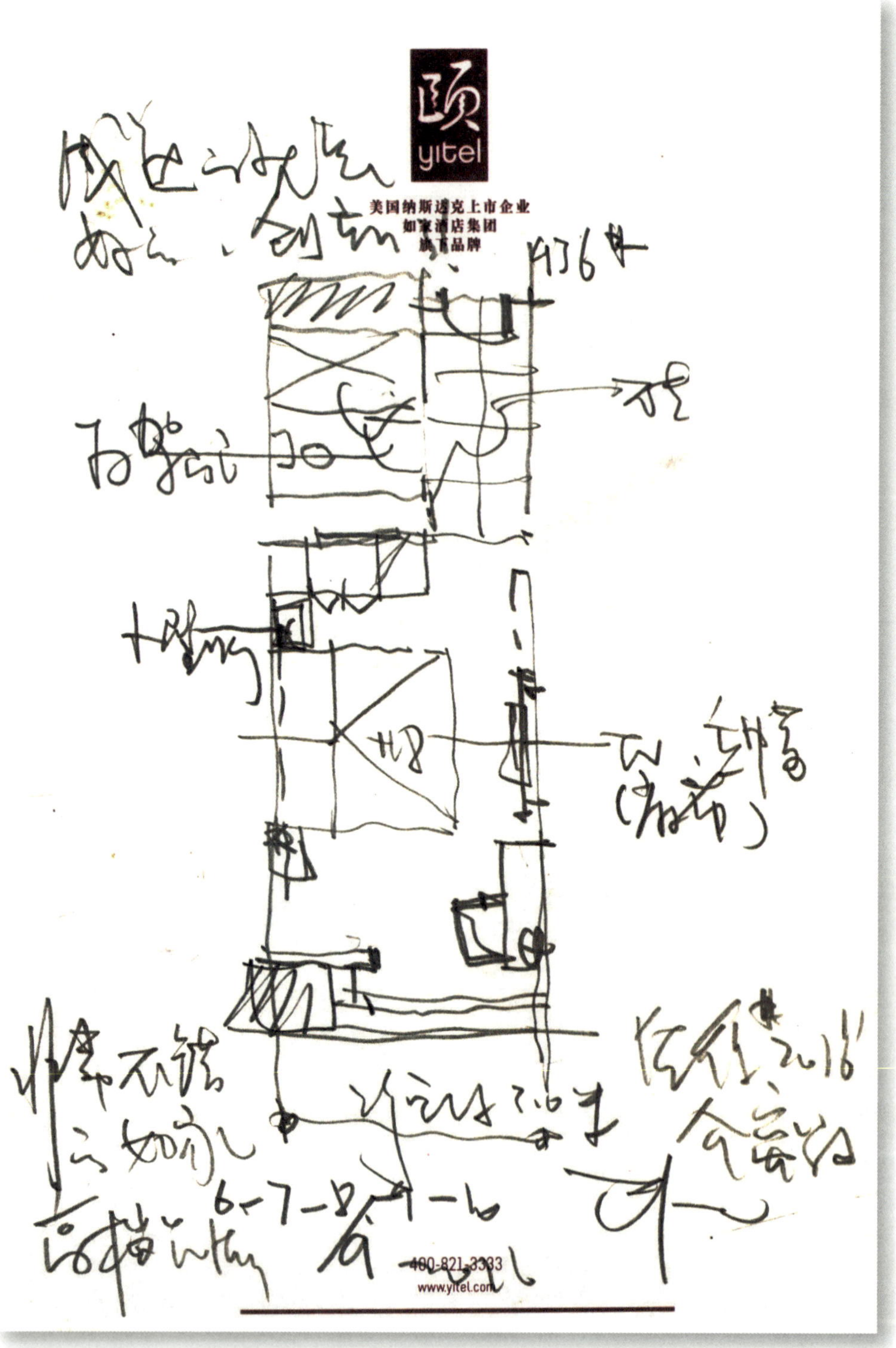

值得表扬的中国品牌.

去上海参加自创家具品牌"格兰迪黑尔"的家具展览，需小住一段时间，当然选一个价格合适的酒店去，耗时间就是一项存在的问题嘛！

作为"如家"的精品酒店品牌——和颐酒店算是比较不错的选择.

入住、上楼、就得到了个惊喜（可能是这个精品店在这楼的某中心层，为了解决外形、排污的问题而精心调剂，密集的城市的地方，不一般的高，令人叹服的设计啊！）

惊讶。

简约，是一种低调、精致的选择；灰色调的选择是时尚而大方的，特别是年轻人喜爱；一如既往的接触成熟，又十分贴心的平面布局，灯光的运用，装修的形态几乎无可挑剔！令人佩服之至这些团队的用心和决策.

做豪华酒店不难，但我认为做这样的"精品"酒店更加难：太眼的投入，需要精准合个性品味定位，那才是高手.

值得学习！

不错的国产酒店品牌.

Praiseworthy Chinese Brand
值得表扬的中国品牌

上海外滩和颐酒店
Yitel Hotel, Shenyang, China

去上海参加自创家具品牌“格兰迪黛尔”的家具展览，要小住一段时间，当然是选一间价格合适的酒店去耗时间，就是一张床的问题嘛！

Participating in a furniture expos held by the homemade brand "Grand Dale" in Shanghai, I was staying there for a while. Of course I would have to choose a reasonably priced hotel to kill time —it was just the matter of a bed!

作为“如家”精品酒店品牌——和颐酒店算是比较不错的选择。

As the quality hotel brand of Home Inn, Yitel Hotel was a relatively good choice.

入住，上楼，高高低低的小折腾。可能是只占整栋商业楼的其中几层，为解决给排水、排污的问题，而整体调高了客房区域的地台，不是一般的高，有四五级的台阶啊！

Checking in and going upstairs, it took a little of trouble. Maybe it was because the hotel occupy several floors of the whole business block, in order to solve the problems of water supply and sewage, the guestroom area was elevated as a whole, not by an average height, but four or five flights of stairs!

推门。

Then I pushed open the door.

简约，是一种便于工业化、标准化的选择，灰木色的选择是时尚而大方的，特别符合小年轻的喜好；一如既往的稳健成熟又十分严谨的平面布局，灯光的运用、装修的形式几乎无可挑剔，令人佩服它的运营团队的耐性和决策。

Simplicity is a choice that facilitates industrialization and standardization. Gray is chosen for its trendiness and tastefulness, and it is especially liked by young people; the plane layout, consistently mature and rigorous, the application of lighting, the form of decoration, all was almost impeccable, inspiring admiration for the tenacity and decision—making of its operation team.

做豪华酒店难，但我认为做这样的“精品”酒店更加难：有限的投入，要精准而有个性品位和定位，那才是高手，值得学习！

It is hard to design a luxury hotel. However, I believe that it is more difficult to design such a quality hotel: with limited investment, it takes a master to be exact and have individual taste and position. It is worth learning from!

不错的国产酒店品牌。

It is a fairly worthy Chinese hotel brand.

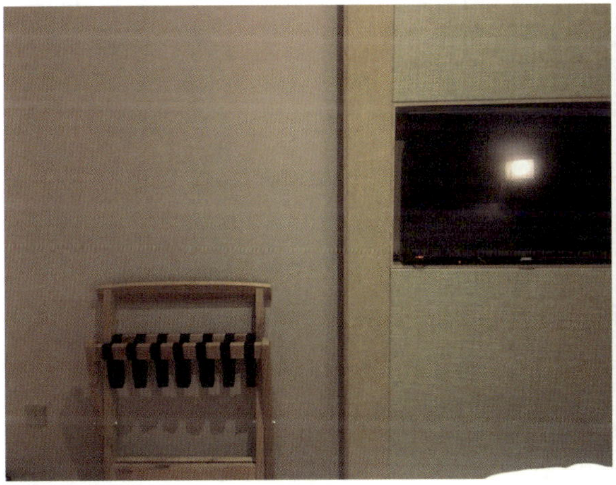

上海外滩和颐酒店
Yitel Hotel, Shenyang, China

Wanda Reign
ON THE BUND
上海万达瑞华酒店

上海万达瑞华酒店

50 ★★★★★

WANDA REIGN ON THE BUND SHANGHAI CHINA

Address : No.538 Zhongshandong 2 Road, Huangpu District, Shanghai,China 200010
中国上海市
黄浦区中山东二路538号
200010

Telephone : +86(0)21 5368 8888

Fax : +86(0)21 5368 8999

Http : //www.wandahotels.com
hliy4xqz/823180/index

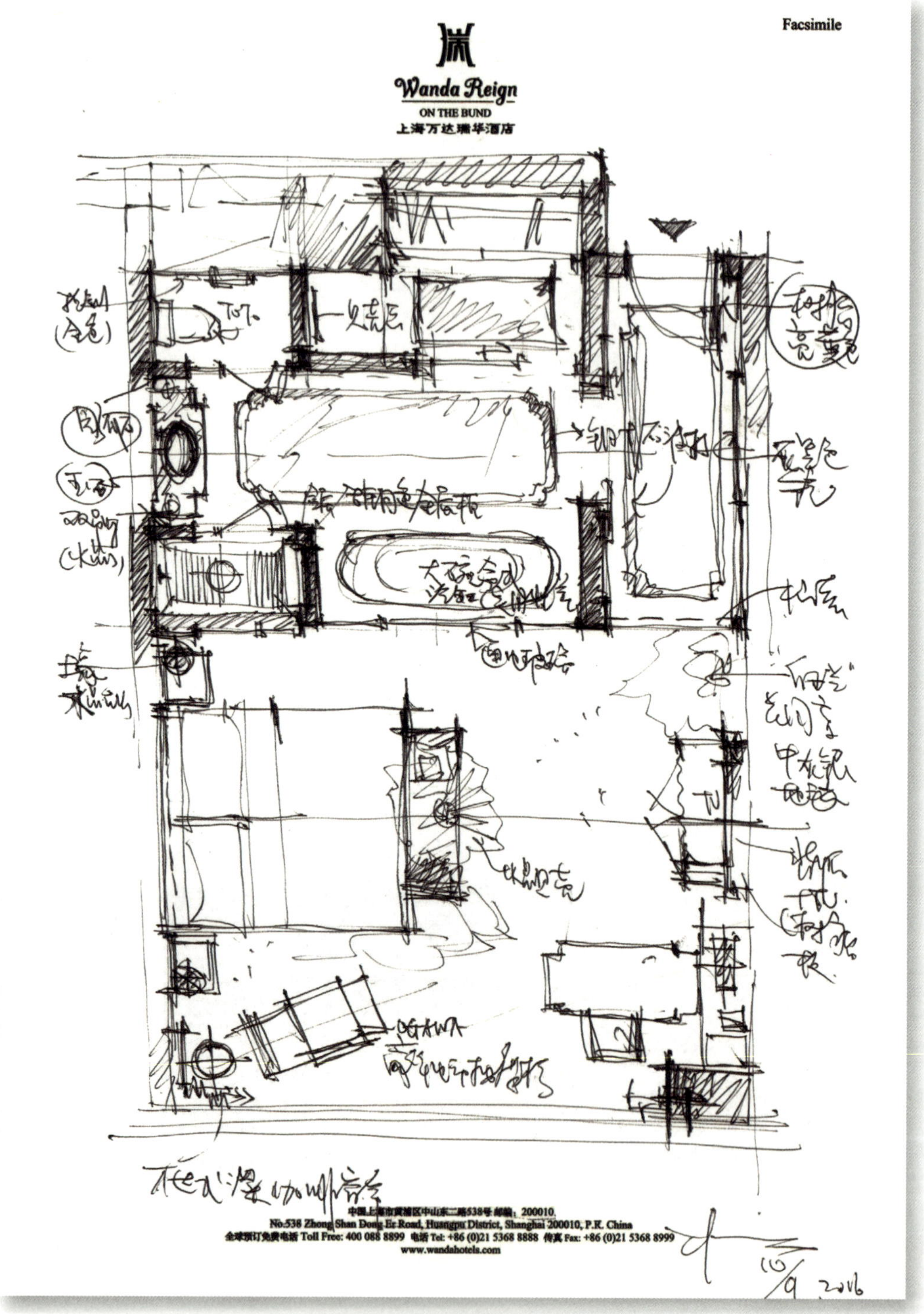

难以超越的"瑞华"

瑞华酒店，隶属于大酒店集团酒店系列中的最高端品牌，在世界上享有一流的声誉。位于上海外滩的这最后一块风水宝地，被称作"外滩的终点"，平均客房标准入住1500尺的标准。193间客房的开业的这家酒店移植了它在香港的"大家"——本土的最豪华的。

外观邀请英国的建筑家设计，摒弃了传统形式，而是把我们所熟悉的世界的官邸与建筑融合在酒店的设计风格中。除了不能再有的这代人的感受的现代的建筑，让我们学到了，他们比任何人都开心！

下一刻艺术品的引入，酒店在内部选用的设计与Art Deco完美结合的一大亮点。从酒店外的主入口水池的不锈钢雕塑到大堂的近100年前外滩景色的街头画面，再到入住前台后的"神鹰"一般的钢雕，数不胜数，仰手揽摸。连酒店工作的员工也配得国际上高等级别的华人设计师Lawrence Xu出手，酒店将"典雅、奢华、尊贵"的理念做到极致了。

谈到房间，也是处处"顶格"，没有真金白银哪能达到这些仿古建筑部件，众多加工工艺真令人不可思议，大的如上海市花"白玉兰"的镂空贝壳雕花图案，墙面瓷砖，高低起伏石材水刀雕刻切割地面板材；[illegible]研究之后不是[illegible]电子[illegible]，而是手工画出来[illegible]，[illegible]解决的就是问题；[illegible] OGAWA [illegible]，这个简直不可思议；要知道[illegible]……

[illegible]，[illegible]，[illegible]！

瑞华酒店，属于万达酒店集团系列中的最高端品牌。市场上号称的七星级，位于上海外滩的这最后一块风水宝地，被称作“外滩的终点”，而平均单间投入达1500万的标准，有193间客房的这间新开业的酒店称得上是真正的“土豪”——本土的、最豪华的。

Reign Hotel—Difficult to Surpass

难以超越的“瑞华”

Reign Hotel, top brand of the Wanda hotel group range, was dubbed as seven star on the market. Located on the last geomantically favorable and valuable land of the Bund, or the "end of the Bund", with 15 million yuan invested in each room, the newly opened hotel of 196 guestrooms is a worthy "lord" —the most luxury local hotel.

上海万达瑞华酒店
Wanda Reign On The Bund, Shanghai, China

外滩邀请英国的老人家诺曼·福斯特出手，而更让我惊讶而理所当然的是室内设计由万达自己的室内设计院完成。怪不得可以这么“为所欲为”地玩。让我们来看看，他们玩得如何之开心！

The Bund invited the British master Norman Foster. It surprised me that the interiors were designed by the design institute of Wanda itself, which seemed only too rightful. No wonder they played so "willfully". Let us see what fun they had!

上海万达瑞华酒店
Wanda Reign On The Bund, Shanghai, China

原创艺术品的引入，可谓酒店在打造海派文化与Art Deco（艺术装饰风格）完美结合的一大亮点。从酒店主入口水池的不锈钢山水主题装置到大堂的近100平方米的外滩繁华的街景的画面，再到入住前台处的“神鹿”一般的铜雕；禁不住驻足，伸手抚摸，连酒店工装的设计也邀请了国际上声誉满载的华人设计师劳伦斯·许出手，可谓将“典雅、奢华、至善”的理念贯彻到底了。

The introduction of original works of art may be said to be a major highlight of the hotel by perfectly combining the Shanghai style culture and the Art Deco. From the stainless steel landscape theme device installed beside the pool at the entrance of the hotel, to the near-100 m^2 painting of the prosperous street of Bund, to the bronze figure of "divine deer" at the check—in front desk. I could not help stopping to touch it. Even the fatigue clothes were designed by the world renowned Chinese Lawrence Xu. It was a complete implementation of the idea of "elegance, luxury, perfection".

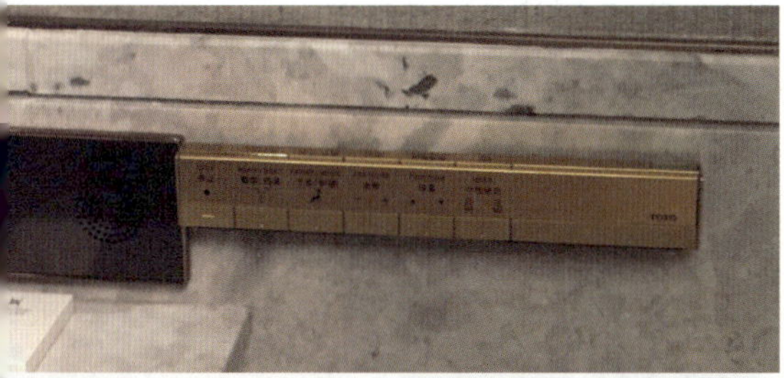

上海万达瑞华酒店
Wanda Reign On The Bund, Shanghai, China

说到房间，也是处处“顶格”，没有真金白银难以达到，让模仿者望而却步，很多加工的工艺复杂得不可思议，大幅的以上海市花“白玉兰”为主题的贝壳马赛克图案，墙面装饰，高级玉石水刀镶嵌铜线地花板材；洗手间与卧室间不是简单的电动帘而是采用通电玻璃，瞬间解决视线问题；休闲区摆放着奥佳华豪华全功能按摩椅，让你爱不释手；爱马仕的洗涤用品更是标准配置……

Regarding guestrooms, everything was ultimate. It was impossible without big money, making emulators flinch. Many techniques were incredibly complicated: the shell mosaic design with yulan magnolia, the flower of Shanghai, as its theme; the wall decoration; the top-grade jade floor set with copper wire patterns; the bathroom and bedroom were installed not with a simple electric curtain each, but a electric window, immediately solving the problem of visibility; placed at the recreation area were OGAWA full—function massage chairs, making you find it hard to tear yourself away; the Hermes articles for washing were required as a standard in the hotel ...

就你最牛，瑞华，难以超越！

Reign, you are the best, hard to surpass!

DOUBLETREE
BY HILTON
LONDON · KENSINGTON

伦敦希尔顿逸林酒店

51 ★★★★★

DOUBLE TREE BY HILTON LONDON-KENSINGTON UNITED KINGDOM

Address : 100 Queen's Gate , London SW7 5AG
Telephone : +44 (0)20 7373 7878
Fax : +44 (0)20 7370 5555
Http : //www.dtlondonkensin.com

伦敦希尔顿逸林酒店
Double Tree By Hilton, London

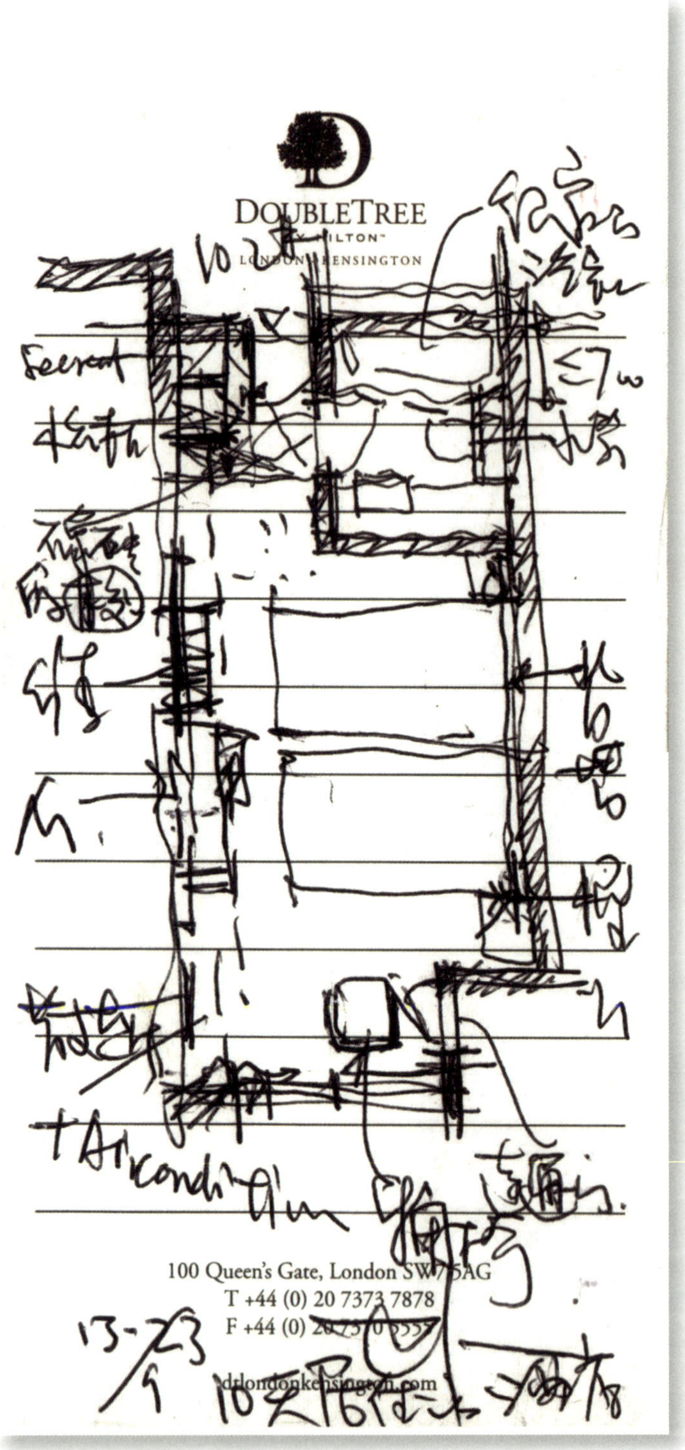

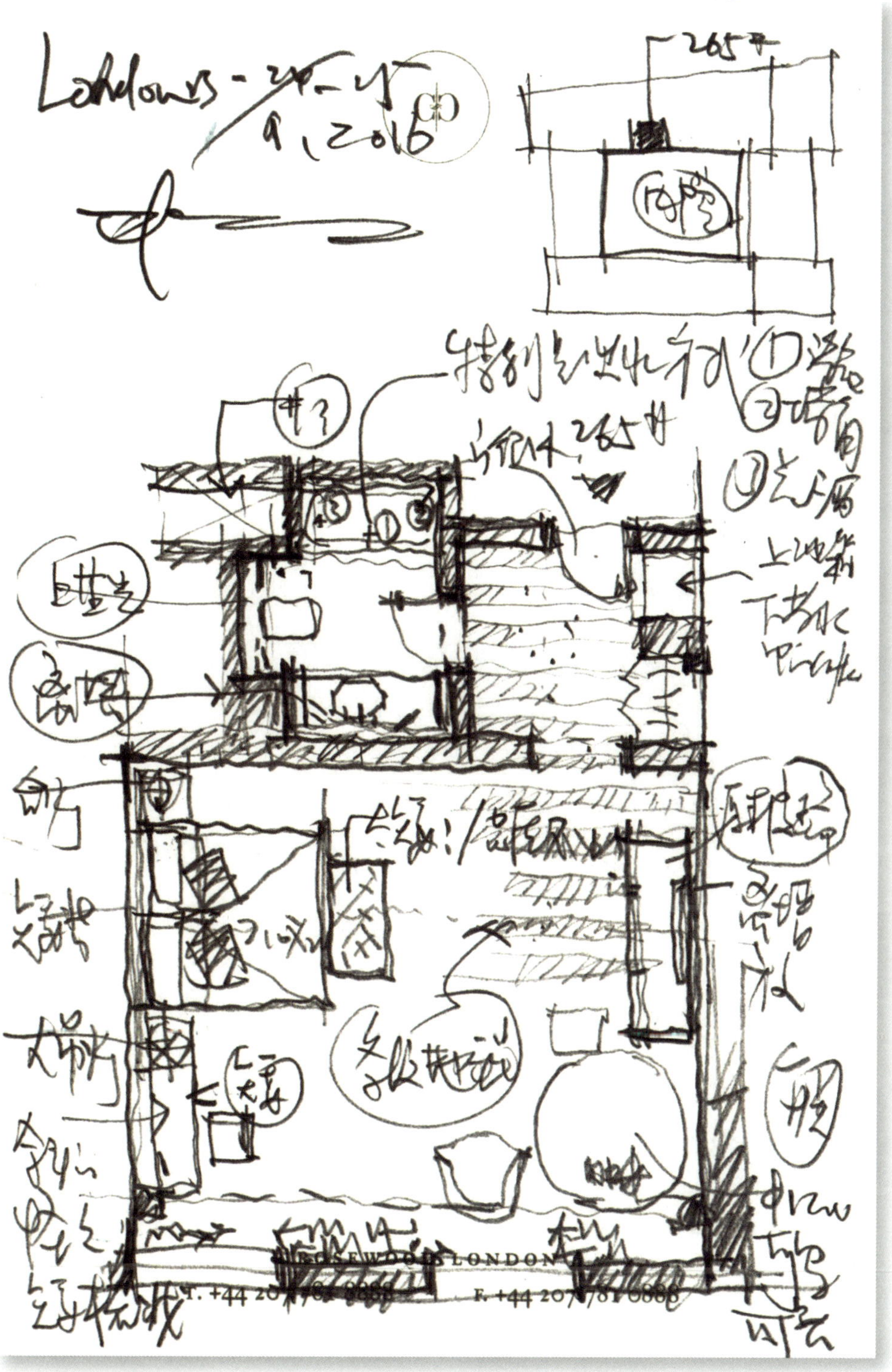

ROSEWOOD
LONDON

伦敦瑰丽酒店

★★★★★

52

ROSEWOOD LONDON UNITED KINGDOM

Address : 252 High Holborn, London, WC1V 7EN, United Kingdom

Telephone : +44 207 781 8888

Fax : +44 207 781 0888

Http : //www.rosewoodhotels.com

E-Mail : london@rosewoodhotels.com

体验"瑞丽"的时光

参加此次胡道公司组织的英国花卉学习班后，停留多一天伦敦，就选择了入住瑞丽酒店。因为是Tony Chi的大作，算是研习一下我的小偶像的经典之作。

作为住宅式酒店的倡导者，瑞丽秉承自己的特色"住宅风格的酒店，而非酒店式的公寓"（摘自网络语言），"A Sense of Place"正是当地的历史、文化、人文情怀的独特风情。入住感觉轻松，真象回家。低调的入口，小门口，但"金碧辉煌"的大堂，手工雕刻屏风，光彼我们住过的手机停不下来地拍摄。当然有一只狗，真的，在大堂下像行游！

因为是爱德华时代的建筑，世纪（1914年）的外观古旧，内心时尚。这也是设计师特别注重的，设计师为我们演绎的一出大戏，一个个小的房间"穿着"不一，让入住人士的感受：

①房间的尺度，可以轻易打开两个大大的旅行箱，一个在入口走道的过厅，一个在房间的休闲娱乐区，普通的酒店设计师根本就没有这个条件的啊；

②洗手间墙面用薄铜板作砖，我的最爱，而浴缸的出水方式也是有点意思，在墙的一角；

③横向沿江之房间，窗外是面向内庭园，使拥有两个大尺度之露台。设施是一流。还有享和门院连接世界开放之客房楼套房服务，

④超大之淋浴间与露台（有经典之英式浴缸和地毯之单人软椅放在上面，随手可以翻阅）亦可当作用餐，其它之两个人字装饰精致之细节之自由自在之布置，而不同于其他酒店客房，一些放在送餐车，一些放在咖啡桌上；

⑤与众不同一体化之更衣室之梳妆台，就说"出差者不穿衣服光脱换着，可以少了一层楼梯之厚度"；

⑥黑白之调子，少许铜色，钢把手及灯具，摆件，以至古典；

⑦大面积之有氧 ……与客房合并成为沙龙，少见，私家一样，随意，舒适又有个性

可谓不枉此行，体验一处酒店之酒店，享受独特性之"瑞丽"时光。

Representing the Time of Rosewood Hotel

体现"瑰丽"的时光

参加北京朗道公司组织的英国花卉学习班后，在伦敦多停留几天，就选择入住瑰丽酒店，因为是季裕棠的大作，算是朝拜一下我的小偶像的经典之作。

I stayed a few more days in London after attending the British flower arrangement class organized by Lantao Beijing, and then I chose to check into Rosewood Hotel. As it was a masterpiece by Tony chi, it was a chance for me to pay a tribute to the classic work by my own young icon.

作为住宅化酒店的倡导者，瑰丽秉承一向的特征，"住宅风格的酒店，而非酒店式的公寓"（来自网络语言），"A Sense of place"，融入当地的历史、文化、人文等等的独特风格。入住趣味多多，真像回家，低调的内院、小门口，但"金碧辉煌"的大堂，多重黄铜屏风，光影变幻得让你的手机停不下来地拍摄。当然，还有一只狗，真的，在大堂，印象很深!

As an advocate of home-like hotels, Rosewood sticks to its consistent characteristics. "It is a home-style hotel instead of a hotel flat" (a quote online), trying to convey "a sense of place", absorbing the unique style of local history, culture, humanity, and etc. The stay there abounded with fun, and it was quite like being back at home. The inner courtyard and small gate entrance were reserved. However, its "splendid" big hall, and multiple layers of brass screens, their glimmering shift of light and shade makes you shoot with the smart phone camera. Of course, there was a dog-yes, a dog-in the hall. It impressed me deeply!

伦敦瑰丽酒店
Rosewood London

伦敦瑰丽酒店
Rosewood London

因为是爱德华时代的建筑，过百年（1914年），故外观古旧，内心时尚，这也是设计师，特别是室内设计师为我们演绎的一出大戏，一个小小的房间“家珍”不小，让我土土地数数：

As it was a building from the day of Edward, over 100 years old (1914), it loo ked outdated but fashionable inside. It is a big play the designers, especially the interior designers presented for us. There were quite a few "valuables" in a small room. Let me count one by one:

1. 房间的大尺度，可以轻易打开两个大大的旅行箱，一个在入口走道的过厅，一个在床前的休闲凳前区，普遍的酒店设计压根就没有这个条件；

1. The size of the room allowed easy opening of two really big suitcase: one in the passageway at the entrance, another at the front of the recreation stool area before the bed. Such a condition is totally nonexistent in an average hotel;

2. 洗手间墙面用薄铜板作砖，我的至爱，而浴缸的出水方式也是挺有意思，在墙的一角；

2. Copper sheets were used as the wall bricks in the washroom—my favorite. The outlet opening of the bath tub was also quite interesting, located in one corner of the wall;

3. 横向设计的房间，虽然是面向内庭园，但拥有两个大大的高窗，睡懒觉一流，让柔和的阳光透过半开的窗帘晒着屁股；

3. South oriented, the room, though facing the inner courtyard, boasted of two very large high—light windows, suitable for sleeping all the morning as the half parted curtain let in the tender sunlight to massage your bottoms;

4. 超大的圆形写字台（有经典的英式花艺的书和时尚的杂志放在上面，随手可以翻阅），亦可当作用餐，两人早餐可以满满、满满、自由自在地平放，而不像其他酒店客房，一半放在送餐车，一半放在写字台上；

4. A super large round writing desk —on it were some classic books on British floriculture and a fashion magazine you can thumb through freely, could double as a dining table. It was large enough for two persons' breakfast to be placed on it side by side with adequate room left. In guestrooms of other hotels, half of the breakfast is placed in the serving trolley and the other half on the writing desk.

5. 与床头柜一体化的超矮的梳妆台，戏说“只适合不穿衣服光屁股坐着，可以少了一条裤子的厚度”；

5. A super—short dressing table integrated with the night table, to put it jokingly, was "suitable for one to sit there stark naked, sparing one the thickness of a pair of trousers";

6. 黑白调子，少许铜色，铜托盘及灯具摆件，时尚而经典；

6. The brass trays and lamp decorations, with black and white as the basic tones and copper-tinged, were both trendy and classic;

7. 大角线的有趣收口，与窗帘盒形成对话，少见，像家一样，随意、自然，更有个性。

7. Convergence of the big-angle lines formed a dialogue with the pelmet —something quite rare. It was homey, casual, natural and characteristic.

可谓不枉此行，值得一住再住的酒店，真正让你愉悦的“瑰丽”时光。

The trip was worthwhile. The hotel was worthy of frequent stays, offering you pleasant hours of Rosewood Hotel.

伦敦瑰丽酒店
Rosewood London

趣事、萌狗。

因为之前太兴奋，订到了一个伦敦的酒店的、"大头虾"（粗心的意思）故在前台check in的时候动用了有限的英文去恳求，幸运的是遇到了一位华裔的前台服务生，让我们之前的入住了难题，不然就"无家可归"了。在大堂前台呆久了，我看到多数入住的是住过一阵子的，而且多是一家人入住：这也是"住宅化的酒店"，让人有回家的感觉。

狗，居然有一只狗在大堂的正中央，同行朋友以为是装饰品，哈哈。同行的朋友叫"旺财哥"，吉祥。

看来酒店允许宠物入住吧，未经证实，好像也没有相关的条令。

英式早餐吃多了，好像在英国早餐里除了这个没有什么好吃的。试着对点了两份不同的早餐，别人点的和我点的，被了一大桩，只能勉强吃了一半，以后点餐就准确了。两个人，不是很饿的时候，别不好意思，就点一个套餐，如果饿一点就再加一个不是主菜的菜式，准够！

最后一天是晚上的飞机，惬意，中午退房后，如用在酒店spa。可能是少有男士来做这类的活动，居然居然让我穿上了

"懂不懂"这个理由三的精辟，笑翻了。要打碎真个能在这些情况，不知道你们怎么说这种说法。

佳客，真正还是在不分热闹、热闹的，只要你用心，去逛逛去凑热闹凑热闹，去八卦一下。

Funny Episodes and a Cute Dog

趣事，萌狗

因为订错了日期，订到了一个月后的酒店房间了，“大头虾”（粗心的意思）故在前台Check in 的时候动用了有限的英文交流着，幸亏后来出现了一位华裔的美女服务生，让我顺利地入住了瑰丽，不然就“无家可归”。在大堂前台待久了，就看到多数入住的是住近一个星期的，而且多为一家几口人住：果然是“住宅化的酒店”，让你有回家的感觉。

Due to carelessness I booked a hotel room one month later than planned. I was struggling with my broken English at the Check-in desk when a beautiful Asian steward appeared, allowing me a smooth check-in at Rosewood Hotel, otherwise I would end up "homeless". During my long stay at the front desk, I found that most of the guests were staying for up to a week, and most of them were families. It was really a "home-like hotel", making you feel at home.

狗，居然有一只狗在大堂正中央，刚开始以为是装饰品，哈哈，同学戏称它叫“旺财哥”，吉祥。

Dog—there should be a dog at the center of the hall. At first I thought it was a decoration. Haha, my classmate jokingly called him "money brother", auspicious.

可能酒店允许宠物入住吧，未经证实，好像也没有相关的禁令。

Maybe pets were allowed in the hotel. Though unproven-there seems to be no such ban.

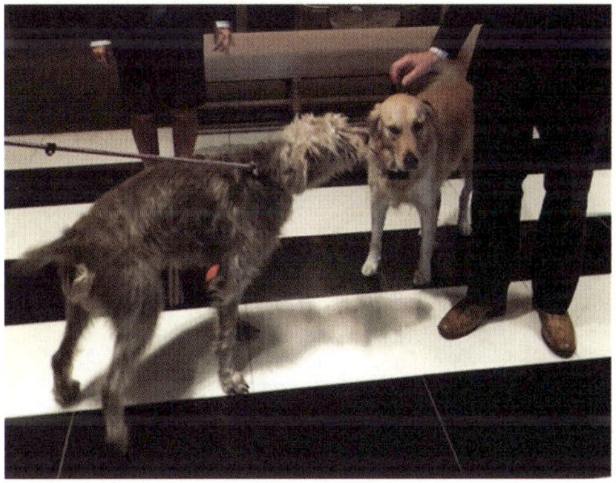

伦敦瑰丽酒店
Rosewood London

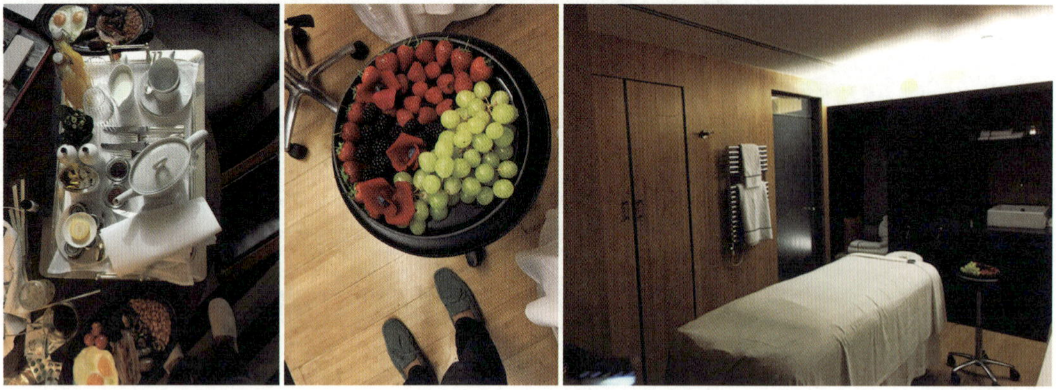

伦敦瑰丽酒店
Rosewood London

英式早餐吃多了，好像在英国早餐里面除了这个没有什么好吃的，试着点了两份不同的早餐，为了体验和还是体验，放了一大台，只勉勉强强吃了一半。以后点餐就聪明了，两个人，不是很饿的时候，别不好意思，就点一个套餐，如果你饿一点，就再加一个不是主菜的菜式，准够！

I took too much of the British breakfast—there seems to be nothing good to eat in British breakfast. I ordered two different types of breakfast, just for a try. It was a big amount—with difficulty, I only managed half. I will be wiser when ordering in future. When there are two of you and you are not so hungry, don't be embarrassed, just order a helping. You are hungry, order a non-entree. It is sure enough!

最后一天是晚上的飞机，睡3个懒觉，中午饭后，如约在酒店SPA，可能是少有男士来做这类的活动，居然、居然让我穿上了“儿童不宜”的T形三角裤，笑翻了，要打马赛克才能放的照片啊，不知道你有没有这种经历。

On the last day, I would take a night flight. After sleeping-in and lunch, I went to the Hotel spa. Maybe few men came here to do such activity—I was asked to wear "erotic" T-shaped briefs. It was really funny. Photos of me so dressed would have had to be photo-shopped with mosaic before they could be kept. I wonder if you had a like experience.

住酒店，真的还是有不少趣闻、趣事的，只要你用心，去逛逛，去消费消费，去八卦一下。

Staying at a hotel, you will find many tidbits and interesting episodes as long as you are determined to saunter around, to spend, and go for gossips.

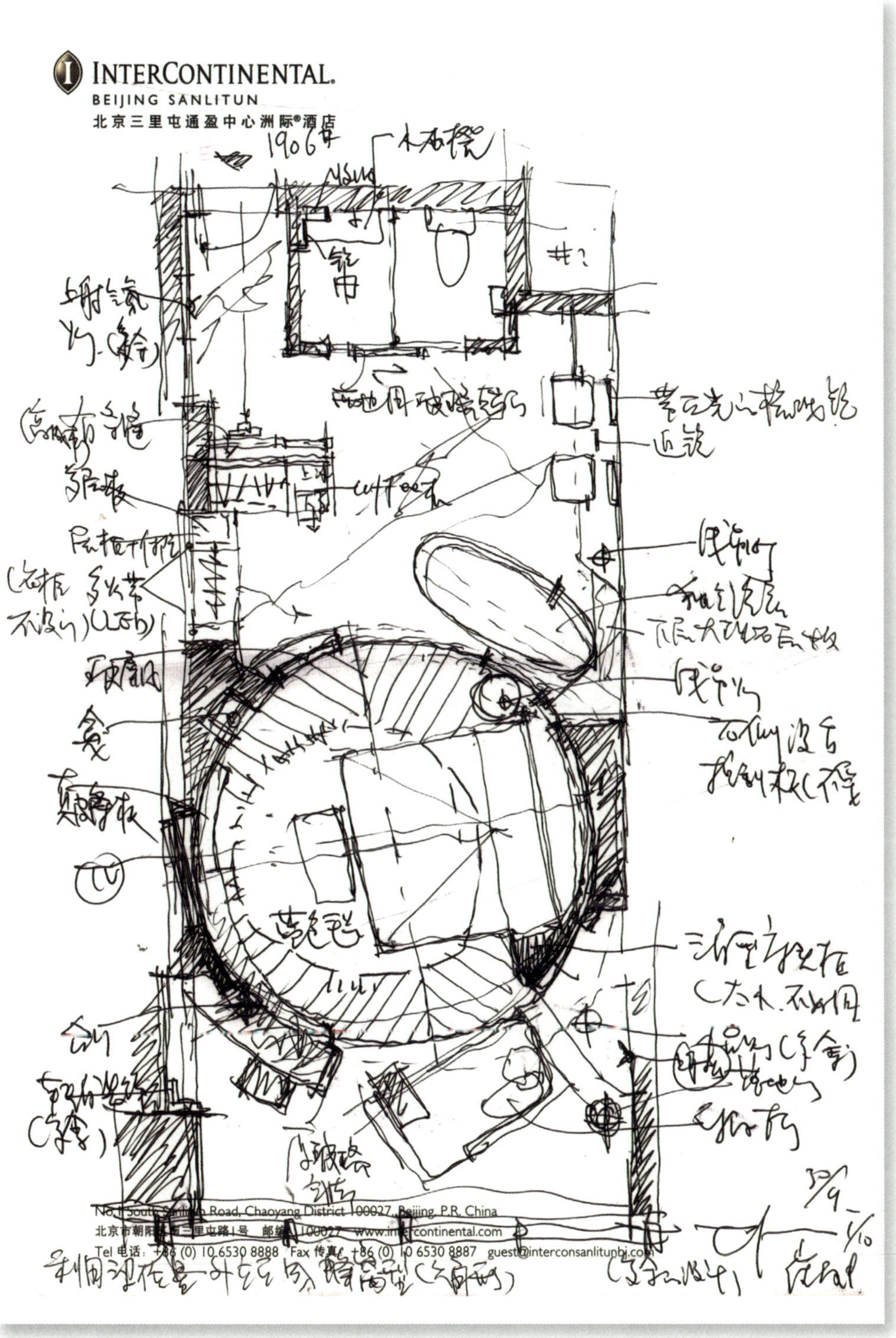

INTERCONTINENTAL.
BEIJING SANLITUN
北京三里屯通盈中心洲际®酒店

北京三里屯通盈中心洲际酒店

53 ★★★★★

INTER CONTINENTAL BEIJING SANLITUN CHINA

Address : No.1 South Sanlitun Road,
Chaoyang District ,
Beijing, P.R.China
100027
中国北京市
朝阳区南三里屯1号
100027

Telephone : +86(0)10 6530 8888
Fax : +86(0)10 6530 8888
Http : //www.intercontinental.com
E-Mail : guest@interconsanlitunbj.com

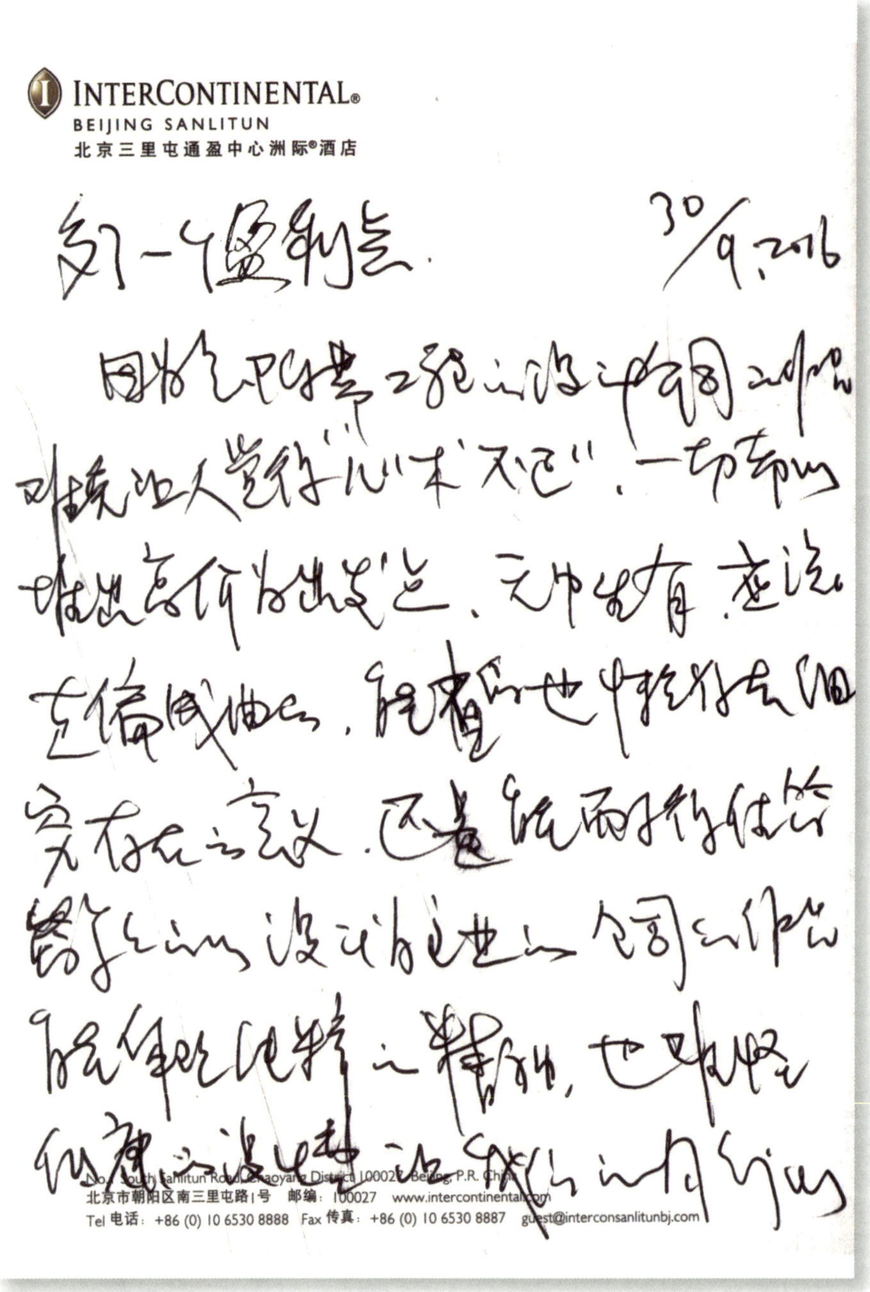

INTERCONTINENTAL®
BEIJING SANLITUN
北京三里屯通盈中心洲际®酒店

[illegible]一个[illegible]　　30/9/2016

[illegible]工程的设计中[illegible]对环境的人觉得"心"和"不心"，一切都[illegible]出发点、[illegible]世中[illegible]存在的意义，还是[illegible]设计[illegible]精神，也[illegible]

No. 1 South Sanlitun Road, Chaoyang District 100027 Beijing, P.R. China
北京市朝阳区南三里屯路1号　邮编：100027　www.intercontinental.com
Tel 电话：+86 (0) 10 6530 8888　Fax 传真：+86 (0) 10 6530 8887　guest@interconsanlitunbj.com

各种设计成为"噱头"，我们仍然
不追求这样的好！
治贵做事之后，30年的演变会
坚持，像我们人的像都不得不如此
入驻刚建成开业的通盈中心洲际酒店，是认真
的人这种要回到设计中活法，还是
成为自己的（可能他们也会一样）
的设计让人们不能够归宿，设计是
不被互动主动的，让人佩服。甚至
是会让人感到设计的同时也能够感
受到"好的氛围"，让世界的设
计更加独特地体现出个人的价值

No.1 South Sanlitun Road, Chaoyang District 100027, Beijing, P.R. China
北京市朝阳区南三里屯路1号 邮编：100027 www.intercontinental.com
Tel 电话：+86 (0)10 6530 8888 Fax 传真：+86 (0)10 6530 8887 guest@interconsanlitunbj.com

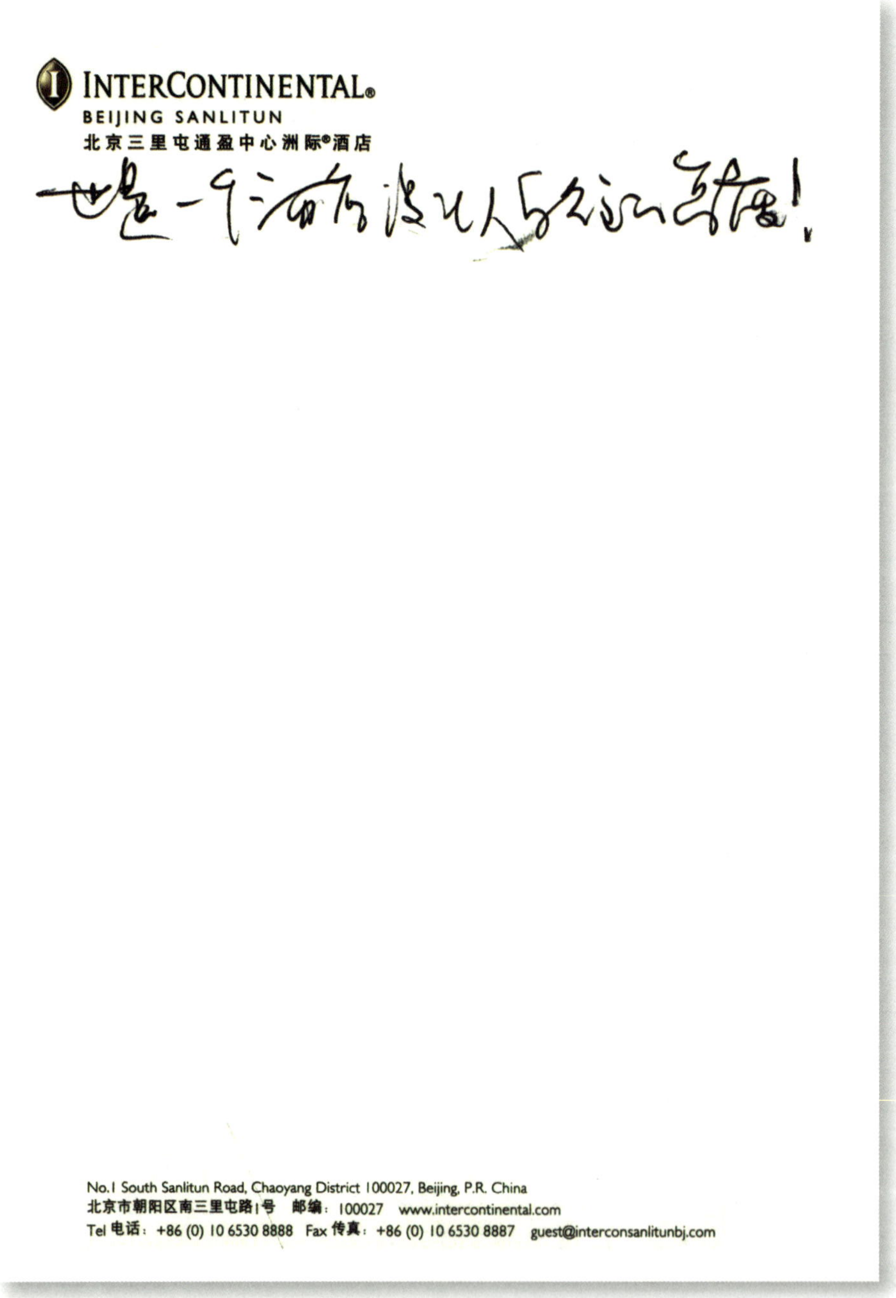

INTERCONTINENTAL®
BEIJING SANLITUN
北京三里屯通盈中心洲际®酒店

也是一个酒店设计人员的家和舞台！

No.1 South Sanlitun Road, Chaoyang District 100027, Beijing, P.R. China
北京市朝阳区南三里屯路1号　邮编：100027　www.intercontinental.com
Tel 电话：+86 (0) 10 6530 8888　Fax 传真：+86 (0) 10 6530 8887　guest@interconsanlitunbj.com

One More Profit -Making Point
多了一个盈利点

北京三里屯通盈中心洲际酒店
Inter Continental, Beijing, China

因为是附带工程的设计公司的作品，难免让人觉得“心术不正”，一切都以堆出高价为出发点，无中生有，应该直，偏成曲的，能省的也懒得去细究存在的意义，还是能耐得住简简单单的设计为主业的公司的作品，能体现纯粹的精神，也难怪低廉的设计费让我们的同行以各种方式去“维生”，我们可以不这样生活就好了！

As it is a work by a design company with projects, it feels inevitably “improper”, as everything is launched for high prices, made out of nothing, distorted or twisted. We don't trouble ourselves to pore over the meaning of their existence. It takes design-specializing companies insisting on simplicity to embody the spirit of purity. It is no wonder that such low designing fees have compelled people in our trade to make a “bare existence”. If only we were not living such a life!

沾崔健的福，30年的演唱会，坚持、偏执的人好像都不得“好结果”，入住这间刚开两个月的洲际酒店，专门入住。这种变通的设计活法，还是成为在国内（可能国际也会一样）做设计的人的不错归宿，设计与工程互动、互利，让人佩服，其实关键还是这家设计公司的业绩让它可以“为所欲为”，让业主与酒店管理方信任和依赖，这个很重要，也是一个酒店设计人向往的高度！

Thanks to Cui Jian, who was having his 30th anniversary concert (it seems that persistent and headstrong people never fare well), I especially checked into this international hotel opened just two months before. Such flexible life is a good destination for designers in China (maybe it also applies internationally). It is admirable that design interacts with the project and they are mutually beneficial. In fact, the key lies in the performance of the design company which makes it do as it wishes, trusted and relied on by property owners and the hotel management. It is very important, a height aspired to by hotel designers!

Sheraton

美国洛杉矶喜来登酒店

54 ★★★★

SHERATON GATEWAY LOS ANGELES AMERICAN

Address : 711 South Hope Street
Los Angeles, California
90017 United States

Telephone : +(1)(213) 488-3500

Http : //www.starwoodhotels.com/sheraton

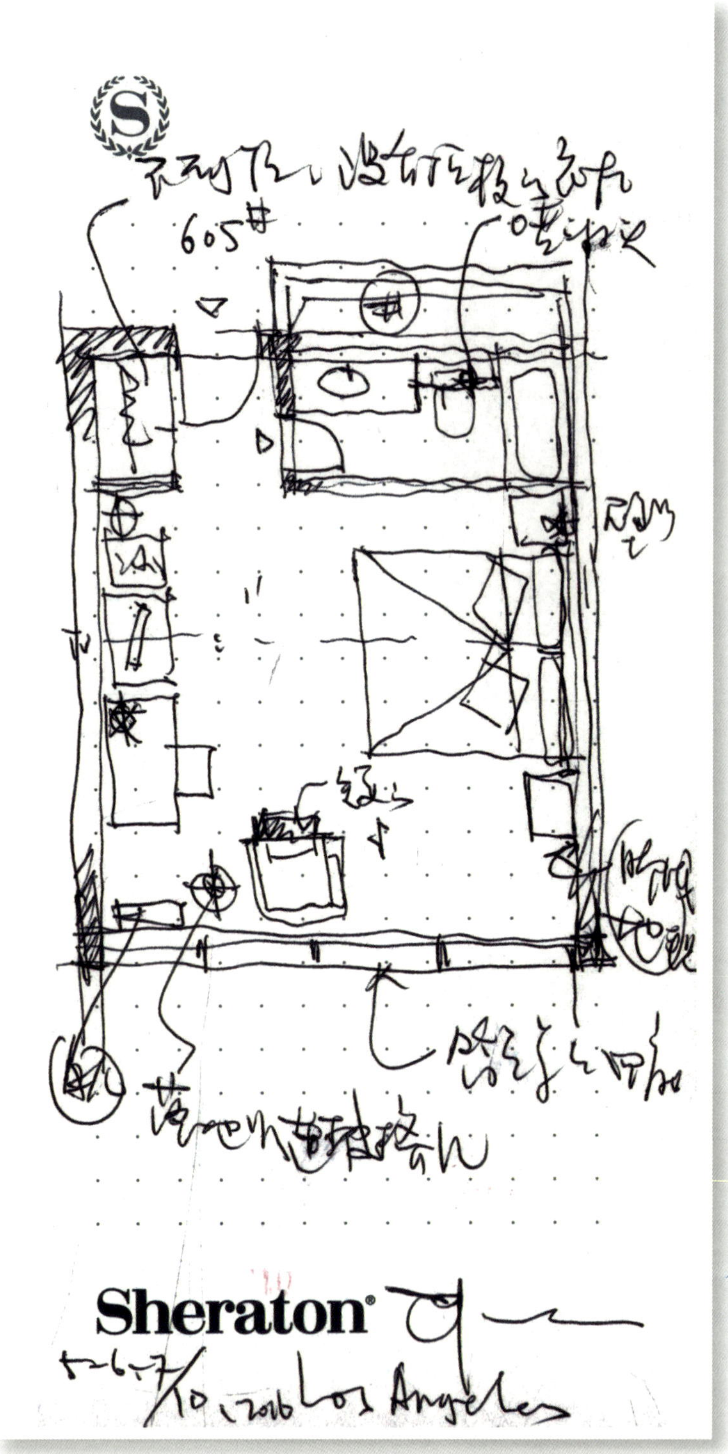

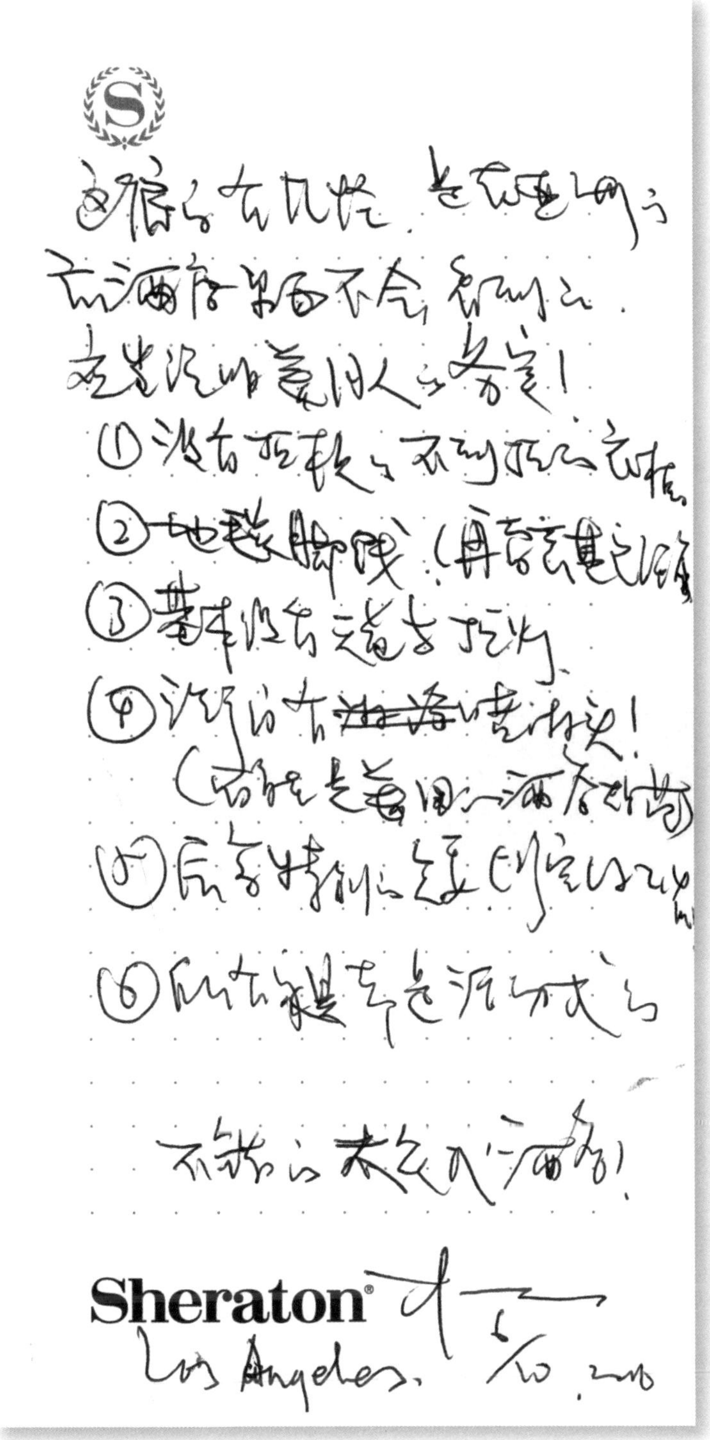
Sheraton
Los Angeles. 6/10, 2010

Several Queer Things in This Room

这个房间有几怪

这个房间有几怪，是在亚洲的新酒店里面不会看到的，应当说明美国人的务实！

This room had several queer things that cannot be found in newly opened hotels in Asia, attesting to American pragmatism.

美国洛杉矶喜来登酒店

Sheraton Gateway, Los Angeles, American

1. 没有顶板的不到顶的衣柜；

1. Roofless non-full-length wardrobe ;

2. 地毯脚线(再留意其他酒店)；

2. Carpet skirting (look at other hotels) ;

3. 基本没有天花与顶灯；

3. Basically no ceiling light or overhead light ;

4. 洗手间有喷淋头（可能是美国酒店规范）；

4. Shower nozzle in the washroom (maybe it is a standard for American hotels) ;

5. 层高特别矮（净空约2.4米）；

5. Extremely small floor height (about 2.4 meters) ;

6. 所有家具是活动式的。

6. All furniture movable .

不错的大众式酒店！

It is a quite good hotel for the average customer!

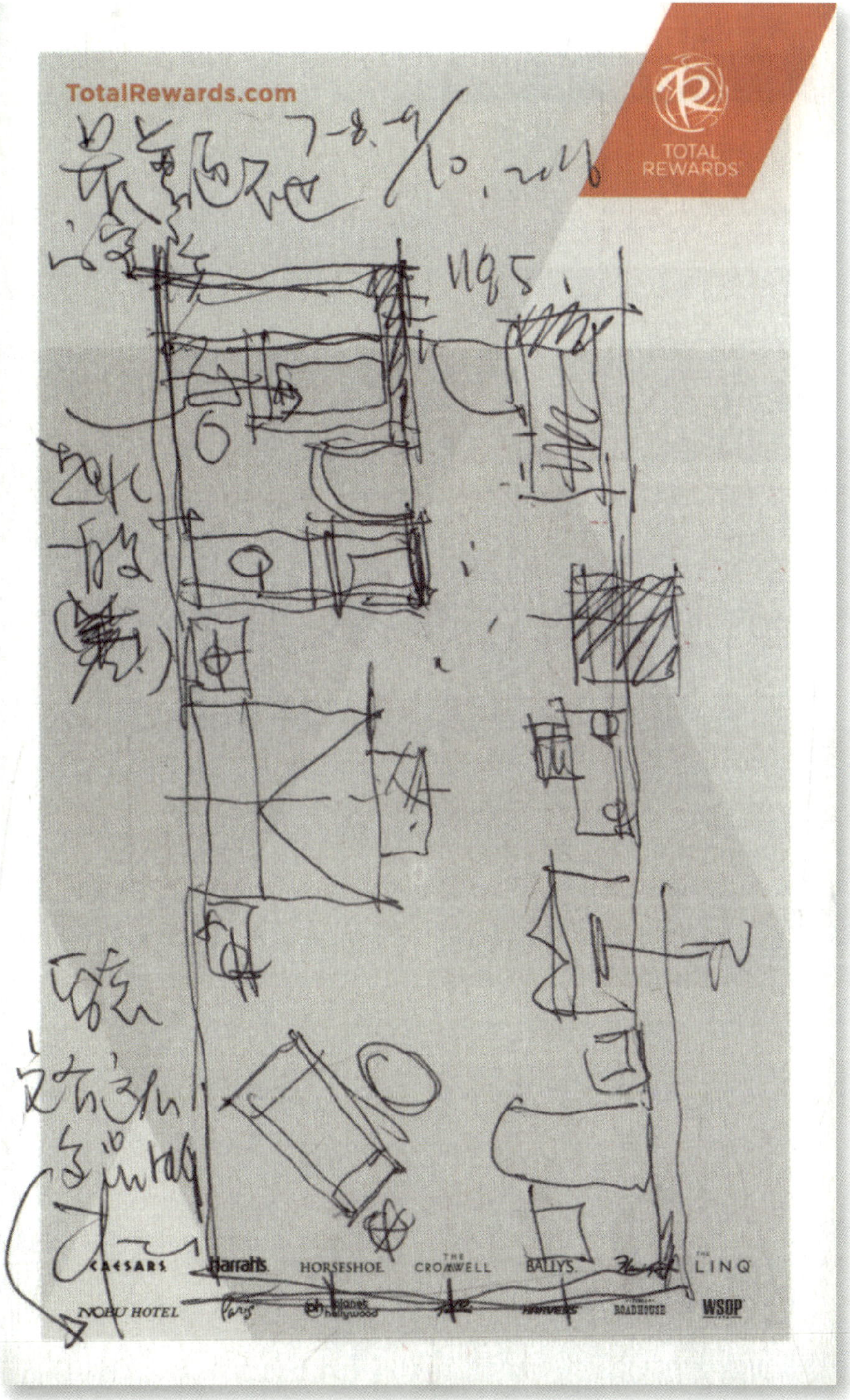

美国拉斯维加斯百利度假酒店

55 ★★★★★

BALLY'S CENTER OF THE ACTION LAS VEGAS

Address : 3645 Las Vegas Boulevard South

Telephone : +(877) 603-4390

Http : //www.caesars.com/ballys-las-vegas

美国拉斯维加斯百利度假酒店
Bally's Center Of The Action, Las Vegas

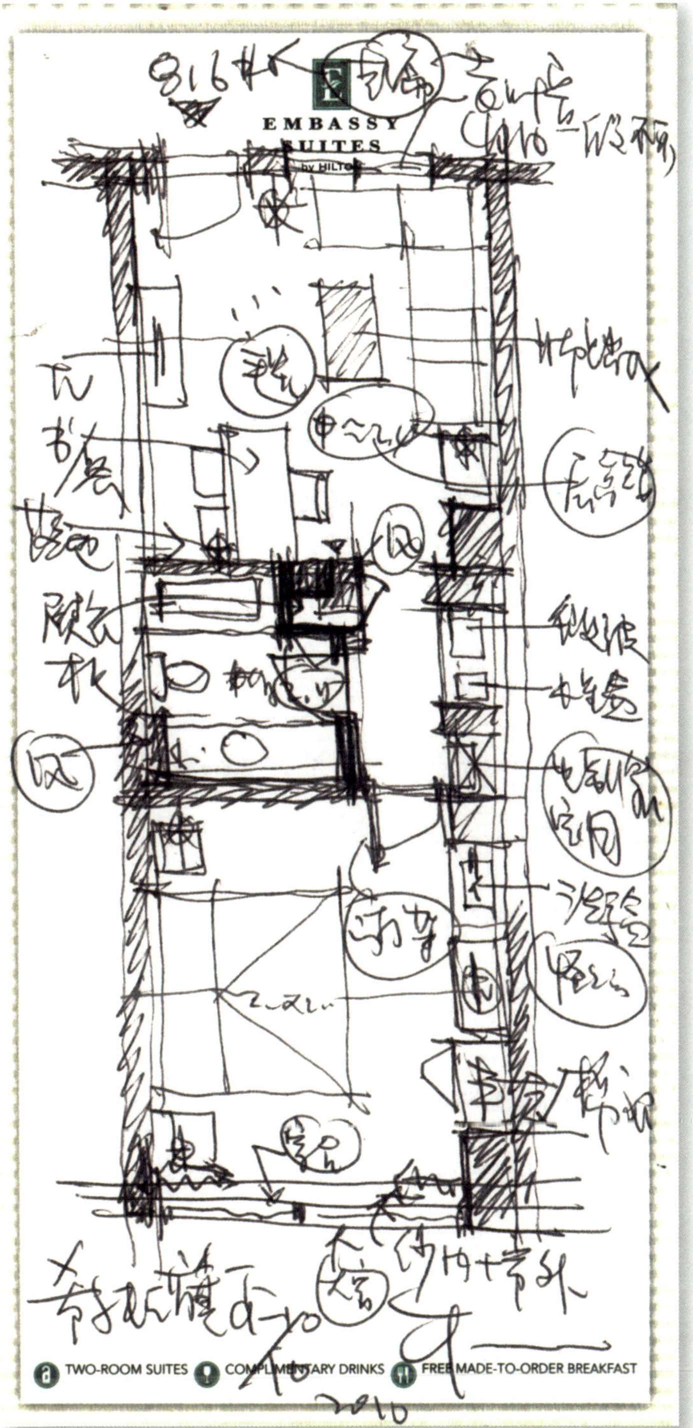

EMBASSY SUITES
by HILTON

美国洛杉矶希尔顿尊盛酒店

56 ★★★★★

EMBASSY SUITES BY HILTON LOS ANGELES GLENDALE

Address : 800 North Central Avenue, Glendale,California, 91203, USA

Telephone : +1-818-550-0828

Fax : +1-818-550-1289

Http : //zh.hiltonworldwide.com/portfolio/embassy-suites/

用一天体验一个旅客的生活。

美国行的第二站，入住Hilton酒店集团旗下的酒店式公寓（也称公寓式酒店）品牌"EMBASSY SUITES"

我们住了去芝加哥，地段繁华的市中心处，先用一个晚饭前和大家一起跟着去前台后面巡视了一遍：前台、中庭、走道区、自助餐厅、会客室等。最后就是那个房间，房客在这里可以得到休憩解决洗衣、用水、通风、采光、合理、舒适的布局，接着，我用一个旅客的角色去"打扰"这个房间，用一次的体验来找自己别人住一个房间的感觉。

情况，我的大旅行箱不知道在哪里打开好，只能放在客厅的电视柜前的空地上，沙发很大，茶几很矮，连"茶几"都嫌矮，够不着。沙发的座椅太深了，会掉到地上；还会可能是那就吃饭之用，太大了，妨碍了客厅的感观。但是人性化地在前边去厅的墙面开了一个带通道的可调节的隐藏门窗，可以享用到自然光，中部有细节创造的设计，茶水吧配有微波炉，可以简单煮食，让旅客能长住，有意思的是地方是在房间内也没有一个洗衣机，这个在

我使出我全身的力[illegible]往日[illegible]，[illegible]一切全在于这

不合之感觉，且是[illegible]也之，难免说谎。一切[illegible]一气呵成

到底，投入些火，撑坏，这样的[illegible]和绝对冷[illegible]

之指之整半生般，方缝，不得多日居住；[illegible]大家，[illegible]

[illegible]，[illegible]是[illegible]眼之。到西部[illegible]。

今到用了三分之冷水口，煮水在锅[illegible]，用了两次之牙，[illegible]

[illegible]了一次。煮水完毕，[illegible]地[illegible]了……

果然，用了一天一夜，我们终于做了一个自以为精彩的[illegible]

式会宴。

Experiencing a Week's Life in a Single Day

用一天体验一个星期的生活

美国游的芝加哥一站，入住希尔顿酒店集团旗下的酒店式公寓（或称公寓式酒店）品牌“EMBASSY SUITES”。

At Chicago, I checked into a hotel flat (or flat hotel) of the brand Embassy Suites under Hilton Hotel Group.

我住在位于走道的尽端，比较私密的小走廊处，先用建筑技术人员的眼光去前前后后地“巡”了一遍：前厅，中部是湿区后勤配套空间，最里面就是卧室了，层高矮，净高约2.4米，多管井解决给、排水，送风及排风，合理、老到的布局。接着，就用一个旅客的角度去“折腾”这个房间了，用一次的体验来模拟别人住一个星期的感觉。

My guestroom was at the end of the corridor, a relatively little one — a place of privacy. First I made a survey of the room with my eyes of a building technician: the anteroom at the front, the wet area with logistic provisions at the middle, and the bedroom at the innermost. It was 2.4 m in height between floors.The multiple tube well solved the issues of provision and discharge of water and air — the layout was reasonable and skillful. Then as a guest, I put the room to the"test": trying to simulate a week-stay experience in a day.

惨了，我的大旅行箱不知道在哪里打开为好，只能在客厅电视柜前的空地上。沙发很大，茶几很矮，连“葛优躺”都嫌够不着，且沙发的坐垫太滑了，会掉到地上；写字台可能是兼顾吃饭之用，太大了，妨碍了客厅的感观；倒是人性化地在靠近走廊的墙面开了一个带活动可调节功能的白色百叶窗，可引入柔和的自然光；中部有细节舒适的洗手间；茶水吧配有微波炉，可洗可基本煮食，让你安心长住；有趣味的地方是在房间内也设有一个洗手盆，这个在我住过或公司设计过的项目中都比较罕见。一则会有干湿不分的感觉，且是地毯的地面，难免污染；二则多了一套给排水系统，投入增大，需谨慎，这样 的配套相对就会影响衣柜的整体长度，太短，不够多日居住。整体的大窗，城市夜景观，早上晨曦观感都是非常舒服的。开了两部电视，分别用了三处的给水口，煮水在茶水吧，刷了两次牙，狠狠地洗了两次澡，热水充足，终于安安乐乐地躺在床上了……

Terrible — I did not know where to open my suitcase. I had to do it in the space before the TV cabinet in the living room. The sofa was so big and the tea table was so short that even Ge You (176cm tall) could not reach it lying on the sofa. The sofa cushion, so slippery, would fall on the ground.The writing desk, perhaps designed to double as a dining table, was so big as to interfere with the whole appearance of the living room. It was user-friendly to have a movable and adjustable white louver window in the wall near the corridor, admitting soft natural light. At the middle was a elaborate and cozy washroom. The tea bar was equipped with a microwave oven, with which you can cook basic food — making you settled for a long stay. The interesting part was that there was a hand basin even in the room — something rather rare with the projects I had stayed in or my firm had designed. For one thing, it got the line between the dry and wet area blurred, not to say that the carpeted floor was inevitably subject to pollution. For another, it meant an extra set of supply/discharge system, adding to the input and care. Such a design would, to some degree, affect the whole length of the wardrobe — if too short, it would not suffice for a stay of days. The full-length window gave a quite comfortable view of the nightly city and morning sun. I turned on two TV sets and used the waste supply at three inlets; boiled water in the tea bar; brushed the teeth twice; bathed twice to my full satisfaction — the hot water supply was more than adequate. Finally, I lay on the bed, settled and happy.

当然，用了一天一夜，狠狠的体验了一个这么不错的酒店式公寓。

Of course, I spent a day and a night indulging in experiencing such a rather nice hotel flat.

DESTINATION
KOHLER

美国威斯康星州科勒美国俱乐部

57 ★★★★

KOHLER DESTINATION RESORT SPA GOLF WISCONSIN AMERICAN

Address : The American Club
419 Highland Drive
Kohler, WI 53044
Telephone : + 800-344-2838
Http : //www.americanclubresort.com

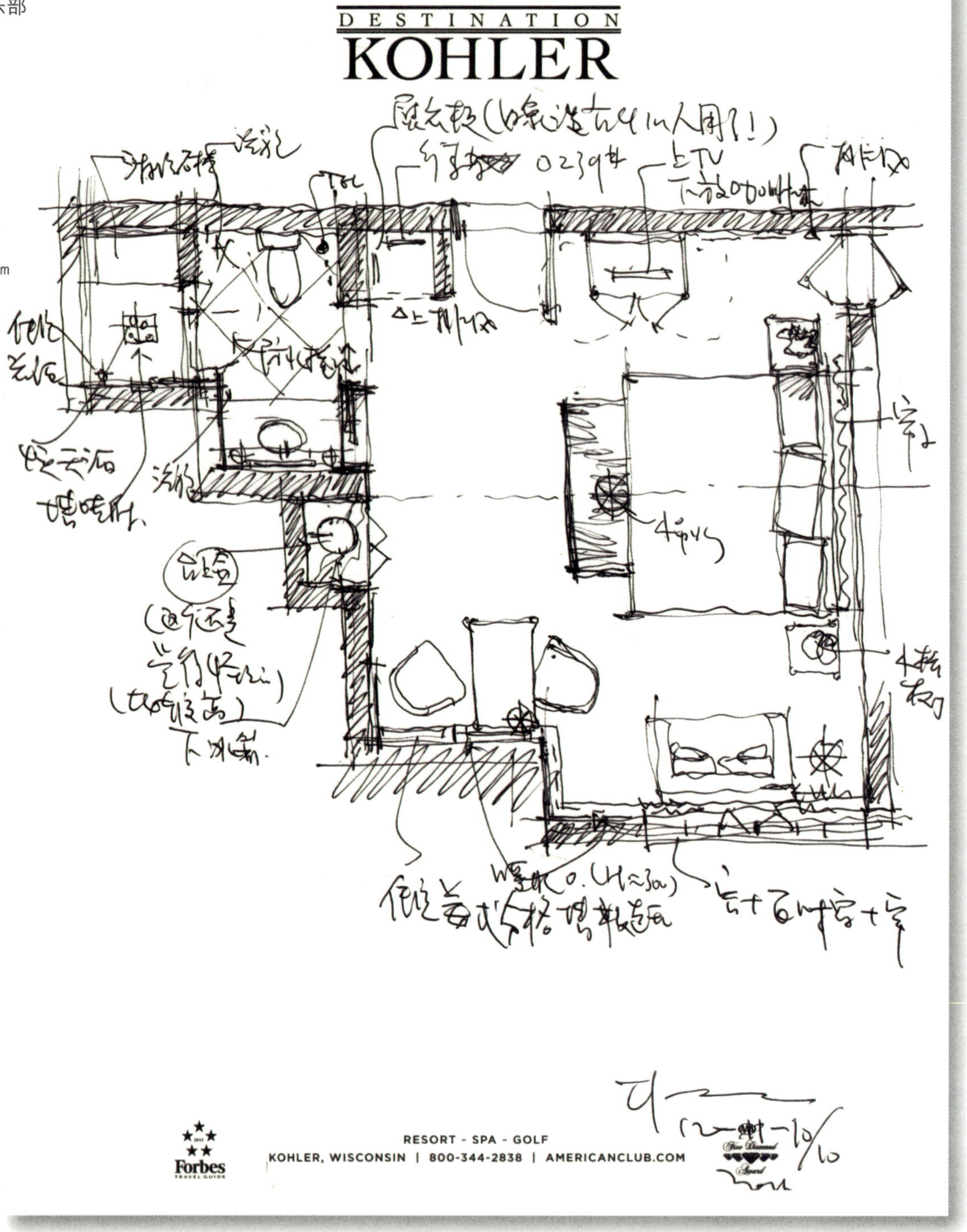

姜、还是老的辣。

再次入住这个古老的酒店(俱乐部)，凭拿到手的房门钥匙，走走找找，好像绕了不少时间才找到房间(每一间房都有一个名字，这次是 John Brown)开门。哈哈，还好，跟上一次住的不同的房型，而更加仔细地去体味"程序"酒店的厉害。

第一次入住的记录详见(《住哪?》之 P150)

这里感觉灯光控制的合理性，而且相当的模式化的趣味。当然，开关是非常非常的古老而经典，都全部用在英式酒店里少见的浴缸和洗脸台的墙面开关，特别适合年纪大一些的住客，而不是一味追求时髦的 PAD 控制台。而更多的冷热水更是传统一样，冷热调节还是不可少的！

例如当年的设计师，洗手间等最好而有趣，因为天冷，就放到了宿舍了单个小时的热水澡(是不是旅游分记录了)我不止一次谈起……酒店客房的热水供给(水量：水压)在这里是非常古老而有力的三种不同的出水方式：传统花洒、顶花洒(四头)、墙面长身按摩喷头，可以一齐使用哟！如果水压和水量不够，那这一切就只能是一种装饰，而沐浴后会很惬人的舒爽样子，让你可以静静地长时间泡在水中。相信这样对来说对于高个子后的美式帅哥是非常有帮助的。

姜，还是老的辣。

魅力，不减当年，更加让人迷恋（美的享受快感啊！）

沐浴的时候，习惯看看用品的说明，用"blue"开发的这套沐浴用品：浅蓝色的洗发水（shampoo）、粉蓝的Condition护发素和天空蓝的Gel沐浴液，这一切都会让人对蓝天所有着不同的遐想。牛！

Age Means More Taste
姜,还是老的辣

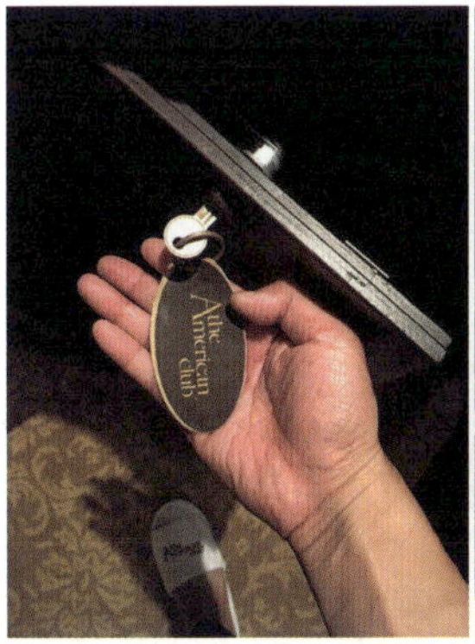

美国威斯康星州科勒美国俱乐部
Kohler Destination Resort Spa Golf ,Wisconsin

再次入住这么古老的酒店（俱乐部），自己拿行李，高高低低，走走找找，好像花了不少时间才找到房间（每一间房都有一个名字，这次是John Brown）。开门，哈哈，还好，跟上一次住的不同的房型，可以更加仔细地去体验“科勒酒店”的厉害。

Once again I checked into this so ancient hotel (club). Carrying the luggage by myself, going up and down, looking out for the room all the way, after quite a while, I finally found it (each room had its own name. This time it was John Brown). I opened the door. Haha,OK, it was a type of room different from that of last time, allowing me a more careful study of the edge of Kohler Hotel.

第一次入住经历详见（《住哪？2》P150）

See Where to Stay? 2, P150, for my first stay here.

首先感觉到灯光控制的合理性，而且有相当多的模式与趣味。当然，开关是非常非常的古老而经典，部分采用在其他酒店已少见的滑动调光的墙面开关，特别适合年纪大一些的朋友，而不是一味追求时髦的PAD控制台，而更多的细节更是值得一提，绘制平面是个可少的。

First of all, I was impressed with the sensible design of the lighting control, offering many modes and great fun. Of course, instead of using the trendy Pad control board, it used switches quite quaint and classic, some being wall-mounted switches which adjust light by sliding, rarely seen in other hotels today but quite suitable for older people. More details were more worthy of a mention.So it could not do without a sketch plan.

佩服当年的设计师，洗手间紧凑而有趣，因为头痛，就放肆地享受了半个小时的热水澡（是不是有点过分浪费了）。我不止一次说过，酒店客房的热水很重要（水量、水压），在这里是非常有说服力的：三种不同的出水方式：传统花洒、顶花洒（四头）、墙面喷射按摩喷头，可以一起使用啊！如果水压和水量不够，那这一切就只能是一种装饰，而淋浴间有很怡人的石头椅子，让你可肆意地长时间地洗热水澡，相信这样对来这里打高尔夫后的美女帅哥是非常有吸引力的。

I really admire designers of that time who had designed a washroom so compact and interesting. Suffering from a headache, I indulged myself in a half-hour hot bath(I wonder if it was a little too wasteful).I have mentioned it again and again that the hot bath water (volume and pressure)in the guestroom of a hotel is very important. It was quite convincing here: three types of showering was available: the traditional shower faucet, the ceiling shower faucet with four heads, and wall-mounted massage shower faucet. And you can use all the three at once! Without adequate water pressure and volume, all this would have been a mere form of decoration. Besides, the cozy stone chair in the shower compartment allow you an extravagantly long hot water bath. I believe that it is quite appealing to handsome men and beautiful women who come to play golf here.

姜，还是老的辣。

Age means more taste.

魅力，不减当年，更加让人迷恋（爱的高级版）啊！

Instead of being reduced in charm, it is more charming (more than lovable)!

沐浴的时候，习惯看看用品的说明，用“Blue”开发的这套洗涤用品：浅蓝透明的洗头水，粉蓝色的护发素和天空蓝的沐浴液，这一切都会让人对蓝天、白云有着不同的遐想，牛！

While bathing I habitually had a look at the direction for using the washing articles: this set developed by Blue came with light blue crystal shampoo, pink blue air conditioner and sky blue liquid soap fires different kinds of imagination of the blue sky and white clouds. Great!

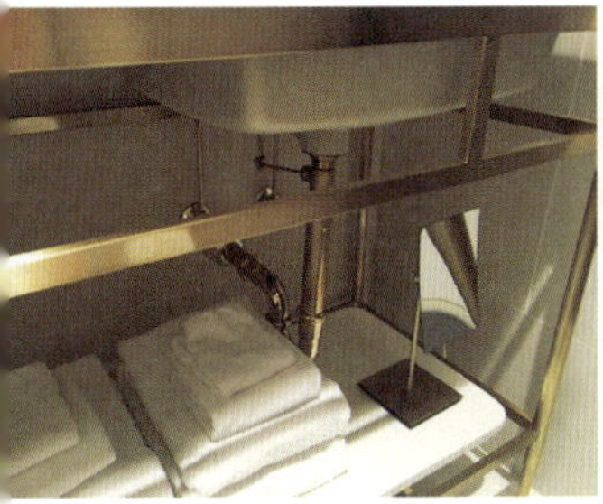

美国纽约11霍华德酒店

58 ★★★★★

11 HOWARD NEW YORK AMERICAN

Address : 11 Howard Street
New York, New York 10013
United States

Telephone : (1)(212) 235-1111

Http : //www.starwoodhotels.com/

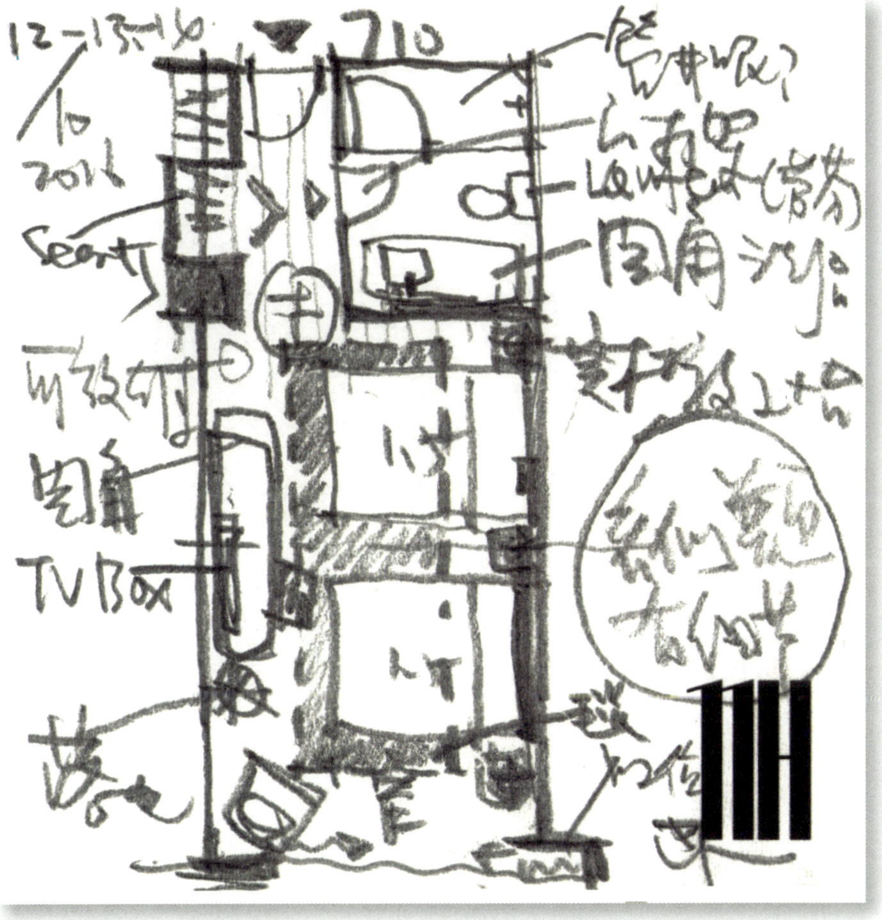

美国纽约11霍华德酒店

11 Howarad, New York, American

美国纽约巴卡拉酒店及公寓

59 ★★★

BACCARAT HOTEL & RESIDENCES NEW YORK AMERICAN

Address : 28 West 53rd Street, New York City, NY 10019

Telephone : +00 1 844-294-1764

Http : //www.baccarathotels.com/

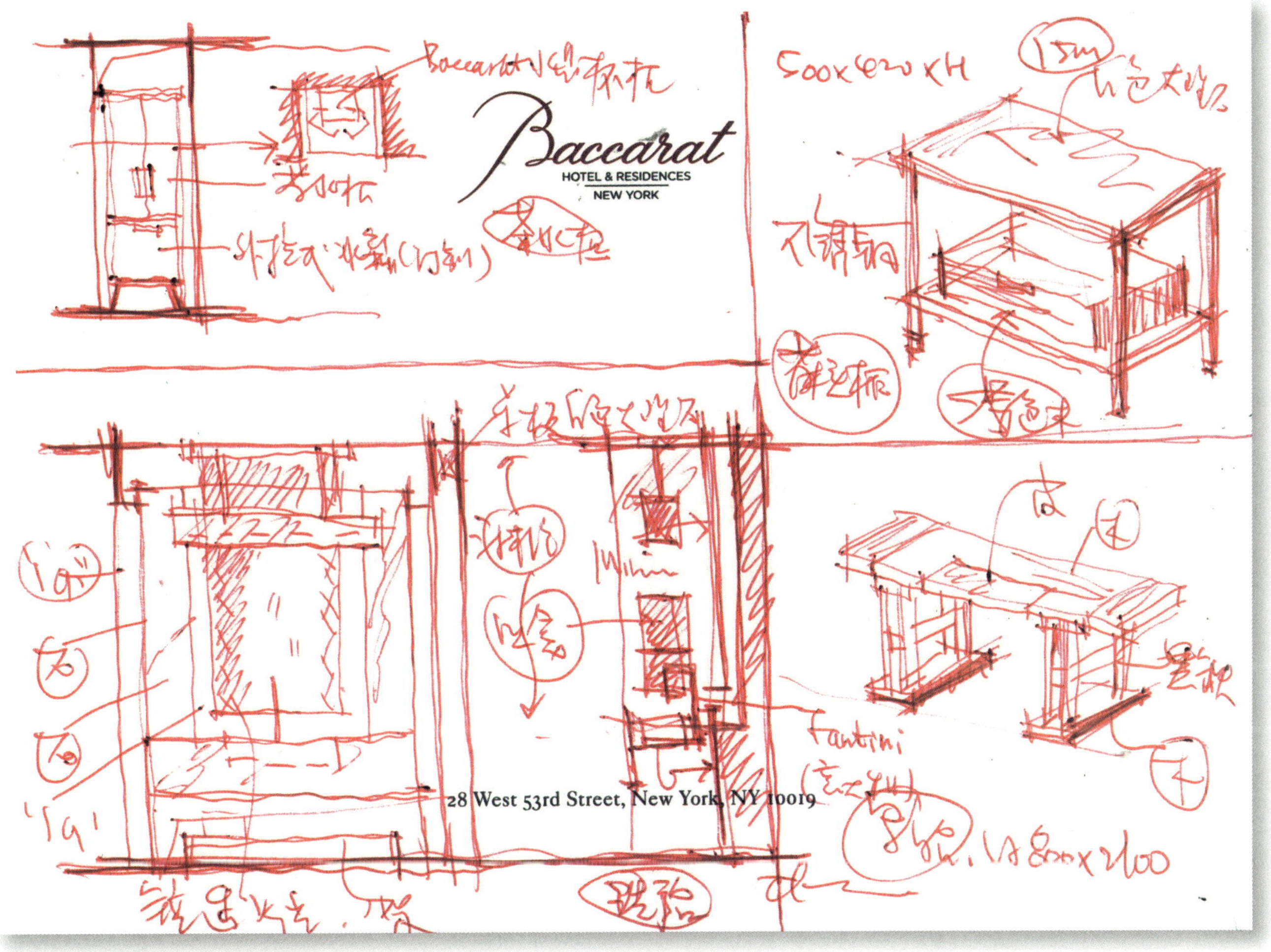
Baccarat
HOTEL & RESIDENCES
NEW YORK
28 West 53rd Street, New York, NY 10019

入住。

选择从艺术学院到酒店的停留，会让可以去看一下老街区，也可以顺便重温一下上一次到过的地方。

心仪已久的巴卡拉酒店，之前从相关的渠道已看到的资料和介绍，推开门时，小激动了一番，倒两天的11 Howard精品酒店，从Chinatown出发打车到位于MoMA对面的这个耀眼的水晶之城。全玻璃水晶的外墙安全地融入了传统建筑的酒店与街道艺术的思维，让内外的视线更能通过光影的折射而不能洞悉到我像真的整个可以接受这个概念的世界，我是要感谢的。

进入酒店首层大厅，水晶吊灯晶莹如火炉，灯光随音乐舞蹈，厚重的黑色墙面让水晶精显其璀璨。

大堂位于二层的高大空间，真心佩服能在这样的旧建筑内有如此优势的层高的地方，让完全合适于巴卡拉的大型水晶大吊灯长到近7、8米高的吊架，相当震撼！

入住的仪式也太牛了，巴卡拉水晶高脚杯斟满冰镇香槟，在经典的沙发上慢慢享受中午的休闲片刻。

路上的堵车、绕路（大家算计我们，常走大道收拾）的闷气一下子就消了。同行的驾驶员总是咱们俩，偷两天，在这里享用两天的精美、健康的格智早餐，一天晚上的红葡萄酒牛扒大餐和一次晚上的红葡萄酒小聚会，令我不想选择这家价格不菲的全球的酒店。

全程是快乐相伴！

迷恋是一种境界，就像幸福是快乐的感觉一样。相信人性也是接纳的感觉，就像是爱情的感觉一样，是一种迷恋。

Ecstasy
入迷

纽约巴卡拉酒店及公寓

Baccarat Hotel & Residences, New York, American

选择从芝加哥飞到纽约停留，希望可以去看一下老同学，也可以故地重温一下上一次到过的地方。

I chose to fly from Chicago to New York for a stay, where I hoped I could see my old classmate and revisit places I had visited last time.

心仪已久的巴卡拉酒店，之前从相关的渠道已看到的资料很多很多。提前订好了，小激动了一番，住两天的11Howard 精品酒店。从唐人街坐出租车到位于MoMA对面的这个神秘的水晶之城，全玻璃水晶的外墙完全摒弃了传统位于旺地的酒店对街道景观渴求的思维，让内外的视线只能通过光影的折射而不能清晰成像，真的佩服可以接受这个概念的业主。我是来朝圣的。

I booked a room with Baccarat Hotel I had been admiring and had read a lot about. With some excitement, I stayed two days in 11Howard boutique hotel before taxiing from Chinatown to this mysterious Crystal City. Its outer wall is all made of crystal, a total abandonment of the traditional idea that hotels in a prosperous spot yearn for good street landscapes. Instead, it makes for no clear images by allowing only refraction of light. My sincere admiration goes to this property owner who could accept such an idea. I had come as a pilgrim.

进入酒店首层过厅，水晶数码墙如大火炉，灯光随音乐舞蹈，厚重的环境色调让水晶更显其璀璨。

The first floor passage of the hotel had crystal digital walls like big burners, with lights dancing to music. The solidness of the environmental hues set off the brilliance of the crystal.

大堂位于二层的高大空间，真心搞不明白为什么这样的旧建筑物内有如此优势的层高的地方，让它充分展示巴克拉的大型水晶大吊灯及到顶近7、8米高的陈列架，相当震撼！

The hall was located in a lofty space. I really have no idea why there was such a tall space in such an old building, allowing a full show of the gigantic Baccarat crystal ceiling lights and the 7-8 meter high display stands touching the ceiling-quite a shocker!

入住的仪式也太牛了，巴克拉水晶高脚杯盛满了冰镇的香槟，在经典的沙发上慢慢享受中午的偷闲片刻。路上的堵车、绕路（太多单行线了，第五大道附近）的闷气一下就消了。同层的餐厅亦是吸引，住了两天，在这里享用了两天的精美、健康的极简早餐， 一天晚上的红葡萄酒、牛扒大餐和一天晚上的白葡萄酒小聚会，令我不枉选择这家价格不菲的全水晶酒店。

The ritual of check-in was great—with Baccarat goblets overflowing with champagne on rocks, I took my time enjoyed the leisurely moments of the noon sitting on a classic sofa—my sullenness and frustration at the traffic congestion, detours (there were too many one—way routes near 5th Avenue) dissipated at once. The dinning room on the same was equally appealing. During my two-day stay here, I spent two morning enjoying fine, wholesome, extremely simple breakfast, one night relishing red wine and grilled fillet steak, and one night reveling at a small white wine party. All this made it worthwhile for me to choose this all crystal hotel that was not so inexpensive.

全程是水晶相伴！迷恋是一种境界，就像幸福是快乐的高级版一样，相信入住巴卡拉的感觉就像爱情的高级版一样， 是一种迷恋。

All the way I was accompanied by crystal! Ecstasy is an elevated state of mind, just as happiness is the advanced edition of enjoyment. I believe that the feeling of one staying at Baccarat is the advanced version of love, it is a kind of ecstasy.

New York American Travels

美国旅游之纽约

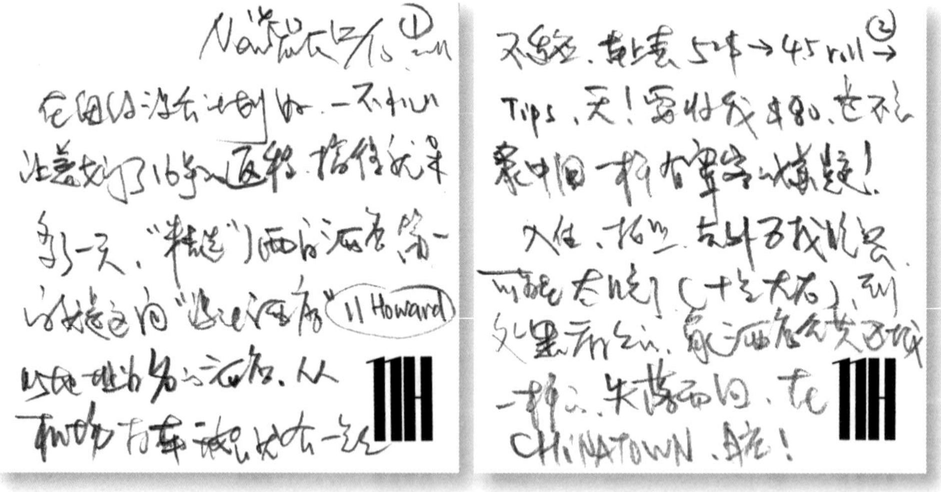

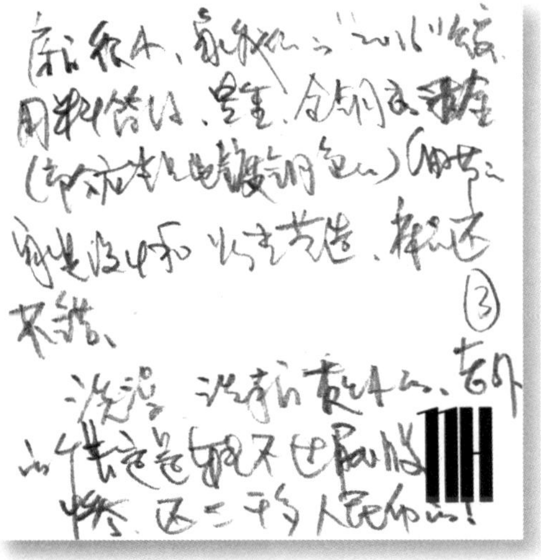

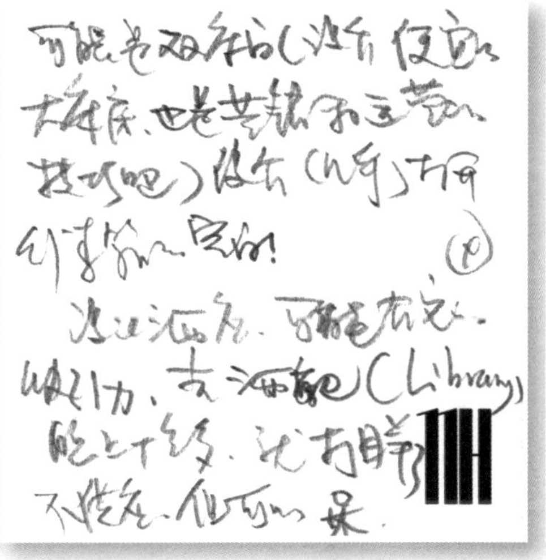

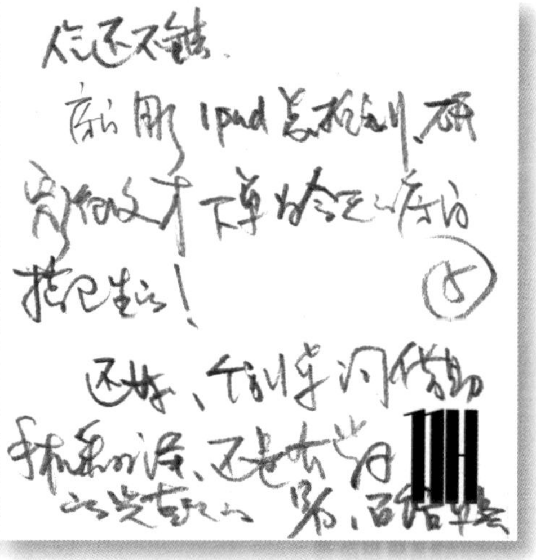

Diary A / Day 1 in New York / Breakfast at the Noodle Restaurant / October 12, 2016

日记A（2016年10月12日面馆早餐）纽约第一天

在纽约没有计划好，一不小心订了16号的返程，就多待了一天，“精选”了两间酒店，第一间就是这间“设计酒店”，11 Howard。以地址为名的酒店，从机场打车开始就有一点点不适应，车上表$52+4.5roll+小费，天！要收我$80，是不是像在中国一样有宰客的嫌疑！

As my stay in New York had been poorly planned, my girl colleague mistakenly booked the return ticket for October 16, making me stay in New York for an extra day. I "handpicked" two hotels. The first was this "design hotel", 11 Howard, a hotel named after an address. From the moment I took a taxi at the airport, I felt a little bit out of place — the taximeter showed $52+4.5roll+Tips. God! Charging me $80! I suspected I was overcharged as back in China.

入住，拍照，去外面找晚餐，可能太晚了（十点左右），到处黑麻麻的，像酒店公共区域一样，失落而回，在唐人街,脏！

After checking in, taking shots, I went out to look for our supper. Maybe it was too late (around 10 PM), it was pitch dark everywhere, like the public area of the hotel. The Chinatown was really dirty. Disappointed, I returned.

房子很小，像我们的“2016”公寓，用料简约、贵重，全铜的五金（部分应当是电镀铜色的）细节的家具设计和灯光营造，床品还不错。

The house was quite small, like our own"2016" flat. The materials were simple and valuable, all bronze hardware (some were electronically coppered). The furniture was elaborately designed and the lighting thoughtfully created. And the bedding was alright.

洗澡，洗手间挺小的，老外的个头肯定是转不过屁股，惨，还三千多人民币！

The shower compartment and the bathroom were rather small. A foreigner would have found it impossible to turn around. Terrible — it cost me more than 3,000 yuan!

可能是双床间（没有便宜的大床房，也是营销和运营的技巧吧），没有（几乎）打开行李箱的空间！

Perhaps it was a double room (there was no queen size bed room — maybe it was also a marketing and operation skill), as there was (almost) no room to open my suitcase!

设计酒店，可能有它的吸引力，去酒吧（Library），晚上十点多就打烊了，不供应，但可以待。

The design hotel may have its own appeal. At past 10 PM, the bar Liberty was about to close for the night. Unable to order anymore, you could stay there for a while.

人气还不错。

The ambiance was quite good.

房间用了ipad总控制，研究了很久才下单为今天的房间搞卫生！

The room used an ipad as the master control. It took me a long time before I knew how to order the room cleaning cleaning service for today!

还好，个别单词借助手机翻译，还是有学习的兴趣的。

Ok, I had learned some of the words by using the mobile phone translator. I still have the interest in learning.

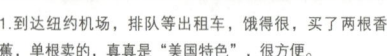

1.到达纽约机场，排队等出租车，饿得很，买了两根香蕉，单根卖的，真真是“美国特色”，很方便。

At New York Airport, waiting in line for a taxi on an empty stomach, I have bought two bananas. It sell by the piece—very "American" and convenient.

2.酒店在唐人街附近，街道脏兮兮的，墙面和招牌也是乱乱的，大苹果纽约也是有千面的。

The hotel is near Chinatown. The streets are dirty, the walls and signs disorderly. The Big Apple is also multifaceted.

3.可能是精品酒店，没有预备你是大行李箱，一打开就没有路可走了。洗手间及其中的淋浴间，以我这身材也觉得窄，“大块头”就惨了！

Maybe the quality hotel did not expected my luggage is so big. Once open, it leaves no room to pass it. The washroom and shower compartment are too cramped even for a person my size; it will be terrible for a big fellow!

4.床头的ipad总控，智能化的投入也是不错的，只是英文一般般的我很艰难才完成预约下单第二天搞卫生。

Place on the night desk is an iPad general control—a very good input in smartness. However, as my English is just so so, it is with great difficulty that I have managed to book the room-cleaning service for next day.

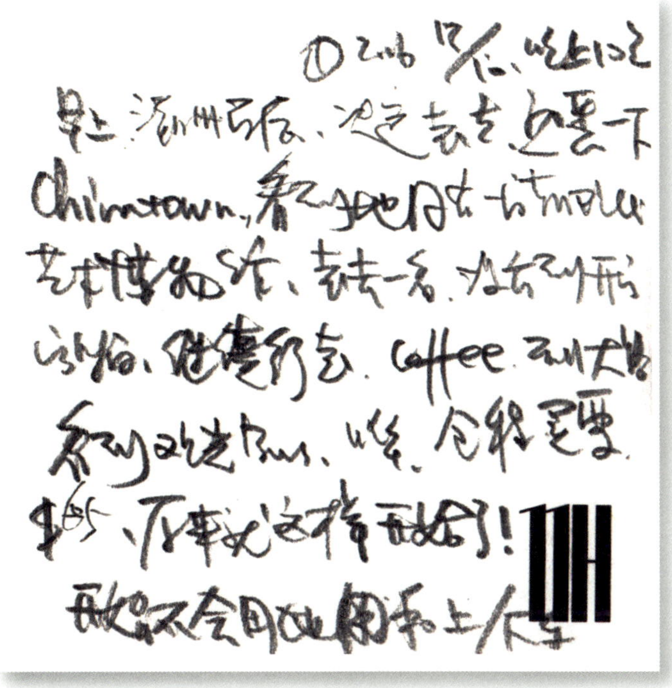
2016
Chinatown
Coffee
Bus

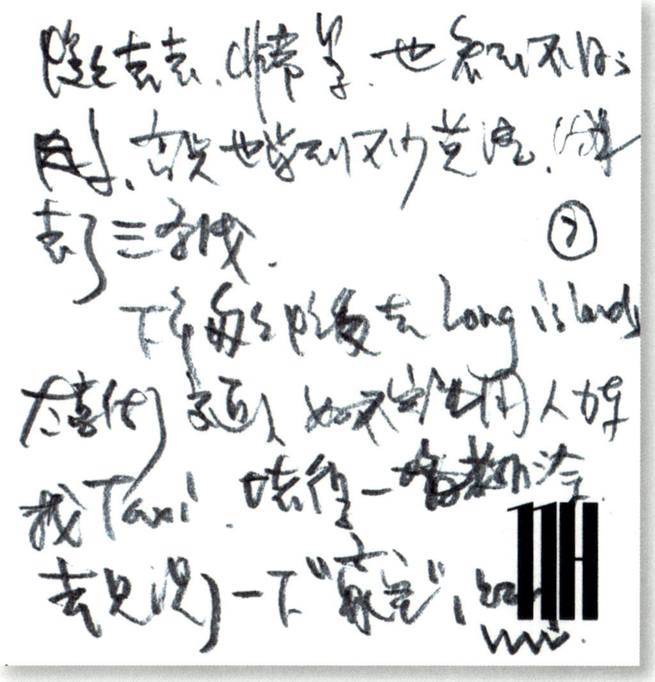
Long Island
Taxi

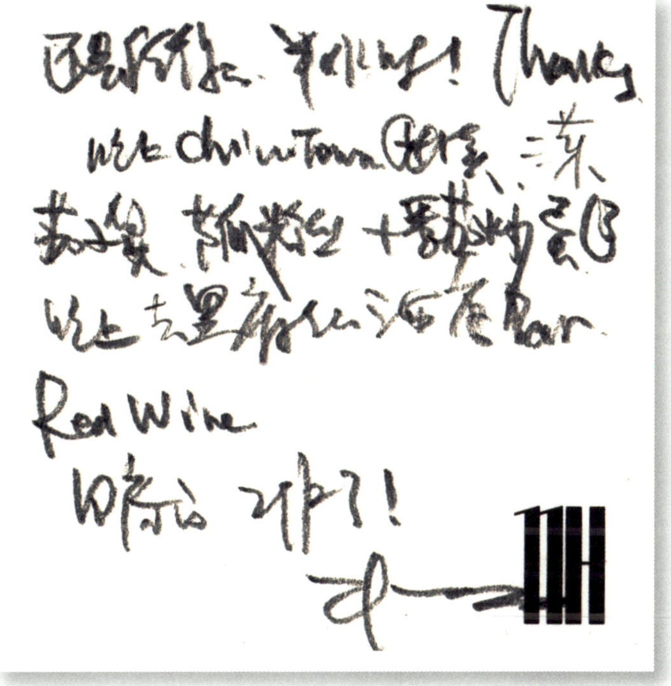
Thanks
晚上 Chinatown
Bar
Red Wine

Diary B / Day 2 in New York / Night October 12, 2016

日记 B（2016年10月12日晚上）纽约第二天

早上，吃过潮州面后，决定走走，熟悉一下唐人街，看到地图有一间新现代艺术博物馆，走去一看，没有到开门的时间，继续行走，喝杯咖啡，到大路看到观光巴士，哗！全程买要$65，于是就这样开始了！

In the morning, after eating Chaozhou noodle, I decided to get around to learn something about Chinatown. Seeing a new MOMA on the map, I went to visit it, only to find that it was not yet the time to open. I moved on, had a cup of coffee, and caught sight of a tour bus on the road. Wow, the full trip would cost me $65. And it began that way!

开始不会用地图和上下车，慢慢走走，非常累，也看到不同的风景，高兴，也学到不少英语，结果走了三条线。

At the beginning I didn't know how to use the map or how to get on or off the bus. It crawled along at a snail's pace —so tiring. Nevertheless, I was glad that I had different views of the city and that I had learned much English. I ended up traveling three lines.

下午匆匆忙忙去长岛，太高估了交通了，好不容易用人力车找出租车，堵得一塌糊涂，去见识了一下“豪宅”。

In the afternoon, hurried to Long Island. I had overvalued the traffic—only with great difficulty did I manage to find a Taxi by hiring a rickshaw, only to find that we were caught in a terrible jam. Finally I visited a "luxury residence".

还是值得的半小时！ Thanks.

Still, the half an hour so spent was worthwhile. Thanks.

晚上唐人街继续，三菜：茄子煲+节瓜粉丝+番茄炒蛋，晚上去黑麻麻的酒店酒吧，喝杯红酒。

In the evening I continued my tour of Chinatown. I had three dishes: the potted deep fried eggplant, Chieh-qua mung-bean noodle and tomato omelet. Then I went to the dim bar for a glass of red wine.

回房间，工作了！

Finally, I returned to my guestroom to get down to work!

1.看到小公园的老人在围着下象棋，真的以为是在广州呢，移民干什么啊!

The sight of the old gentlemen sitting around the table outside playing chess makes me think it was Guangzhou. What's the purpose of immigration?

2.非常正宗的潮州面馆，服务热情，生意好。

Quite authentic Chaozhou noodle restaurant with good service and brisk business .

3.尝试坐旅游观光车，可以自由自在，上上落落，到处看看，也可以学习英语听听解说。

Take the tour bus to get around for sightseeing and to learn English by listening to the narration.

4.一天下来，可以看到大半个纽约的重要景点，确实方便。

You can see more than half of New York in one day. It is really convenient.

5.下雨天，找出租车，花了20美元体验了一个外国人的黄包车。赚钱不容易。

It was raining. I spent 20 dollars experiencing a foreign rickshaw. It is not easy to make money.

6.去长岛的路上，几乎都是蚂蚁般的前行，在高速公路也不例外，近两小时的车程，佩服住在郊区的美女帅哥们。

On the way to Long Island, even on the expressway, the traffic is almost crawling along. It is near-two-hour trip. I really admire people living in the suburbs.

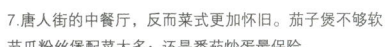

7.唐人街的中餐厅，反而菜式更加怀旧。茄子煲不够软；节瓜粉丝煲配菜太多；还是番茄炒蛋最保险。

The dishes in Chinese restaurants in the Chinatown make people homesick. The stone potted deep fried eggplant is not soft enough; potted Chieh-qua mungbean noodle has too many fixings. It is safest to order the tomato omelet.

8.走了一天的路，酒店酒吧的一杯红酒最舒服。

After a day's travel, it is most comfortable to have a cup of red wine at the bar of the hotel.

最后一天的纽约，离开的时候，回想前两天没记的笔记。 0:25 ~~[illegible]~~ 16/10 2016 ①

早上在11 Howard 吃完早餐后 Check out and Taxi 到第五大道的 Baccarat

住两天。当时不知此酒店，同时也不知是一个近年开业的

一个近十年的新酒店，Hms 查看路线很详！

终于到酒店里面一瞬间，我才一直认为前门（或者是换了）一个门的所谓

的大堂）lobby，check in 的时候也有一种气氛，到了二楼，之后每一件

东西 glass 都像是水晶一样，都是 Baccarat 的 Kims 的，大家的不眨眼

是太过分也是讲究，大大的窗口看到几十万（看 Goodman 的

专柜上面的）但挡不住了一两个人使用而让人觉得是一个好的

地方去两天吧。入住也是很愉快的。

20 West 53rd Street, New York, NY 10019

我们一边房间的窗外就看到的是对面的 MoMA 艺术博物馆（现代艺术）

Baccarat
HOTEL & RESIDENCES
NEW YORK

Day 3 in New York

纽约第三天

最后一天的日记，属于回忆的，因为偷懒了，两天没动笔了。

I wrote the diary of the last day from memory as I had not written any for two days because of laziness.

10月14日早上在11 Howard 早餐休闲后办理退房手续，乘出租车到市中心的巴卡拉。堵车很要命，为了少走几步路，车因为单行线又多走了近半个小时，也多了近十美元的车费，所以走路最经济了。

October 14

After breakfast and a little rest at 11 Howard, I checked out and taxied to Baccarat at the city center. There was a terrible traffic jam. I was unwilling to walk; the taxi used an extra of nearly half an hour because it was a one-way street, meaning that I had to pay an extra of nearly ten dollars for the fare. So walking is the most economical way to get around.

看来这儿的酒店生意一般般，我十一点就有房了（或许是提前了一个月订房的好处吧）在大堂办理入住 的时候就有一杯香槟了，别忘了，这里每一件貌似玻璃的东西可都是巴卡拉水晶的，大堂的大吊灯更是太过分地肆意可见，大大的器皿动辄几十万（今天在Goodman专柜印记了价格并收了一两个自己使用而让自己贵族了一瞬间），让我土豪两天吧，入住过程是愉快的。

It seemed that the business here was just so so as I had already got my room by 11, or it was because I had booked the room a month in advance. While checking in at the lobby, I had already got a glass of champagne. Don't forget it — every glass-like article here was made of Baccarat crystal. The electroliers in the hall, especially, were exceedingly glaring. The massive wares were worth tens of thousands of dollars (today on the counter of Goodman I witnessed their prices and collected two pieces for personal use, making me a nobleman immediately). Let me be free with my money for two days. The check-in process was pleasant.

拍了一通房间的设计，就去旁边的MoMa艺术醺醉（就是醉）了半个下午，可以用手机连WiFi的收听方式，还是第一次体会，不错了！下午柏悦酒店的一杯白葡萄+一个汤，忘记了晚上吃了什么！老年痴呆症了！

After taking photos of the guestroom design, I went to the nearby MOMA for art exposure for half the afternoon. Here you could tune in with the mobile phone connected to WiFi — my first experience. Good! In the afternoon I drank a cup of white wine and had a soup, but I forget what I had for supper — a sign of dementia!

Baccarat
HOTEL & RESIDENCES
NEW YORK

[illegible]下午，可以用手机连WiFi，[illegible]，不错！[illegible]
[illegible]一样的[illegible]十一个[illegible]
[illegible]吃了[illegible]！ 一起[illegible]来[illegible]！ ②

[illegible]，上午[illegible]了600[illegible]大衣，也给Jules[illegible]，[illegible]
[illegible]，[illegible]了一圈 Central Park，[illegible]/[illegible]，[illegible]
[illegible]，[illegible]不[illegible]人[illegible]，[illegible]
回[illegible]（[illegible]），大[illegible]
[illegible]一[illegible]，也[illegible]，[illegible]，一[illegible]，一[illegible]，两[illegible]
[illegible]，也[illegible]！

[illegible]！[illegible]，[illegible]图书馆，再[illegible]
[illegible]"[illegible]"[illegible] 20 West 53rd Street, New York, NY 10019 [illegible]，
[illegible] Baccarat [illegible]，[illegible]。

我们又一Uncle Jack's Steakhouse。后来这里的[illegible]
"~~[illegible]~~要大，我们[illegible]"

那些，再加一些[illegible]的酒在这里，也很便宜。

[illegible]，我们都来[illegible]，[illegible]，喝到了！[illegible]，吃饱了

大家都[illegible]一下，[illegible]！

[illegible]回到了！

16/10，上午10:04

Day 4 in New York

纽约第四天

10月15日，上午睡大懒觉，好好地享受了6000多元一天的大床，也好挑挑毛病，这样才有资格。认认真真去了一圈纽约中央公园，近24000步的纪录，领略纽约人的休闲生活，公园众生相，不同层次和人群的生活，风景一流。回程看到了古根海姆博物馆（没有进去），大都会就匆匆忙忙看了一个小时。晚上酒店大餐，一汤、一牛扒、两杯香槟$175，可谓奢侈！

October 15

I slept in, fully enjoying the big bed costing me more than 6,000 yuan a day, a chance to find flaws. Only by so doing was I physically qualified for the upcoming walk of near 24,000 steps around the Central Park, appreciating the recreation and life of the New York people from different walks of life. The views were excellent. On my way back, I saw the Guggenheim Museum but did not enter. I spent one hour having a cursory tour of the big capital. Then I spent 175 dollars eating a big supper at the hotel: a soup, a helping of grilled fillet steak, and two cups of champagne — really lavish!

把酒店的平面画好，近中午！继续出发去旁边的公共图书馆，再走路去吃上海菜“鹿鸣春”。第五大道的逛逛生活，小购买，包括巴卡拉水晶，回酒店小休一会，继续去去停停。

When I finished sketching the hotel plans, it was already near noon! Then I went on to the nearby public library, before walking to eat the Shanghai dish Great soup dumplings. Then I did a little shopping on the Fifth Avenue, including the Baccarat crystals before returning to the hotel for a little rest. Then the round trip continued.

找牛扒——Uncle Jack's Steakhouse，居然碰到戴昆同志，“世界很大，我们很近”。

While looking for the grilled fillet steak, I happened to meet comrade Dai Kun at Uncle Jack's Steakhouse. "The world is big and we are close to each other", as they say.

聊聊天，开心一晚上，来参观酒店房间，继续聊。

We enjoyed chatting the whole evening. Then he came over to visit my guestroom. Our chatting continued.

晚上十点，就是自己的时间，洗澡，写日记，看电视，明天还有半天可以享受一下！结束美国之行！

By 10 PM I kept the time to myself, bathing, writing the diary, and watching TV. I still had half a day to enjoy the next day before the trip in the US. Ended!

安安心心回国了！

Then I would return to my homeland, fully relieved.

1.巴卡拉酒店的看似玻璃的东西都是这一品牌出品的真真正正的顶级水晶，非常璀璨。

The glass-like articles in Barkla Hotel, all authentic top-grade crystals crafted by this brand, glisten brilliantly.

2.什么是艺术？我说“看不懂的就叫艺术”。一片红，五条不同大小、颜色各异的线，就是大师的作品。果然非常的MoMA(当代艺术博物馆)。

What is art? My answer: “It’s what you cannot appreciate”. An eyeful of red and five lines of various sizes and colors—it’s a master piece. It is really quite MoMA (Museum of Modern Art).

3.西式下午茶时间的柏悦酒店，非常有人气，西餐的摆盘也非常有美感，黑鱼子龙虾汤，味浓口甘，配上香槟，绝佳！

Park Hyatt sees brisk business at the Western afternoon tea time.The arrangement of the Western-style foods is quite pleasant-looking. The soup of caviar malossol and lobster, highly seasoned and sweet, is a perfect match with champagne!

4.专心走了一圈中央公园，阳光明媚，绿树成荫，各种艺术表演相当有水平，走走停停，乐在其中。

On an attentive tour of Central Park,sunny and lined with shady trees, I see various art performances of fairly high levels. Sauntering along on and off, I enjoy it.

5.公园有很多艺术家摆摊，因为是一个华人画家，就驻足近大半个小时看看他的“工作”，外国小孩在妈妈的陪伴下，很配合，一个不错的“客户”。

The Park abounds with street artists. It is because he is a Chinese artist, I have spent nearly half an hour watching him work. The foreign kid, accompanied by his mother, quite cooperative, is a very good “customer”.

6.佩服大苹果纽约城市规划的前瞻性，中央公园的宽阔湖面给城市的倒影真是百看不厌，让我们自愧不如。

The foresight of the urban planning of New York the Big Apple is admirable. The reflection of the city on the broad lake surface of Central Park is really a timeless beauty, putting us to shame.

7.难得晚上还开门的大都会博物馆。与上一次的感觉不同。专门找几个感兴趣的馆走走，太大了，只能走马观花了！

It is rare to see this museum still open in the evening. It feels different from last time. I have to select several venues to visit. It is too big—I have to settle for a cursory tour!

8.小费文化是诚信体系的影子（详见《住哪？2》），一顿酒店大餐，让消费者自己评价服务水平，填上你喜欢的数字。

The tipping culture is the shadow of the credibility system (see Where to Stay? 2 for more). After a big meal in a hotel, the customer is asked to rate the service and write down the figures as they like.

9.特别去看看这个项目：公园大道432号。可惜还没有完工（样板房完成了），还是非常之震撼和完美的，左右紧挨着，施工难度可想而知了。

I am very eager to see this project: 432 Park Avenue. It is shame it is not yet completed (the show house is completed). Still, it is very shocking and perfect. As the blocks are barely apart from each other, the construction work is conceivably very hard.

10.逛逛Goodman商场，专门去看看卖水晶的专柜，看到这个巴卡拉水晶乌龟，就下手让它跟着我回家了！

Sauntering in Goodman shopping plaza, I especially go to visit the counter selling crystals. On seeing this Barkla crystal turtle, I get it and take it home!

11.在最有名的牛扒店，吃上风干35天的熟成牛肉，我说是我这一辈子吃过最好的，没有之一。

Eating cooked beef air-dried 35 days at the most famous grilled fillet steak house—it is the best beef I have eaten so far, second to none.

12.途经百年老店费乐蒙酒店，忍不住再一次进去看看，还是像上一次一样的受欢迎，可能是下午茶时间，排队等位的人好多。

Passing centennial Fairmont Hotel I cannot help entering it for another look. I am welcomed as last time. Perhaps as it is afternoon tea time, many people are waiting for a place.

13.巴卡拉酒店的精美早餐无敌，每一样的出品几近完美。

The delicate breakfast at Barkla Hotel is matchless. Every dish is near perfection.

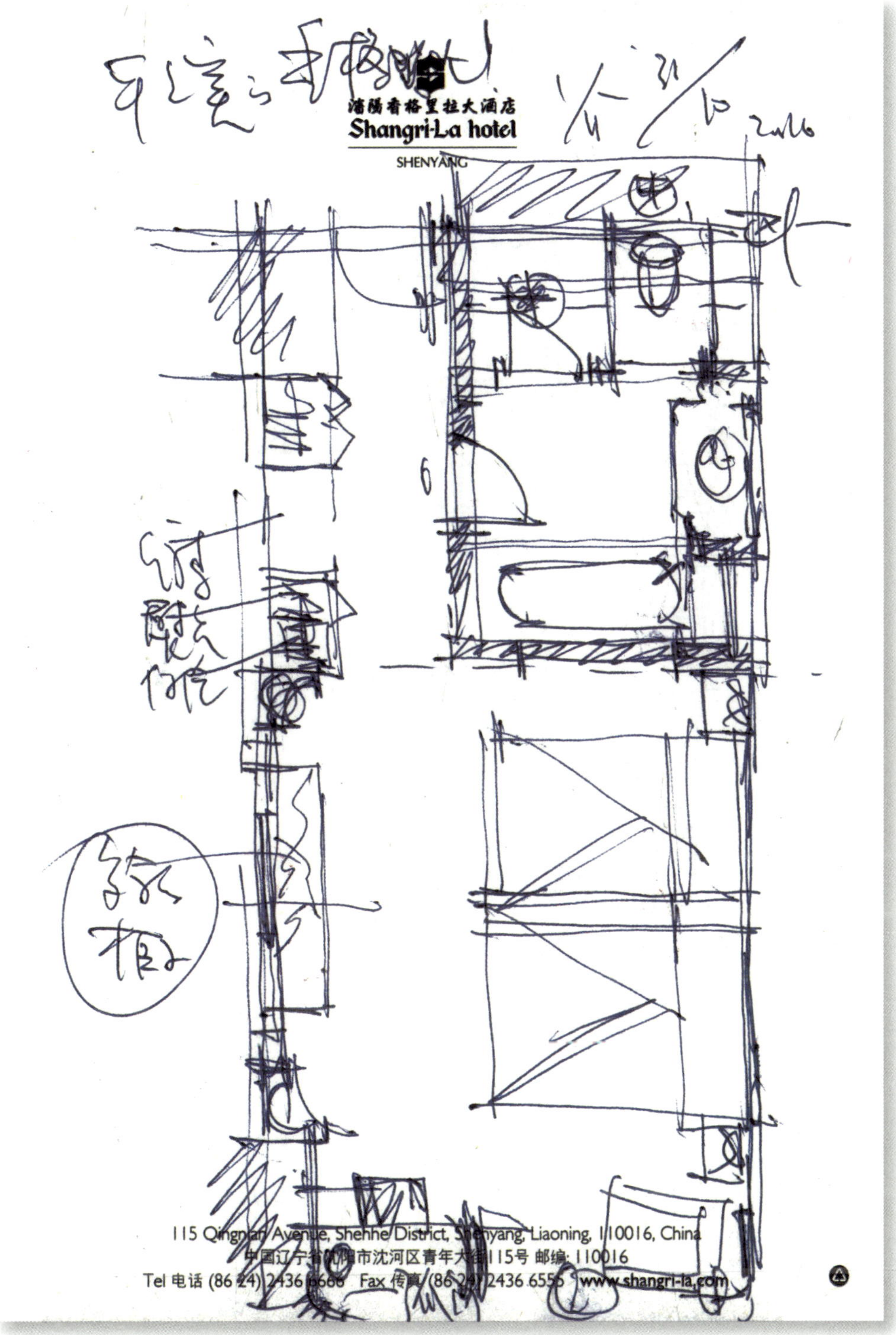

瀋陽香格里拉大酒店
Shangri-La hotel
SHENYANG

沈阳香格里拉大酒店

60 ★★★★★

SHANGRI-LA HOTEL SHENYANG CHINA

Address : 115 Qingnan Avenue, Shenhe District, Shenyang, Liaoning, 110016, China
中国辽宁省沈阳市
沈河区青年大街115号
110016
Telephone : +(8624) 2436 6666
Fax : +(8624) 2436 6555
Http : //www.shangri-La.com

WANDA®
Hotels & Resorts
万达酒店及度假村
兰州万达文华酒店
★★★★★

61

WANDA VISTA LANZHOU CHINA

Address : No.52 Tianshui North Road,Chengguan District, Lanzhou, Gansu Province, 730010,P.R.China
中国甘肃省兰州市城关区天水北路52号
730010
Telephone : +86(0)931 612 8999
Fax : +86(0)931 612 8688
E-mail : wandavista.lanzhou@wandahotels.com
Http : //www.wandahotels.com

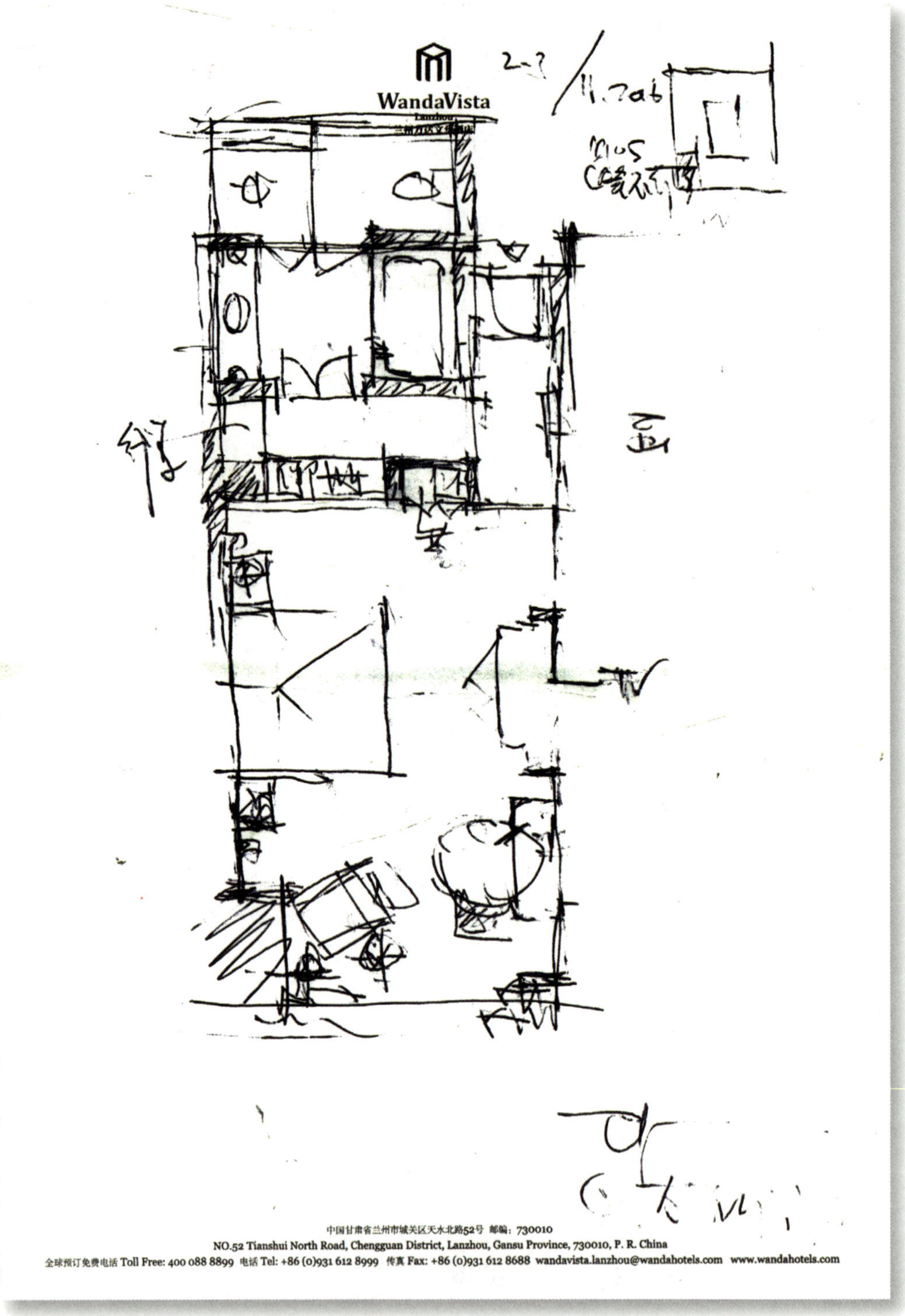

WandaVista
Lanzhou
中国甘肃省兰州市城关区天水北路52号 邮编：730010
NO.52 Tianshui North Road, Chengguan District, Lanzhou, Gansu Province, 730010, P. R. China
全球预订免费电话 Toll Free: 400 088 8899 电话 Tel: +86 (0)931 612 8999 传真 Fax: +86 (0)931 612 8688 wandavista.lanzhou@wandahotels.com www.wandahotels.com

加栋：

因为飞晚了到兰州，天冷，找烤串，喝啤酒，回到酒店已经半夜两点多了，筋疲力尽到前台入住，哇！服务员一个大失误，两个老男人啊，我出差订房偏偏有一个人睡两张床的状态，这次倒刺激。

有劳酒店匆匆忙忙调来折叠床，委屈了我们的总监陈帅哥了，一个我睡两米的大床，他只能……哈哈，很少加班的经历，除了和家人朋友外出旅行的时候才有的。

其实，一番折腾后，还是依旧八样打鼾，倒是房间的细节都没细看，竟然连回的洗手间只能胡乱的使用了一番。

一早，赶时间，用十几分钟来洗澡，气，断断续续的出水，这笔不知道有没有人用过，也算是第一次遇到这种情况。

我去和加栋一样，这也是一种"奇遇"了。

Extra Bed Added

加床

兰州万达文华酒店
Wanda Vista, LanZhou, China

一早赶时间，用十几分钟画平面 ，气，断断续续地出水，这笔不知道有没有人用过，也算是第一次遇到这种事情。

Pressed for time in the morning, I spent more than a dozen minutes drawing plans. Damn, the ink came out off and on. I wondered if anyone had ever used it before me. It was the first time I'd had such an experience.

或者和加床一样，这也是一种“奇遇”了。

Adding an extra bed, it was another kind of "adventure".

因为很晚才到兰州，天冷，吃烤串，喝啤酒，回到酒店已然半夜两点多，筋疲力尽到前台入住，哇！居然是一个大床房，两个老男人啊，平时出差订房偏偏有一个人睡两张床的状态，这次倒刺激。

It was quite late when we arrived at Lanzhou. Since it was cold, we hunted for skewered foods and beer. When we got to check into the hotel, worn out, it was already two the next day. Wow, we were assigned to a double bedroom—we were two men! Usually, when traveling on business, I tend to book double deb rooms for myself alone. It was so sexy this time.

有劳酒店匆匆忙忙调来折叠床，委屈了我们的总监陆帅哥了，一个人我睡2米的大床，他只能……哈哈，很少有加床的经历，只有和家里小朋友一起外出旅行的时候才有的。

The hotel was put into the trouble of fetching a folding bed, and our handsome Director Lu got the shorter end of the stick: I was to occupy the two-meter big bed and he had to … Haha. We had seldom have an extra bed added, except when our little kids were traveling.

其实，一番折腾后，还是依旧入梦打鼾，倒是房间的细节未及细看，宽敞迂回的洗手间只能胡乱地使用一翻。

In fact, after all this trouble, we still fell asleep, dreaming and snoring. The shame was that there was no time to study the details of the room. As for the spacious and roundabout bathroom, we just had a cursory use of it.

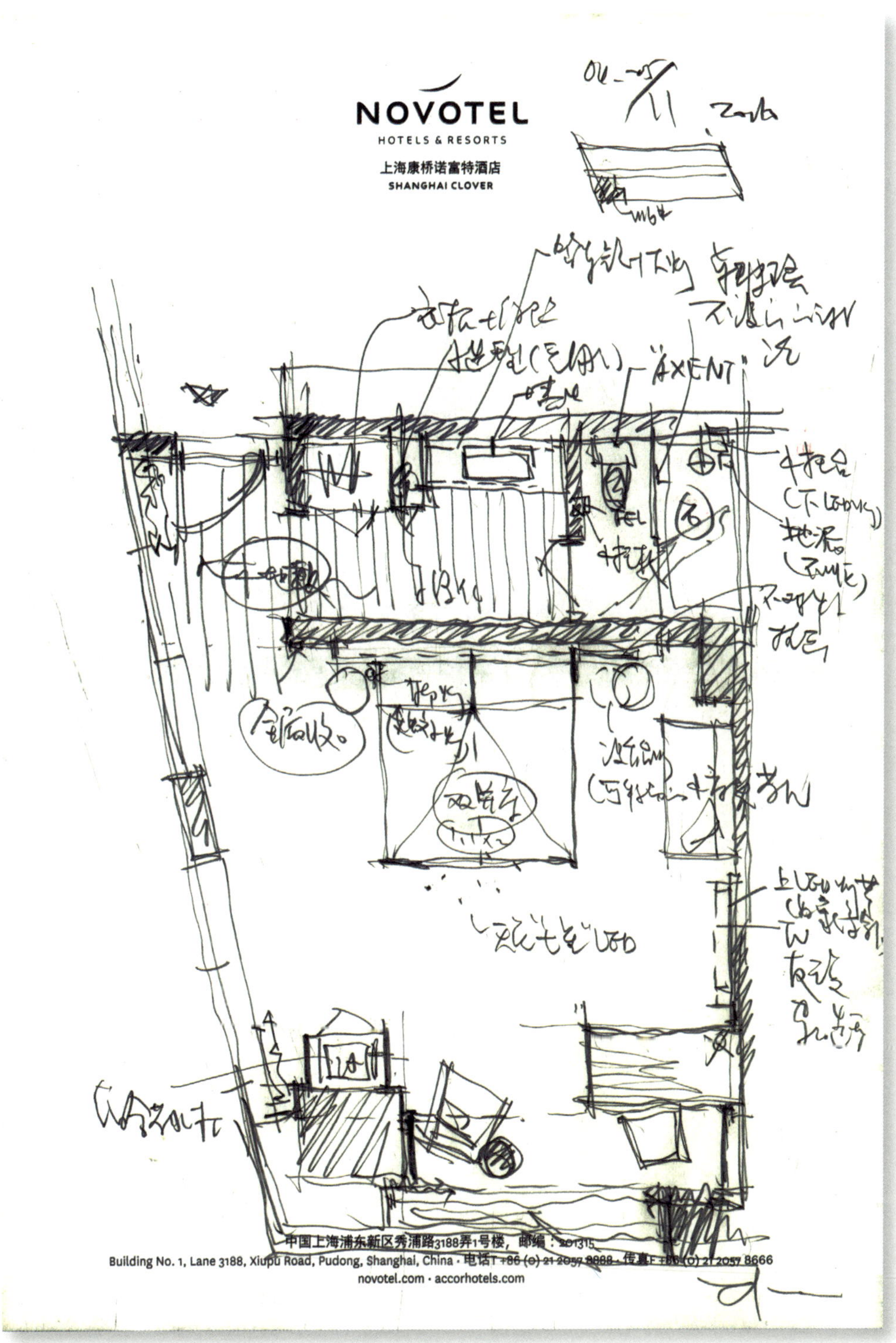

NOVOTEL
HOTELS & RESORTS
上海康桥诺富特酒店
SHANGHAI CLOVER

上海康桥诺富特酒店
★★★★★

62

NOVOTEL HOTELS & RESORTS SHANGHAI CLOVER SHANGHAI CHINA

Address : Building No.1, Lane 3188, Xiupu Road, Pudong, Shanghai, China.
中国上海浦东新区
秀浦路3188弄1号楼，
201315

Telephone : +86(0)21 2057 8888

Fax : +86(0)21 2057 8666

Http ://www.novotel.com
//www.accorhotel.com

超越"天书"

因为服务相关品牌的酒店设计，特地去学习一下同行的设计，地处上海南七宝商圈的这家带宗精酒店还是颇惊喜，从进入门的方式，到大堂区域超高大的空间再到其他功能空间的丽美表现丰富及大规模，同样房间也是极尽"设计感"，真当是这个品牌的高端定义了。这些都已成为品牌"天书"了，我们入住的一间尽端转角的房型，可谓更加不愧是这个品牌的代表作。

但是设计师们的功力还是能体现出来的，平面布局尽用转角空间的超高玻璃幕墙，设计的合适的尺度显得自然，安全舒适，门洞无阻（第一次见到上下的印记），倚靠设计了隐藏柜（包括饰品陈列柜、超级"眼睛"的效果，但也是电视柜立面上的美术。灯光设计可谓"画蛇添足"，也许是对地上人来说，对我家来说可能是乐趣。像在天花上方的美仙"北斗七星"的LED灯具造型。

见仁见智。

Super Tianshu Hotel
超越“天书”

因为做相关品牌的酒店设计，特地去学习一下同行的设计。地处上海迪士尼商圈的这家诺富特酒店还是有惊喜的，从建筑入门的方式，到大堂区域超高大的空间，再到其他功能空间的配套，超级丰富及大规模同样，房间也是极尽“设计感”，应当是这个品牌的新宠儿了，远远超过已有的品牌“天书”了。我挑了一间尽端转角的房型，可谓更加不具备这个品牌的代表性了。

As a hotel designer of the related brand, I especially made a trip to study the design of my fellow designer. Located in the Disney business zone of Shanghai, this Novotel held surprises—from the building entrance to the extra large space of the hall to the equipment of other functional spaces, it boasted richness and largeness. The guestrooms were also full of "designer ideas". It could be said to be a new favorite, overshadowing the already existing brand Tianshu. I chose a corner room at the end, less typical of this brand.

但是设计师的功力还是能体现出来的，平面布局尽用微斜的转角玻璃幕墙，洗手间舒适而尺度紧凑、自然，写字台靠窗，阳光普照（第二天早上正好印证了），倚柱设计茶水柜，结合饰品陈列，起到“吸睛”的效果。倒是电视柜立面过于复杂，灯光过分设计，可谓“画蛇添足”，也许只对专业人士而已，对大家来说可能是乐趣，比如在大床上方的类似“北斗七星”的LED灯具造型。

Still, it showed the skill of the designer. The plane surface was all covered in slightly slanting corner glass curtain wall. The bathroom was comfortable, close-kit and natural. The study was close to the window, letting in adequate sunlight(as proved the next morning). The tea cupboard, together with the decorations displayed, proved to be quite attractive. However, the design of the facade behind the TV bench and the lighting was overdone, and superfluous. Maybe only experts can see it. For the average guests, it could be fun, like the dipper-shaped LED lamp over the large bed.

见仁见智。
Everyone has their own opinion.

上海康桥诺富特酒店
Novotel Hotels & Shanghai Clover, Shanghai, China

宜春迎宾馆

★★★★★

63

YICHUN GUEST HOUSE JIANGXI CHINA

Address : No.669, Luzhou North, Yichun City, Jiangxi, China

江西省宜春市泸洲北路669号 336000

Telephone : +(0795)368 8888

Fax : +(0795)368 8889

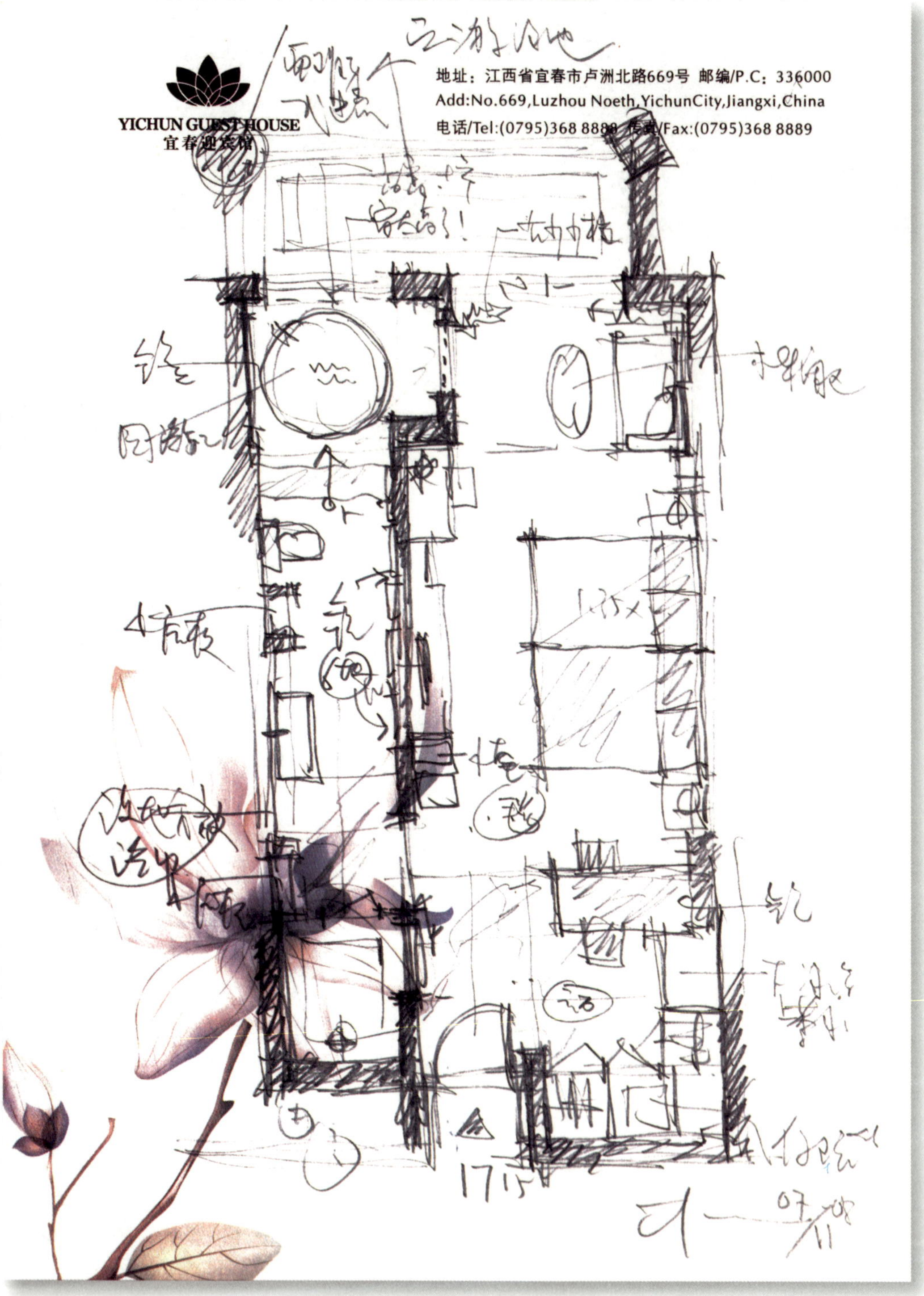

YICHUN GUEST HOUSE
宜春迎宾馆
地址：江西省宜春市卢洲北路669号 邮编/P.C：336000
Add:No.669,Luzhou Noeth,YichunCity,Jiangxi,China
电话/Tel:(0795)368 8888 传真/Fax:(0795)368 8889

怎样洗澡

客人的心情随着旅途长远会变成吃吃喝喝，迷迷糊糊的回到酒店，再好的房间也仿如“虚设”，先睡觉早上再慢慢享受房间的配套。

浙江湖州的客户，入住宜昌最好的酒店，下午洗澡。房间功能齐全，有完善的储物空间，餐厅区；睡眠区，洽谈区，超级大的两张大床人体横在一张床；更有小客厅、观湖游泳池阳台，而最深印象的则是狭长狭长的洗手间，洗漱时被迫采用一整面墙的镜面来扩大空间感，而由于狭窄，毛巾架也没有了正常的位置，小小的薄薄的，不是叠放折好的毛巾、浴巾，倒是一端的淋浴间够大、够深，另一端紧靠的厕所浴缸也都觉得很可惜空间太小，感觉太隐私了。那就早上酒醒起来，洗脸刷牙，再淋浴后泡泡浴缸，完美的“清洁流程”，这样才是住店的最佳的洗浴方式。

宜昌最好的酒店，给我最好的体验。

How to Bathe
怎样洗澡

客户的盛情接待最终都会变成吃吃喝喝，迷迷糊糊地回到酒店，再好的房间也仿如“虚设”，先睡觉，早上再慢慢享受房间的配套。

A client's hospitality will all finally end up with eating and drinking, leaving you back to the hotel dead drunk, making the guestroom seem as if nonexistent, however good it may be. You have to sleep first before taking time to appreciate what is equipped in the room in the morning next day.

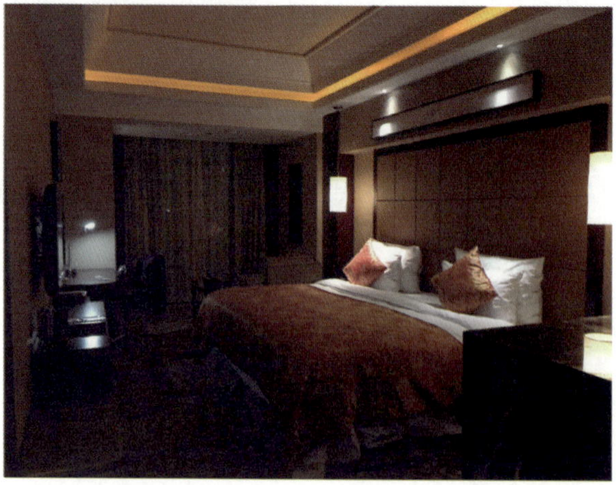

宜春迎宾馆
Yichun Guest House, Jiangxi, China

谢谢我们的客户，入住宜春最好的酒店，印象很深，房间功能超多，有完善的储物空间、餐厅区、睡眠区，哈哈，超级大的两张大单人床拼在一起啦；更有小客厅、观景阳台，而印象最深的则是狭长狭长的洗手间。设计师被迫采用一整面墙的镜面来扩大空间感，而因为狭窄，毛巾架也没有了正常的深度，小小的薄薄的，不足以叠放折好的毛巾、浴巾，倒是一端的淋浴够大、够深， 另一端靠窗有圆形浴缸，也相当气派，只可惜窗台太高，感受“太隐私”了。那就早上酒醒起来，洗脸刷牙，再淋浴后泡泡浴缸，完美的“清洁流程”，这才是住店的最佳的洗澡方式。

Thanks to our client, we was put up in the best hotel in Yichun. I was quite impressed as it had an extra number of functions: a perfect storage area, dining room area, sleeping area—haha, two extra large single beds side by side; a small living room, and a sightseeing balcony. What impressed me most was a long and narrow bathroom. The designer was obliged to use mirror the size of the whole wall for a sense of increased space. As it was too narrow, the towel rack lacked the usual depth, not enough to place the folded towel and bath towel. In contrast, the shower compartment at the end was big and deep enough, and the rounded bathtub near the window was rather imposing. The only shame was that the window sill was too high, feeling "too private". Then, after waking from a drunk sleep, wash your hands and brush teeth, take a shower before immersing yourself in the bathtub—a perfect "cleaning process". And that is just the best way to bathe.

宜春最好的酒店，给我最好的体验。
The best hotel in Yichun gave me the best experience.

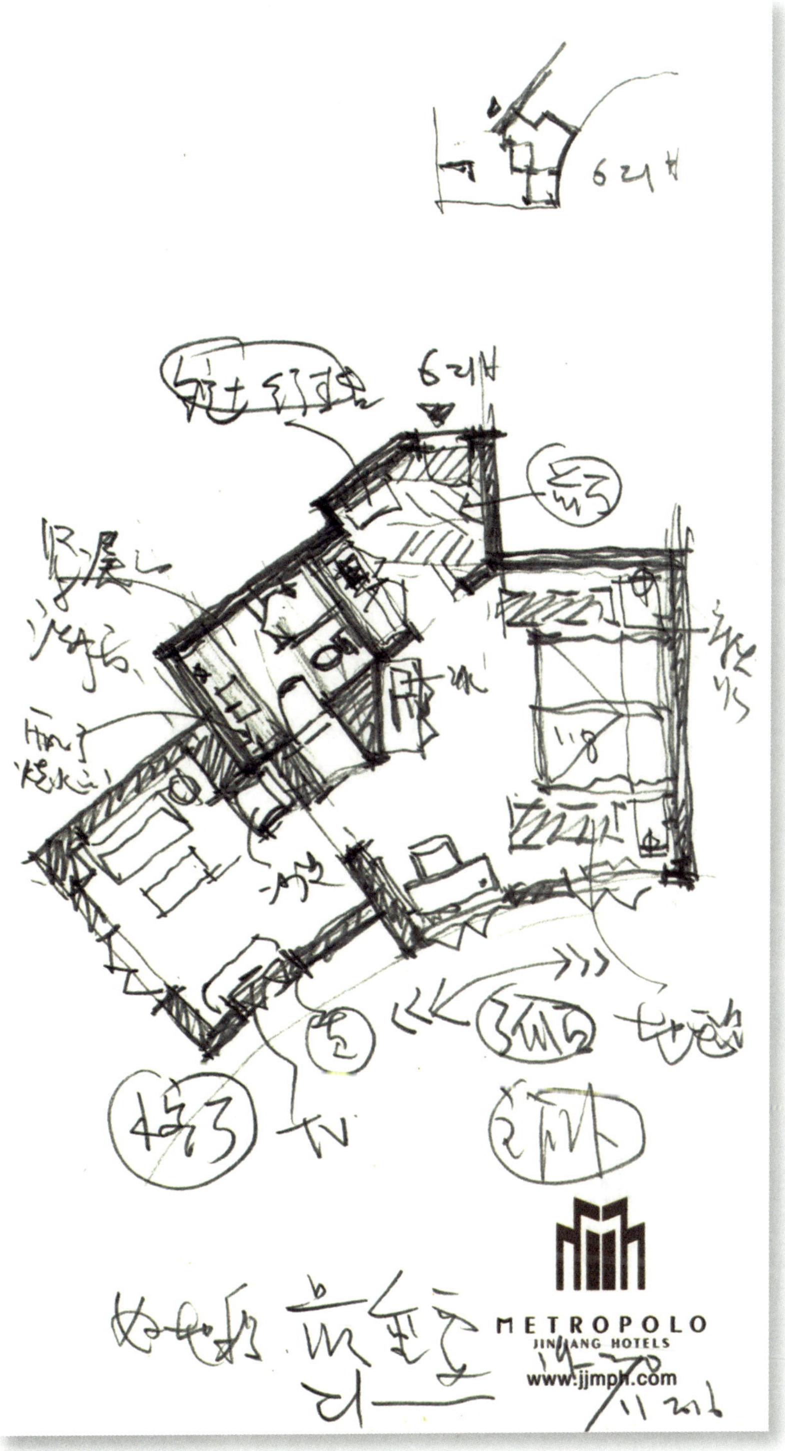

上海锦江都城经典酒店
★★★★★

64

JINJIANG METROPOLO HOTEL CLASSIQ BUND CIRCLE SHANGHAI CHINA

Address : No, 180 Jiangxi Road(M), Huangpu District, Shanghai
中国上海市黄浦区
江西中路180号
Telephone : +(021)6321 3030
Fax : +(021)6329 8622
E-mail : bundcircle@metropolohotels.com
Http ://www.jjmph.com

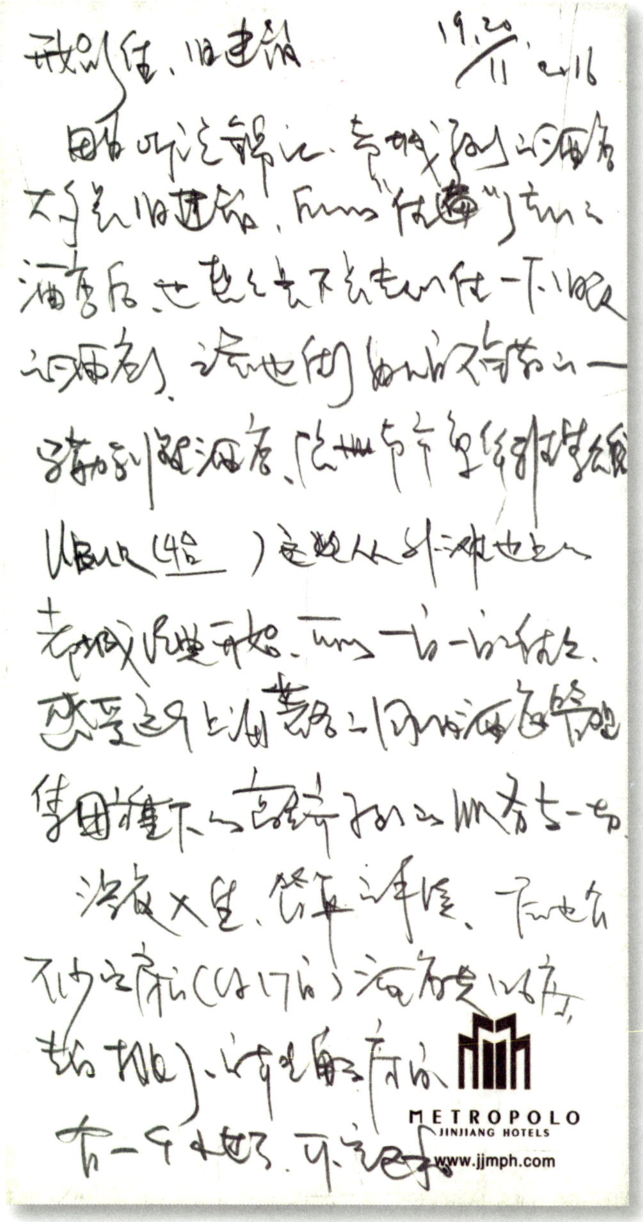
METROPOLO
JINJIANG HOTELS
www.jjmph.com

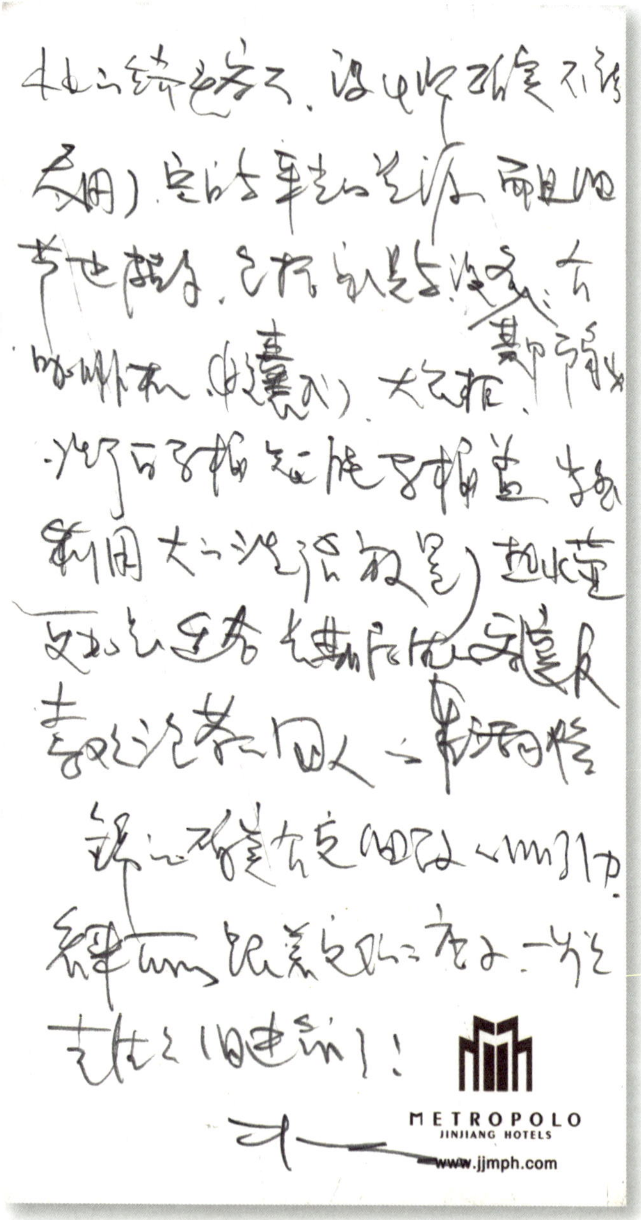
METROPOLO
JINJIANG HOTELS
www.jjmph.com

We Started Living in Old Buildings
开始了住·旧建筑

因为听说锦江都城酒店大多是旧建筑，自己“住遍”了新的酒店后，也想想是不是专心住一下旧改酒店了，之前也住了好几间不错的——马勒别墅酒店、隐居都市繁华雅集公馆、UBRN（艾本）。

Because it's said that Metropolo Hotel is mostly in old buildings, after hading stayed "all around" the new hotels, I also think about the hotels which are converted from old building. Before that I also used to stay in good ones, such as Mahler Villa Hotel, Bamboo Retreat, and UBRN Hotel.

这次从外滩边上的都城经典开始，可以一间一间住，感受这个上海著名的国内酒店管理集团旗下的高端系列的服务与一切。

This time around the bund, I can try the hotel one by one, begin from the Classic Metropolo Hotel, and feel the services and everything from this Shanghai famous domestic hotel management group's high-end series.

锦江都城经典上海新城外滩酒店
Jinjiang Metropolo Hotel Classq Bund Circle

深夜入住，简单的手续，一层也有不少的房间（约17间），酒店共126间房，专门挑了间转角的房间，有一个小过厅、卧室区和小小的端头客厅。设计师确实不错，尽用了空间与采光的资源，而且细节也颇多，包括家具与设备，其中印象有咖啡机（胶囊式）大衣柜，洗手间马桶智能马桶盖，当然利用大的洗手台放置了热水壶更加适合长期居住的宾客及喜欢泡茶的国人的生活习惯。

I checked in late at night, with simple procedure,. There are many rooms on one floor (about 17), and thtal 126 rooms in this hotel. I specially chose single room in the corner, with a small hall, bedroom area and a small living room in the end. The designer is really good, he used up the space and light resources, and also made a lot of details, including furniture and equipments. What impressed me is the capsules coffee machine, large wardrobe, intelligent toilet lid in washingroom, and a hot water pot on the large sink table, what is more suitable for long staying guests and our Chinese who like to make tea.

锦江确实有更细致的吸引力，看来可以跟着它们的店子，一步步去住住旧建筑了！

The Metropolo does have a more detailed appeal, so it seems to be able to follow their store, and take the time to live in the old buildings!

深圳蓝汐精品酒店

★★★★

65

BAY BREEZE SHENZHEN CHINA

Address : 8 Baishi Road East,
Oct Harbour,
Shenzhen, China.
中国广东省深圳市
南山区白石路东
8号欢乐海岸

Telephone : +86(0755) 8615 2533

Fax : +86(0755) 8615 2633

E-mail : wandavista.lanzhou@wandahotels.com

Http : //en.baybreezehotels.com/

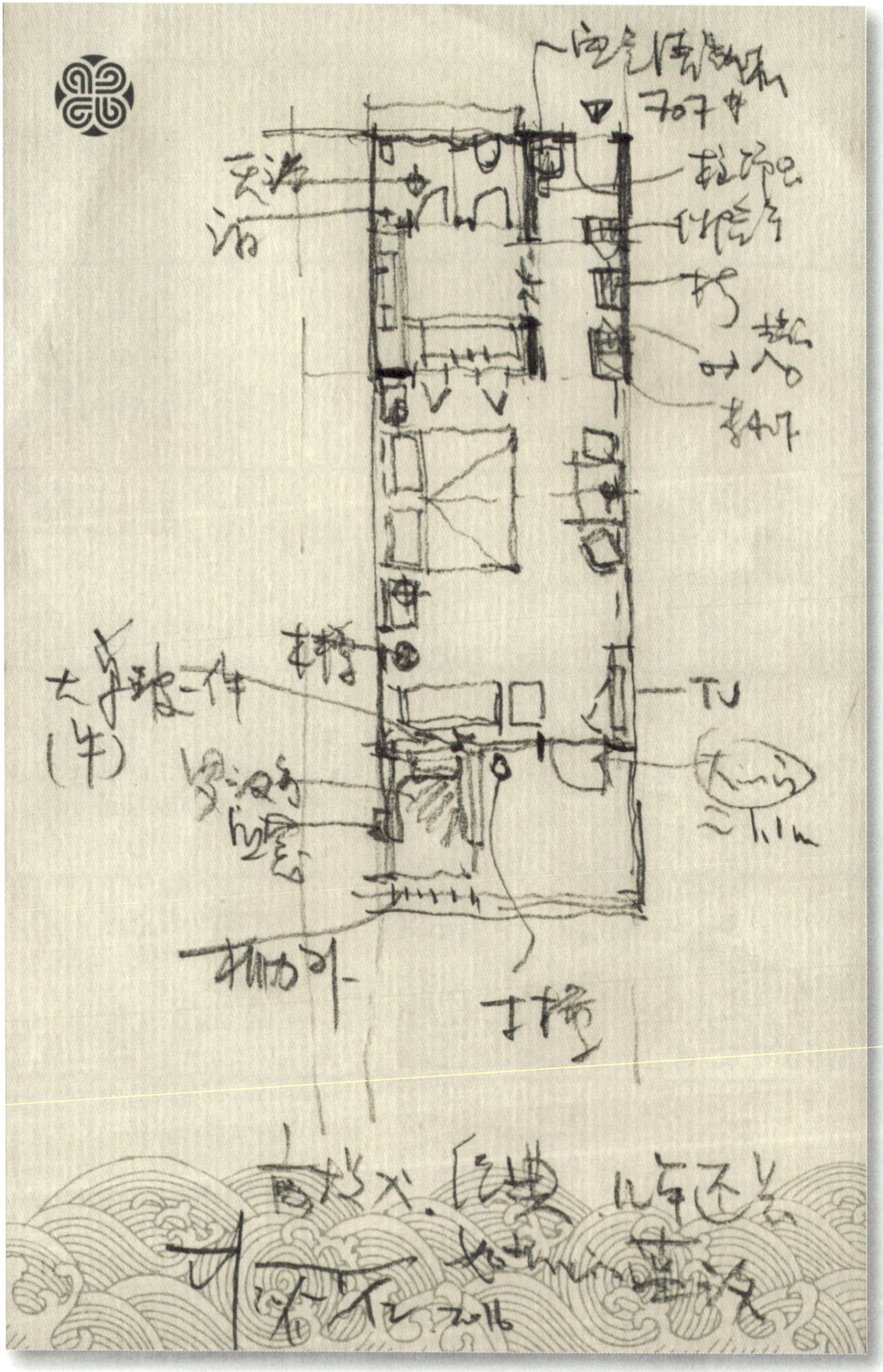

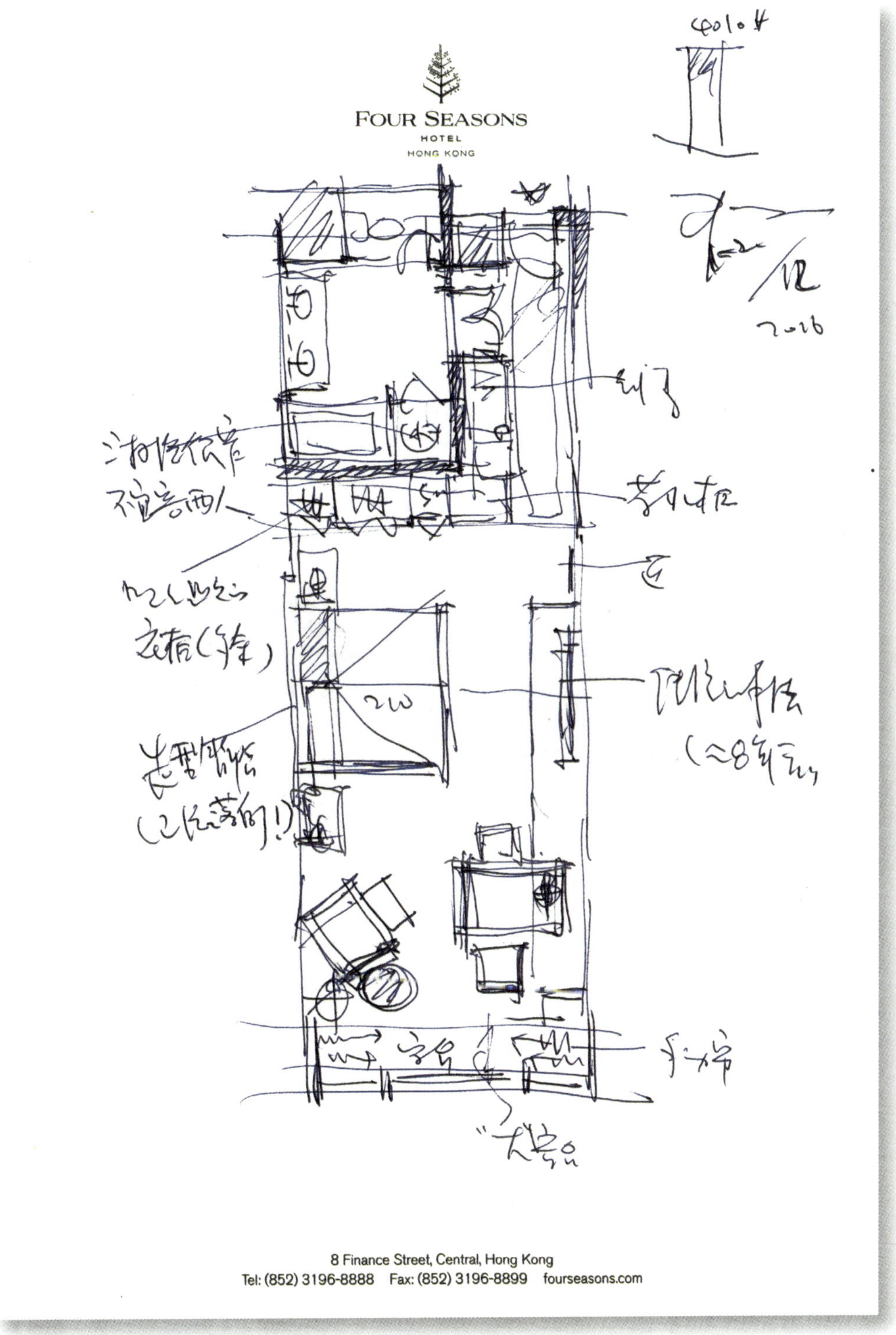

香港四季酒店

★★★★★

66

FOUR SEASONS HOTEL HONGKONG CHINA

Address : 8 Finance Street, Central, Hong Kong
Telephone : +(852) 3196-8888
Fax : +(852) 3196-8899
Http ://www.fourseasons.com

洱海千里走单骑杨丽萍艺术酒店

★★★★

67

QIAN LI ZOU DAN QI YANG LIPING ART HOTEL YUNNAN CHINA

Address : Dali Bai Minority Autonomous Prefecture Of Dali Double Gallery Town Jade Island 大理白族自治州大理市双廊镇玉几岛

Telephone : +86(0872) 2461 401

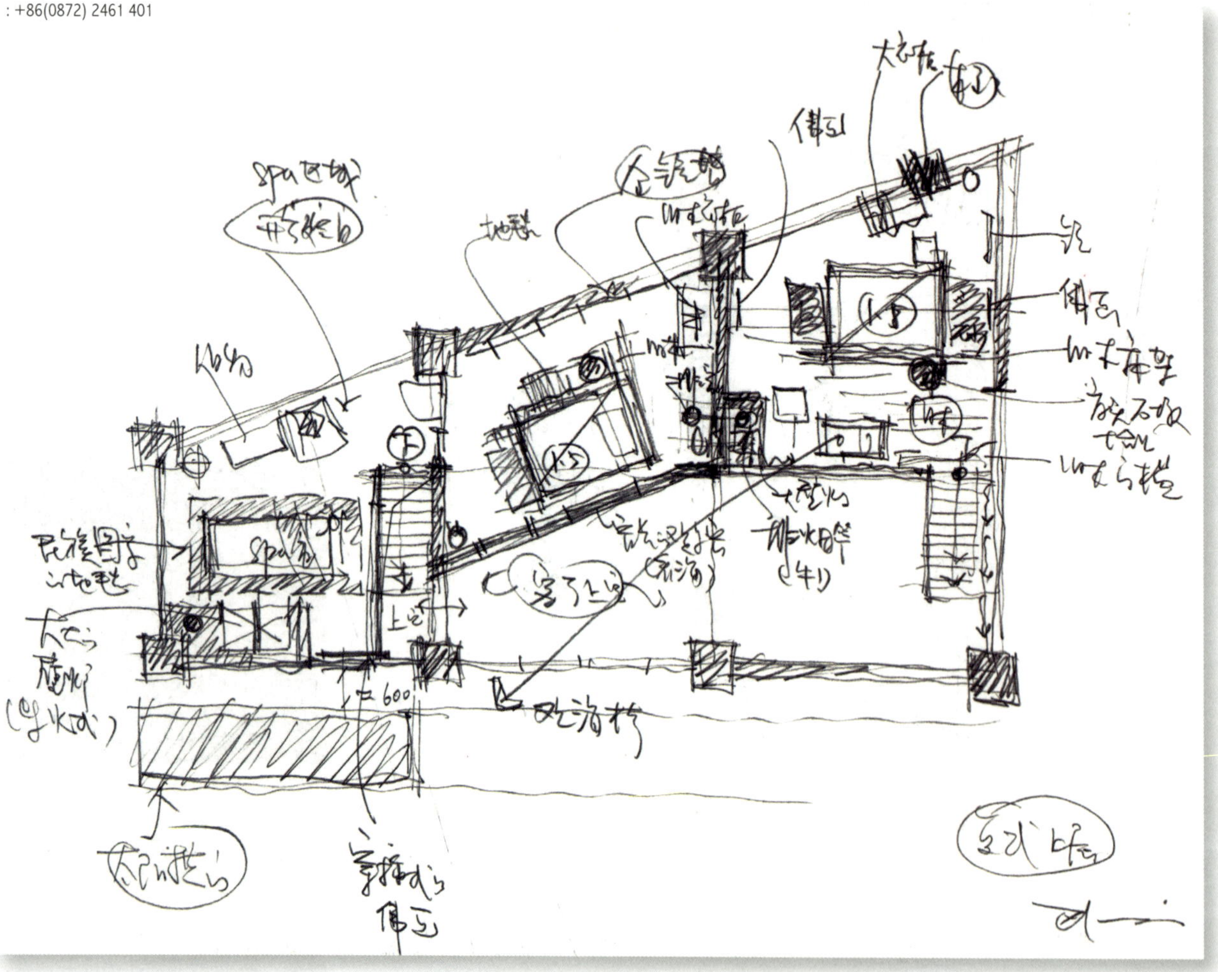

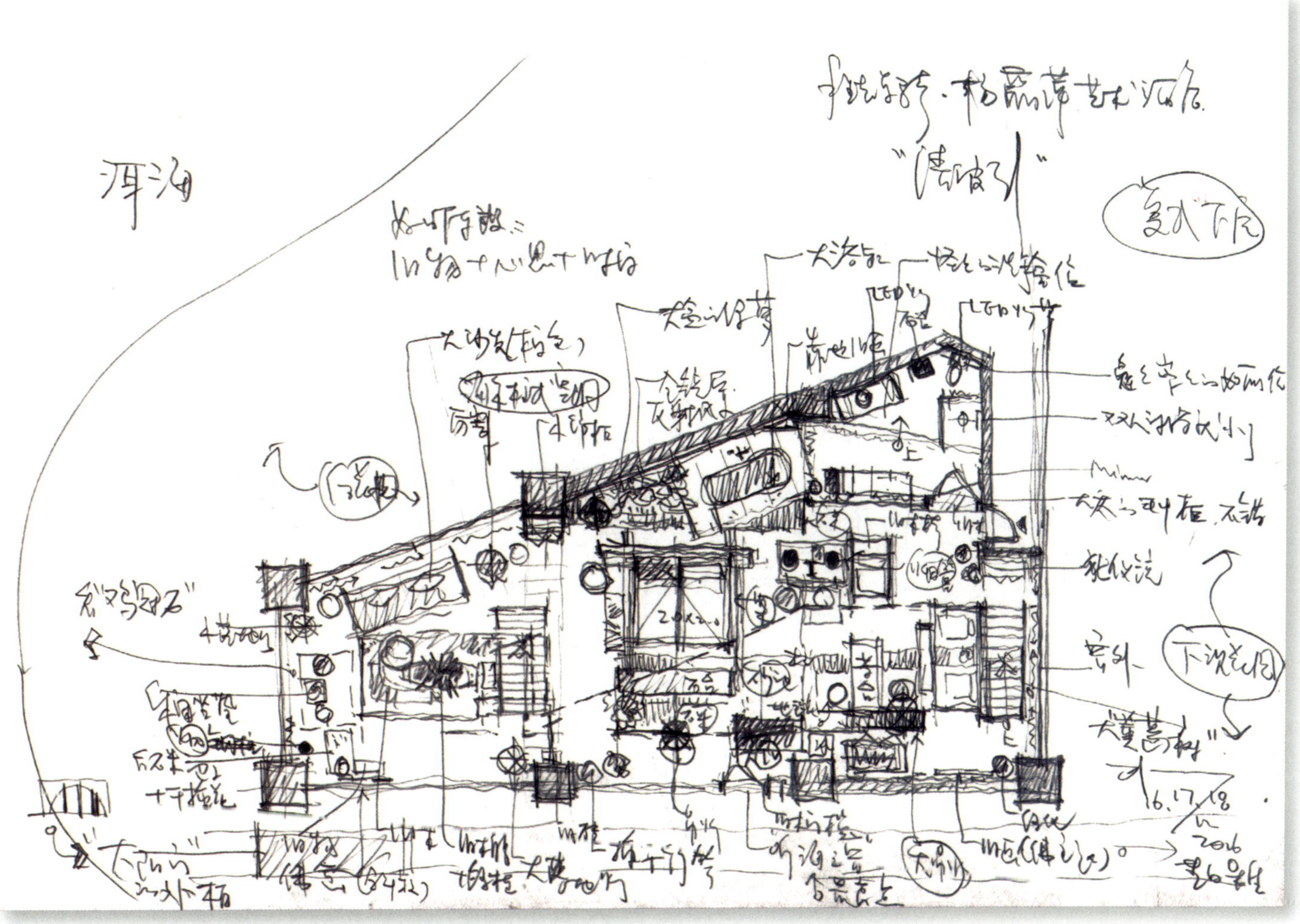
洱海
室外
下沉花园

房挺大，随便用起来看去

生日，找一个地方呆，洱海没有去过，那试一试！找了同事们介绍很久的杨丽萍投资的"千里走单骑艺术酒店"，网上订只知道是复式，公寓式小别墅，湖边，贵，总之，想起来体验一下吧！

离机场非常近，不知道有接送的服务，（可能太贵了，不好意思吧，哈哈）本来入住客人可以从码头由一部大而不讲究的船，从一个小小的码头到达酒店，虽然小了点，第一眼从湖面看到太阳宫（据说是杨丽萍的私宅，太阳宫就是这个酒店）还是非常震撼。

入住，放完毕了下行李，半路，随处看看各种各样的木栏，太美了，慢慢的，细细的品着。走廊，房间里大多，看看，上去，下楼梯，看看书，就是一个适合懒人的地方。看着时间在指间流走，看着太阳沉入苍山，看着湖面霞光渐暗，但入夜后远处城市散点的灯光，看着月亮渐渐挂天上。

呆住了三天，很不情愿的收拾东西时，最后在楼上西边的位建筑多变的布依族民居，上下两层，太多内容了，只能选了这一个木材去写一些描述的文字，以免忘记了需表达的情景，唯独北洋前的两宫（太阳宫与月亮宫）与一序（邻居载着老婆的书序）为整个海南度假作定义了，让这块土地有了艺术的高度和内涵，令人深思。我慢慢地走，慢慢地回味最好的每一分钟，使我多走绕了几次，才依依不舍地上路离开。

一路让你我慢下来的好酒店，每一间房间，房间的每一处，不同的感觉，不同的风景，不同的装饰。

房间太大了，晚上的灯太黄，照明暗了些。

Stay Where You Like in This Big House!

房子这么大，随便哪里待着去！

洱海千里走单骑杨丽萍艺术酒店
Qian Li Zou Dan Qi Yang Li Ping Art Hotel, Yunnan, China

生日，找一个地方呆，洱海没有去过，可以一试！挑了耳闻了很久很久的杨丽萍小姐的“千里走单骑艺术酒店”，网订时只知道是复式，150多平方米，湖边，贵，其他的，去现场体验一下吧！

I was looking for a place for my birthday. As I had never been to Erhai, I might have a try! I had long heard about the Riding Thousand Miles Yang Liping Art Hotel. Yet when I booked it online I knew only that the house has inside stairs, over 150m^2, lies on the lakeside, and is expensive. As for other things, I wanted to experience in person!

离机场非常远， 不知道有接送的服务（可能太贵了不好意思吧，哈哈），原来入住的客人可以从码头由一艘大而不讲究的船，从一个小小的码头到达酒店，虽然只有几百米。第一眼从湖面看太阳宫（月亮宫是杨丽萍的私宅，太阳宫就是这个酒店了），还是非常震撼。

It was far away from the airport. I had not known that the hotel offered to meet and see off guests at the airport (maybe it was because it was too expensive and I felt embarrassed to use the service, haha). It turned out that the would-be guests could take a big yet not so delicate boat to go from a tiny port to the hotel. It was quite shocking to catch the first sight of the Sun Palace from the lake surface—the Moon Palace is the private residence of Yang Liping and the Sun Palace is the hotel.

入住，有免费的下午茶，丰富，阳光照着各种各样的小吃，太美了。慢慢地、细细地嚼着。光着脚，房间里左看看、右看看，上走走、下踱踱，翻翻书，就是一个培养懒人的地方，等着时间在指间流走，等着太阳消失入苍山，等着湖面粼光渐暗，映入很远很朦胧的城市散点的灯光，等着月亮渐挂天上。

Having checked in, I enjoyed the free afternoon tea, which was plentiful. Bathed in the sunlight were various snacks, simply too beautiful. I took time chewing and crunching. Barefooted, I looked around the rooms, going up and down stairs, thumbing through books. It was a place to raise lazy guys, waiting for time to flow through their fingers, waiting for the sun to disappear behind Mount Cang, waiting for the shimmering lake to dim, with reflection of the scattered light from the very remote and city, and waiting for the moon to appear in the sky.

呆住了三天，很不情愿地收拾东西时，才最后画画这复杂多变的平面，上下两层，太多内容了，只能花了近一个小时去多写一些描述的文字，以免忘记了要表达的情景。难怪十几年前的“两宫”（太阳宫与月亮宫）与“一庐”（邻居赵青先生的青庐）为整个洱海度假作定义了，让这块土地有了艺术的高度和内涵，令人佩服。我慢慢地画，慢慢地回味最好的每一分钟，待服务员催促了几次，才依依不舍地上船离开。

I stayed there for three days. At last, when it was time to pack things that I began to draw the plans of the house, complex, full of variation, to which I had grown attached to. As it had two floors and there was too much in it. I could only spend nearly one hour writing down more descriptive words lest I should forget the sights to be expressed. No wonder the "two palaces" (the Sun Palace and the Moon Palace) and the "hut" (the green hut of Mr Zhao Qing, the neighbor of Yang Liping) had defined the Erhai vacation more than a dozen years before, lending height and depth of art to this land—it is really admirable. I drew slowly, appreciating each of the best minutes. Only when urged several times by the waiter did I board the boat to take leave.

一间让我慢下来的好酒店，每一间房间、房间的每一处，不同的感觉，不同的风景，不同的姿态。

It was a good hotel that made me slow down. Each room and each part of each room felt different, gave a different view, and showed a different posture.

房间太大了，待的地方无数，哪里都可以的。

The house was too big. There was too much space to stay, and anyplace would do.

洱海千里走单骑杨丽萍艺术酒店
Qian Li Zou Dan Qi Yang Li Ping Art Hotel, Yunnan, China

兰州金地名庭酒店

★★★★★

68

COURTYARD HOTEL LANZHOU CHINA

Address : 437-451 Jiuquan Road.
Chengguan District
Lanzhou.730030.P.C.
兰州市城关区酒泉路
437号–451号

Telephone : +(0931) 8170000

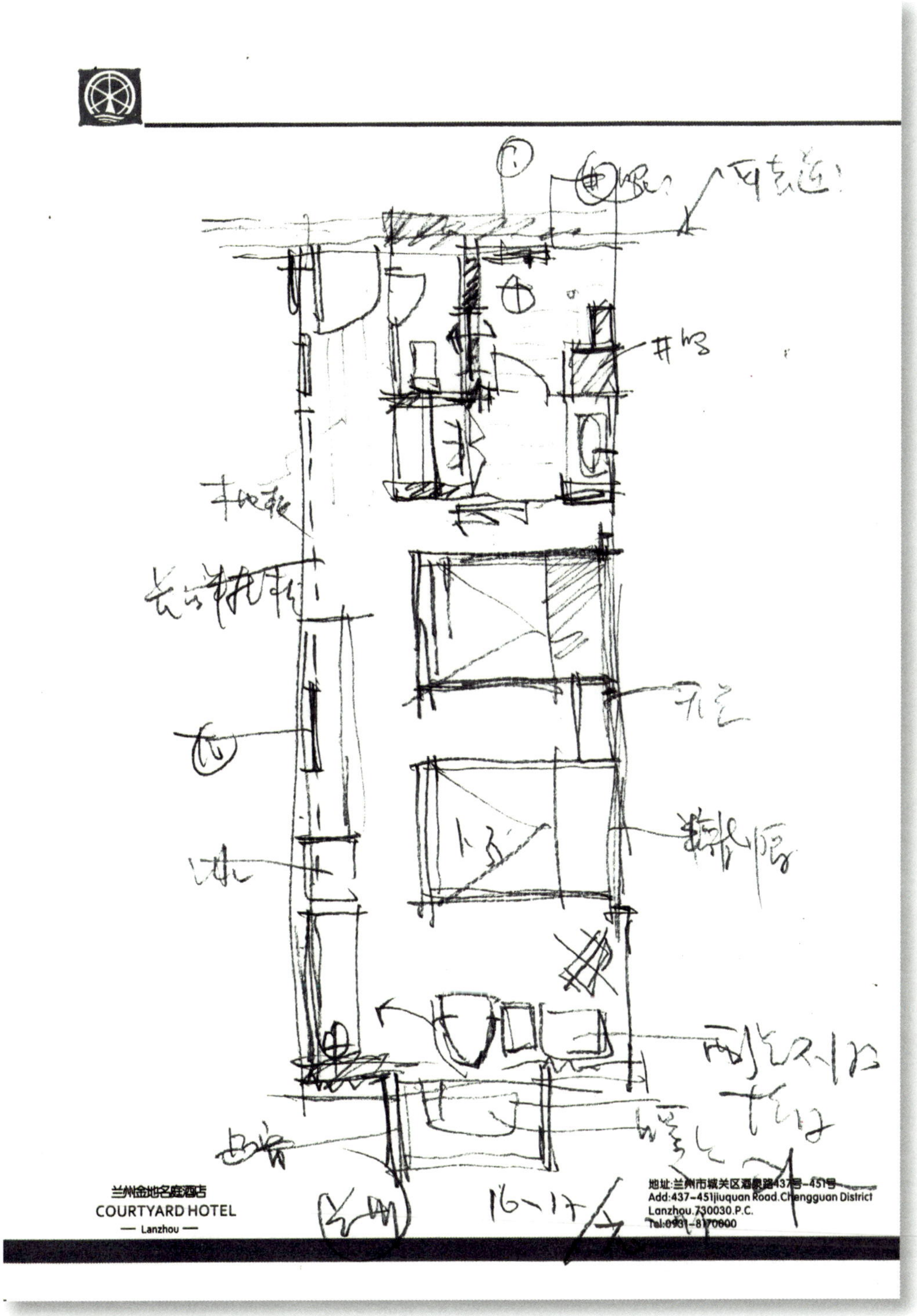

一点点的惊喜已经是不错的结果

让同事选择酒店，一个在楼盘里的精品酒店（金地地产楼下的自营的啊）上五楼 check in 去20楼的房间，打开后，还是有一点小惊喜的！

双人房、单独玄关、"居然"选用了干、湿分离的设计方向，还是有点新意的，而且也做得不错的，两张床之间在走道进出，功能区域接近，就使用区之类睡眠区，声音入侵者、灯光对睡眠影响小；洗、浴再合一而合出的区、出区去洗脸区接合的衣帽区，然后为大的浴室间，非常有意思和有些趣味性。房间所合并用洗浴的灯光，完全不对的，真的有研究的洗浴，体闲区也直接明亮，墙面图案去讨创新的设计和结合，并将它们统一而和谐，当然一两张洗浴的搭配我认为是"大世创意的"，一张偏中式的扶手椅配一张偏欧式的大沙发，有一点怪，也许这个怪也可以算是另一个小惊喜吧。

一间客房，有小惊喜也可，

谢谢惊喜啊！ [illegible]

兰州金地名庭酒店
COURTYARD HOTEL
— Lanzhou —
地址:兰州市城关区酒泉路437号-451号
Add:437-451Jiuquan Road.Chengguan District
Lanzhou.730030.P.C.
Tel:0931-8170000

A Little Bit of Surprise Counts a Not So Bad Result

一点点惊喜已经是不错的结果

让同事选择酒店，一个在楼盘里的精品酒店（金地地产旗下的自营品牌），上五楼办理入住，去20楼的房间，打开门，还是有一点小惊喜的！

According to my colleague's choice, we reached Floor 5 of a quality hotel (a self-run brand under Gemdale) and checked into a room on Floor 20. On entering, we were still greeted with a little bit of surprise!

双人房、单边走道，“居然”选用厕、浴分离的设计布局，还是挺大胆的，而且把握得不错，厕所在走道进出，磨砂玻璃掩门，使用区远离睡眠区，声音及噪音、灯光对睡眠影响小，洗、浴再合一而分前后区：前区是洗漱区及简易衣帽区，尽端为大淋浴间，非常有意思和有肯定性，夜间可以全关闭洗浴区的灯光，完全不影响，真的有研究的深度。写字、休闲区也直接明了，墙面固装与订制活动家具相结合，木饰面统一和谐，当然一两张活动家具的搭配我认为是“太过创意的”，一张偏中式的扶手椅配一张偏欧式的小沙发，有一点点怪，也许这个怪可以算是另一个小惊喜吧。

It was a double room with the passage on one side. Surprisingly, it had the toilet and the bathroom separated—a bold and well-controlled layout design. The toilet, accessible by the passageway, designed with frosted glass hinged door, some distance away from the sleeping area, causing little annoyances of sound, noise and light to the sleeper. The washing area and shower area are next to each other but separated: the forefront was a washing and simple cloak area, and the rear was a big shower compartment—quite interesting and worthy of approval. In the nighttime, all the lights in the washing and shower areas could be turned off without consequences—it deserves a in-depth study. The design of the writing area and the recreation area was also direct and clear cut: the wall fixtures went well with the customized movable furniture, and the wooden finishing was consistent and homogeneous. Of course, there were one or two pieces of movable furniture I think was "too innovative": a quasi-Chinese style armchair was matched with a quasi-European style easy chair—somewhat queer-looking. Maybe that could count as another little surprise.

一间客房有小惊喜也可谓“惊喜”了！

It can be described as "surprising" to find little surprises in a guestroom, little as they are!

兰州金地名庭酒店

Courtyard Hotel, Lanzhou, China

海口朗廷酒店

★★★★★

69

THE LANGHAM HAIKOU CHINA

Address : No.77 Binhai Avenue, Haikou, Hainan 570105, China
中国海南省海口市
滨海大道77号 570105

Telephone : +(86) 898 6696 9777

Fax : +(86) 898 6696 9477

Http ://www.langhamhotels.com/haikou

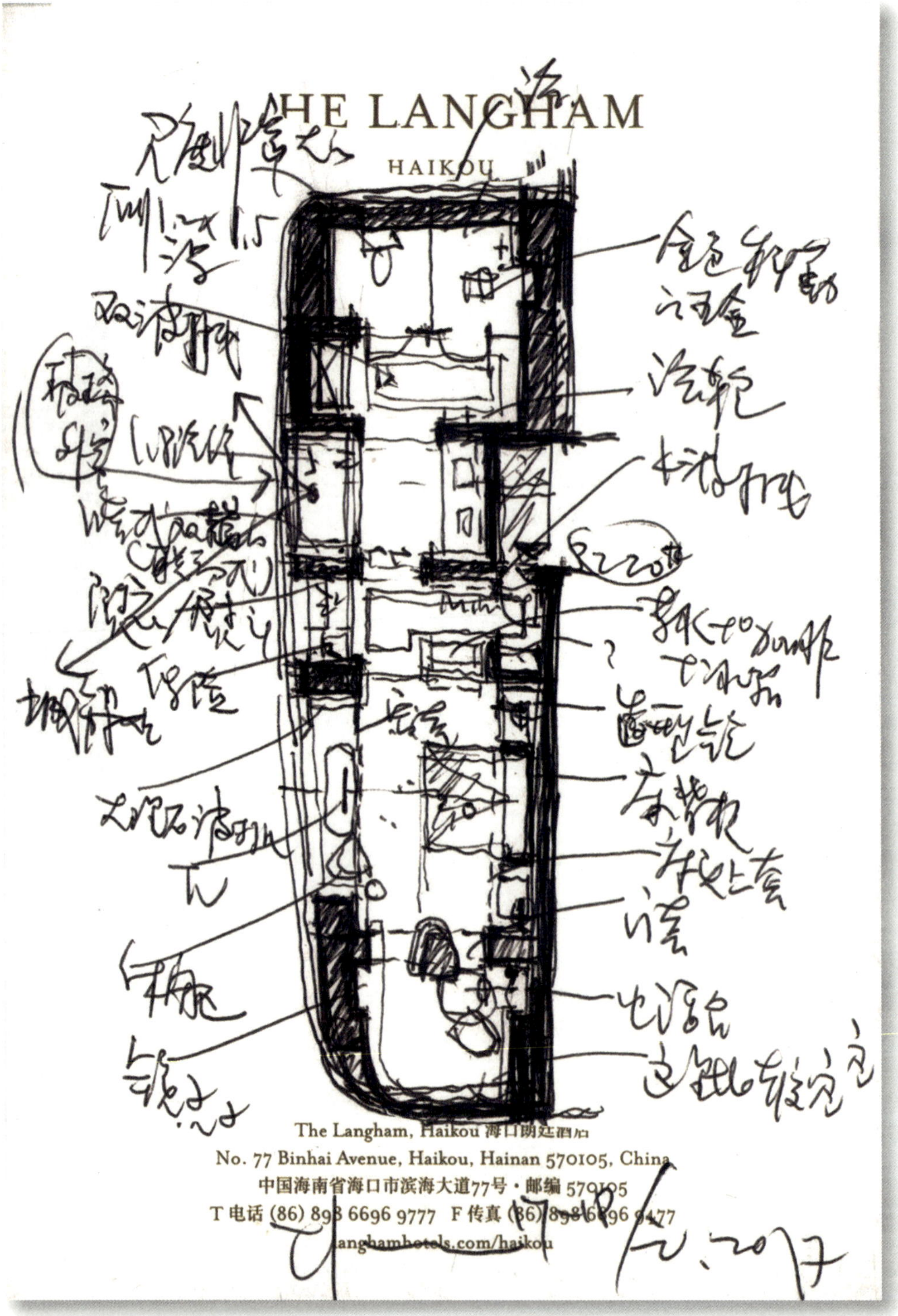

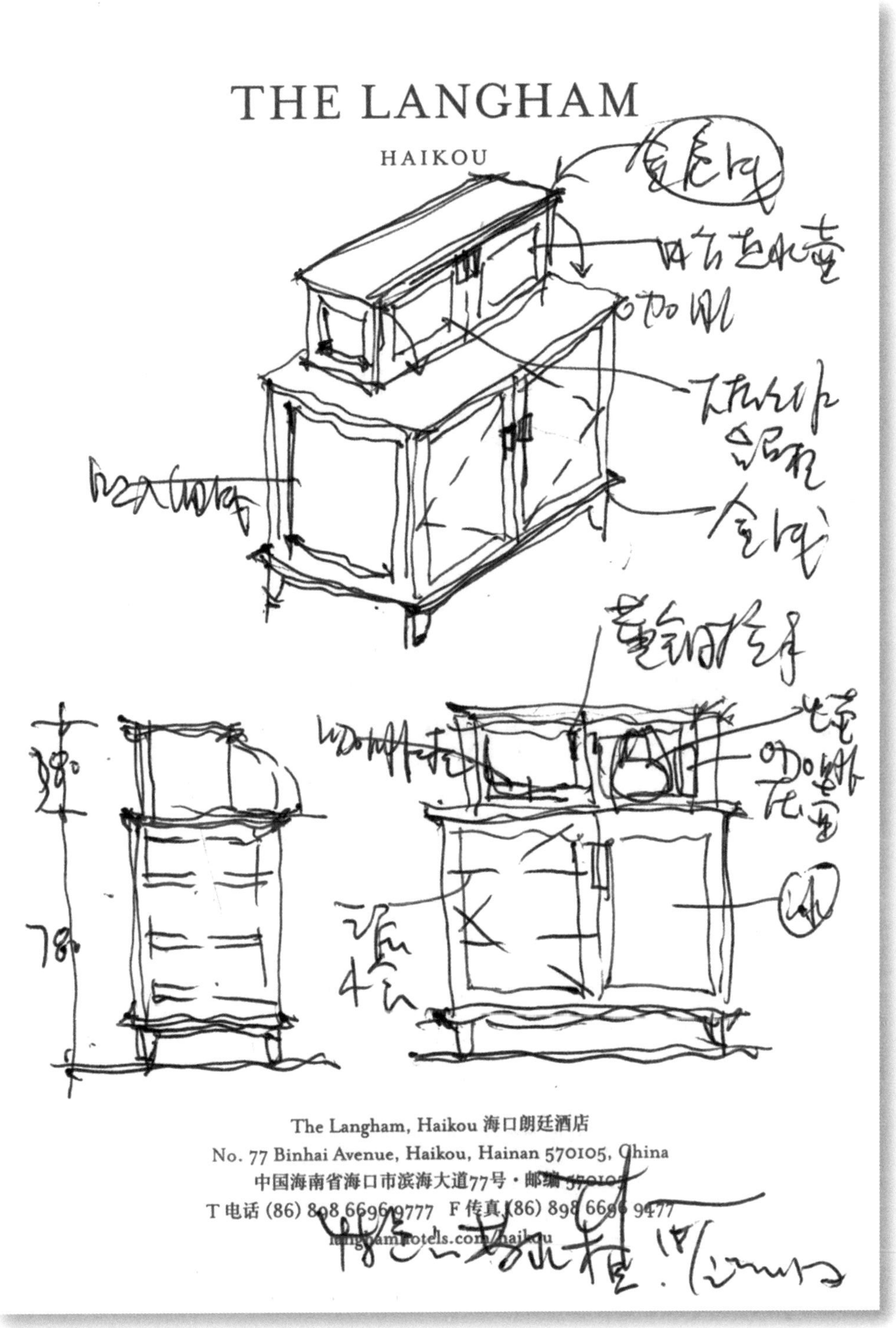
THE LANGHAM
HAIKOU
The Langham, Haikou 海口朗廷酒店
No. 77 Binhai Avenue, Haikou, Hainan 570105, China
中国海南省海口市滨海大道77号 · 邮编 570105
T 电话 (86) 898 6696 9777 F 传真 (86) 898 6696 9477
langhamhotels.com/haikou

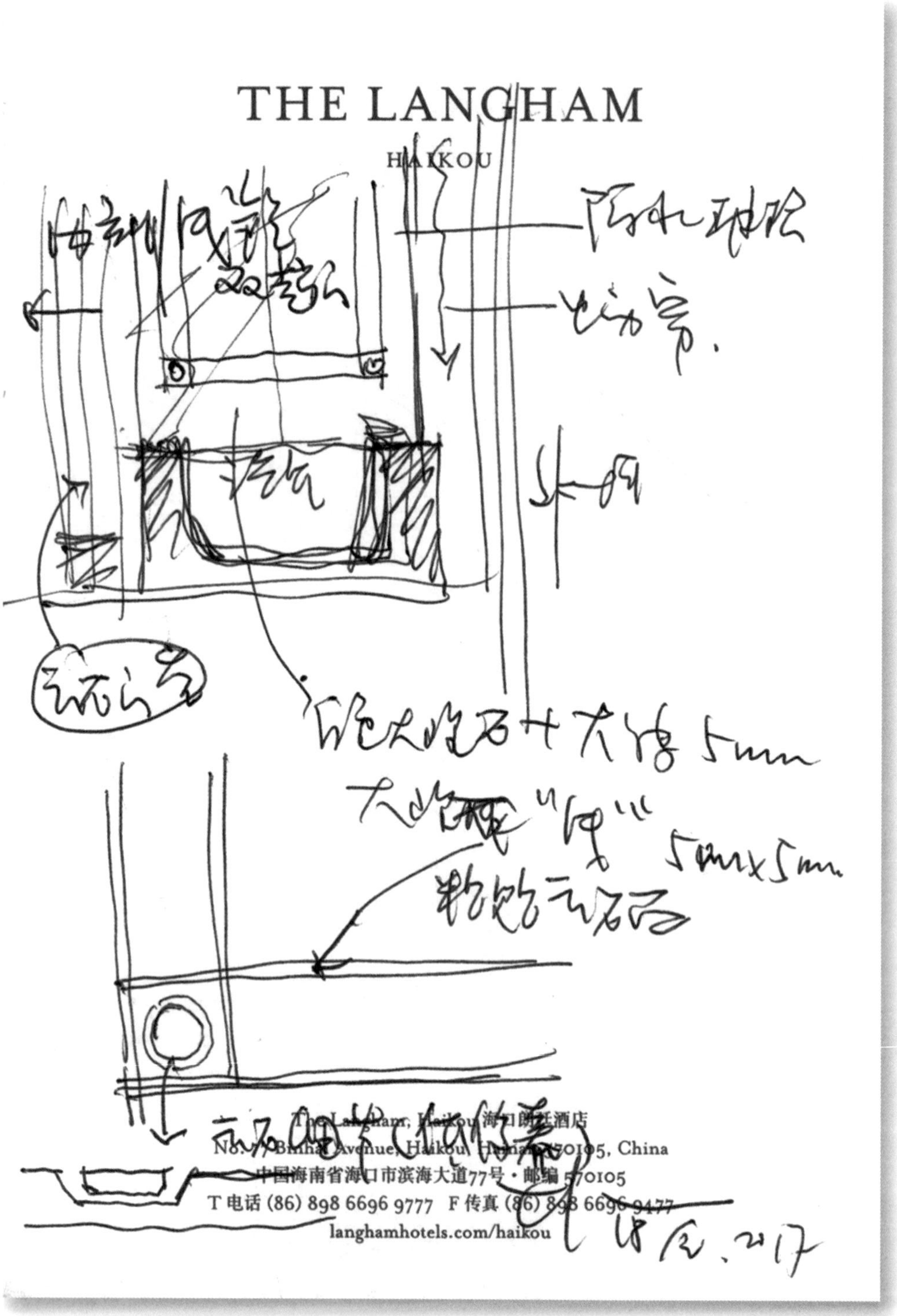
THE LANGHAM
HAIKOU
langhamhotels.com/haikou

跟着网评选酒店

由于太近广州，海口，一般会当天来去或住一下就去另一个城市，这次由于从兰州直接飞过来，提前一天的晚上到达，准备第二天的工作，就住上了。

这几年海口和风一样，一条滨海大道开或将开十间一线国际连锁品牌的我国内最顶级、豪华、海景式的酒店。我，实在患了选择困难症，是选择住的文华海景酒店，还是最新开放的使用的海口丽思卡尔顿酒店，还是全海景的希尔顿酒店，看看网评的推荐，就选定——朗廷酒店

象飞镖形象的塔楼，这造型该别样不俗，酒店的入口设计很传统的气派（如深圳朗廷酒店也是不注重入口的高大上）（详见《住哪了？》2）。标准层一进入室内就感受到某种精致的、内敛的如伦敦七十年代的英伦风范，特意订的稍好一些的房间，但没想到是位于开间较窄的尽端，可惜海景在“飞镖”的另一端，我的房间看到的只是乱乱的城市。

房间内有建筑的创意性，也有其先天的优点，横开展开，

自然而然形成四个区域：仪式入口，进入房间，非常私隐，过了区域是非常之大大的收纳衣柜，梳妆柜也在这个区域，而在睡眠区那一面墙上（地毯地面，舒适温馨）。两米之大床正对着落地之景观玻璃窗，大方明亮，而舍弃传统对称之英式古典造型，而尽端之床头背靠区，在当今可以想象得到那种"园林"——四人也可在此，以娴熟之设计，将不规则之平面得以柔化，尽量地划分，合理。

洗手间是另一大亮点：不单是全部大理石之装饰，更是细节细致细腻和不惜成本的行政套房之金色龙头；大，更是相当的过份，是真正的浪费，前端是无聊的双洗脸盆区域及浴缸，临窗风景；中部是无用的模糊过渡区，尽端才是一个开间一米，进深达四一米至六的如厕及淋浴间，真的是相当之奢侈。

看来同行还是相当前卫，值得作为今后造酒店之重要依据。

我喜欢的经典，英伦格调。

Choose Hotels According to Online Comments

跟着网评选酒店

由于离广州太近，海口一般会当天来去或途经一下就去下一个城市，这次由于从兰州直接飞过来，提前一天的晚上到达，准备第二天的工作，就住上了。

I had been making round trips to Haikou on the same day, or passing it on my way to other cities as it is so near to Guangzhou. This time as I had flown direct from Lanzhou, I had checked in the night before, prepared to work the next day.

这几年海口疯了一样，一条滨海大道开或将开10间一线国际连锁品牌或国内度假、高尔夫、温泉式的酒店。我，突然患了选择困难症，是选神往的观澜温泉酒店，还是最新投入使用的海口丽思卡尔顿酒店，还是全海景的希尔顿酒店，看看网评与推荐，就选了它——朗廷酒店。

These years, Haikou has been developing like crazy. A seafront avenue has been seeing or will be seeing the opening of ten first-line international chain brands or domestic vacation, golf and spa hotels. All of a sudden, I found it difficult to choose: the much-dreamed-of Mission Hills, or the newly put-into-use Ritz-Carlton Haikou, or the full-ocean-view Hilton? Glancing through the online comments and recommendations, I chose the Langham.

海口朗廷酒店

The Langham, Haikou, China

海口朗廷酒店
The Langham, Haikou, China

像飞镖形象的塔楼，远观的识别性不错，酒店的入口没有传统的气派（好像深圳朗廷酒店也是不注重入口的高大上）（详见《住哪？2》），相反，一进入室内就感受到其精致的、内敛的如绅士般的英伦风范，特意订的稍好一些的房间，但没想到是位于开阔景观的尽端，可惜海景在“飞镖”的另一端，我的房间看到的只是乱乱的城市。

The tower, dart-shaped, was easily visible from afar. The entrance of the hotel was not so grand as it is traditionally (it seems that the Langham Shenzhen does not emphasize the grandeur of its entrance, either) (see Where to Stay? 2 for more details). In contrast, once inside the guestroom, I was impressed with the British style, dapper and reserved like a gentle man. I had specially booked a better-than-average room, only to find it was at the end of the open view. It was a shame the oceanic scenery was at the other end of the "dart", and what I could see was just a jumble of the city.

房间内有建筑的制约性，也有其先天的优点，横开展开，自然而然成四个区域：袋式入口，进入房间，非常私隐；过厅区域超常的大大的收纳衣柜，茶水柜也在这个区域，而在睡眠区那一面操作。（地毯地面，有保留意见）

The room had constraints imposed by the building as well as its inherent advantages: it spread out horizontally, forming naturally four areas—the bag-shaped entrance to the room ensured much privacy; located at the hallway was an extra big wardrobe as well as a tea cupboard operable at the sleeping area. (as for the carpeted floor, I have reservations)

2米的大床正对幕墙的景观玻璃窗，大方明亮，配合严谨而优雅的英式立面造型，而尽端的休闲写字区，应当还可以勉勉强强放得下一张“国粹”——四人电动麻将台，以娴熟的门套方式，令不规则的平面得以柔化，老到地划分，合理。

Opposite the two-meter bed was a glass view window on the curtain wall, tasteful and bright, going with a solemn and elegant British style elevation. The writing area at the end barely allowed room for a "national essence"—an electric machiang table for four. With mastery, irregular planes had been skilfully softened and reasonably divided.

洗手间是另一大亮点：不单单全白色大理石的光亮，更是细节、细致、细腻和不惜成本的仿铜五金及龙头；大，更是相当的过分，是真正的浪费，前端是无聊的双洗脸盆区域及浴缸，临窗风景；中部是无用的陈列过渡区，尽端才是一个开间逾1米，进深达1.5米的如厕及淋浴间，真的是相当的享受。

Another highlight was the bathroom. It was not only about the shininess of the all white marble, but more about the detailed, elaborate, fine, exquisite and costly bronze-imitation hardware and taps. The bigness went to the degree of extremity and wastefulness. At the front was a dull double washbasin and a bath tub, and a window with a view; at the middle was an area of excessive display of decorations; at the end was a toilet and shower compartment, over one meter wide and one meter and half deep. It was really enjoyable.

看来网评还是相当靠谱，值得作为今后选酒店的重要依据。
The online comments seemed quite reliable and could be used as references in choosing hotels in future.

我喜欢的经典，英伦白调。
The classic I like—the British white.

房间里吃饭，你会浪费吗？

偶尔也在房间内吃饭，在香港半岛酒店试过一次，印象特深，因为那天我的鞋子露出了脚丫子，而不被允许在"今艾日式"吃早餐，你说气人不气人。女生可以，男生不可以，被逼要回到房间里"撑台脚"（广州话）咯！

这次则是因为回到酒店已经很晚，外面还下雨，一个人，就灰溜溜地想，怎么吃饭啊，"Room Service"，送餐吧！

一个海鲜炒饭，一个水果盆，半小时内送到了。

书台作饭台，开着电视机，看着西片（外国电影）感觉还是不错，关键是可以轻轻松松地也会（儿子会说、哈哈，你懂的，你也可以试试啊！）海鲜饭太大了，水果盘的"内容"也太行(hang)了（太普通了）

结果，浪费度可想而知了，炒饭剩下了一半，水果拣着吃了。撑着肚子睡觉也是一件挺美好的事情，猪一样的生活。

是不是房间点餐都是这样的下场呢？！

房间里吃饭，你会满意吗？！

偶尔也在房间内吃饭，在香港半岛酒店试过一次，印象特深，因为那天我的鞋子露出了脚丫子，而不被允许在“公共区域”吃早餐，你说气人不气人，女生可以，男生不可以，被迫转回到房间里“撑台脚”（广州话），惨！

Occasionally I eat inside the guestroom. I had one such experience in the Peninsula Hong Kong, leaving a deep impression. That day I was forbidden to have breakfast in the "public area" as my toes poked through my shoes. It made me enraged—it was OK for girls to expose their toes, but not men. I was forced to return to eat in the guestroom. What a misery!

这次则是因为到酒店已经很晚，外面微雨，一个人，就灰溜溜地想，怎么吃饭啊，“Room Service”送餐吧！

This time when I arrived at the hotel, it was already very late and it was drizzling outside. Alone, I considered in low spirits how to get my meal. All right, call the Room Service!

一个海鲜炒饭，一个水果盆，半小时内送到了。

Within half an hour, a helping of fried rice with seafood and a plate of fruit was brought before me.

书台作饭台，开着电视机，看着西片（外国电影）感觉应当不错，关键是可以轻轻松松地进食（几乎全裸，哈哈，你懂的，你也可以试试啊！），海鲜饭太大了，水果盆的“内容”也太行了（太普通了）。

Eating at the writing desk and watching Western movies on TV, it felt quite nice. The key point was that I could eat with great ease (almost totally naked. Haha, you know, you can also have a try!).The amount of the fried rice with seafood was too big and the content of the fruit plate too common.

结果，满意度可想而知了，炒饭留下了一半，水果挑着吃了，撑满肚子睡觉也是一件挺美好的事情，猪一样的长肉。

As a result, the degree of satisfaction was conceivable: half of the fried rice was left over and only some of the fruit found their way inside me. It felt good to sleep on a full stomach. And you can gain weight like a pig.

是不是房间的点餐都是这样子的下场呢？！

It was the inevitable consequence of ordering food from Room Service, wasn't it?

长沙瑞吉酒店

★★★★★

70

STREGIS CHANGSHA CHINA

Address : Yunda Central Plaza, No.289,Shawan Road, Yuhua District, Changsha, Hunan Province, P.C.China 410129
中国湖南省长沙市雨花区沙湾路289号运达中央广场，410129
Telephone : +(86) 731 8968 8888
Http ://www.stregis.com/changsha

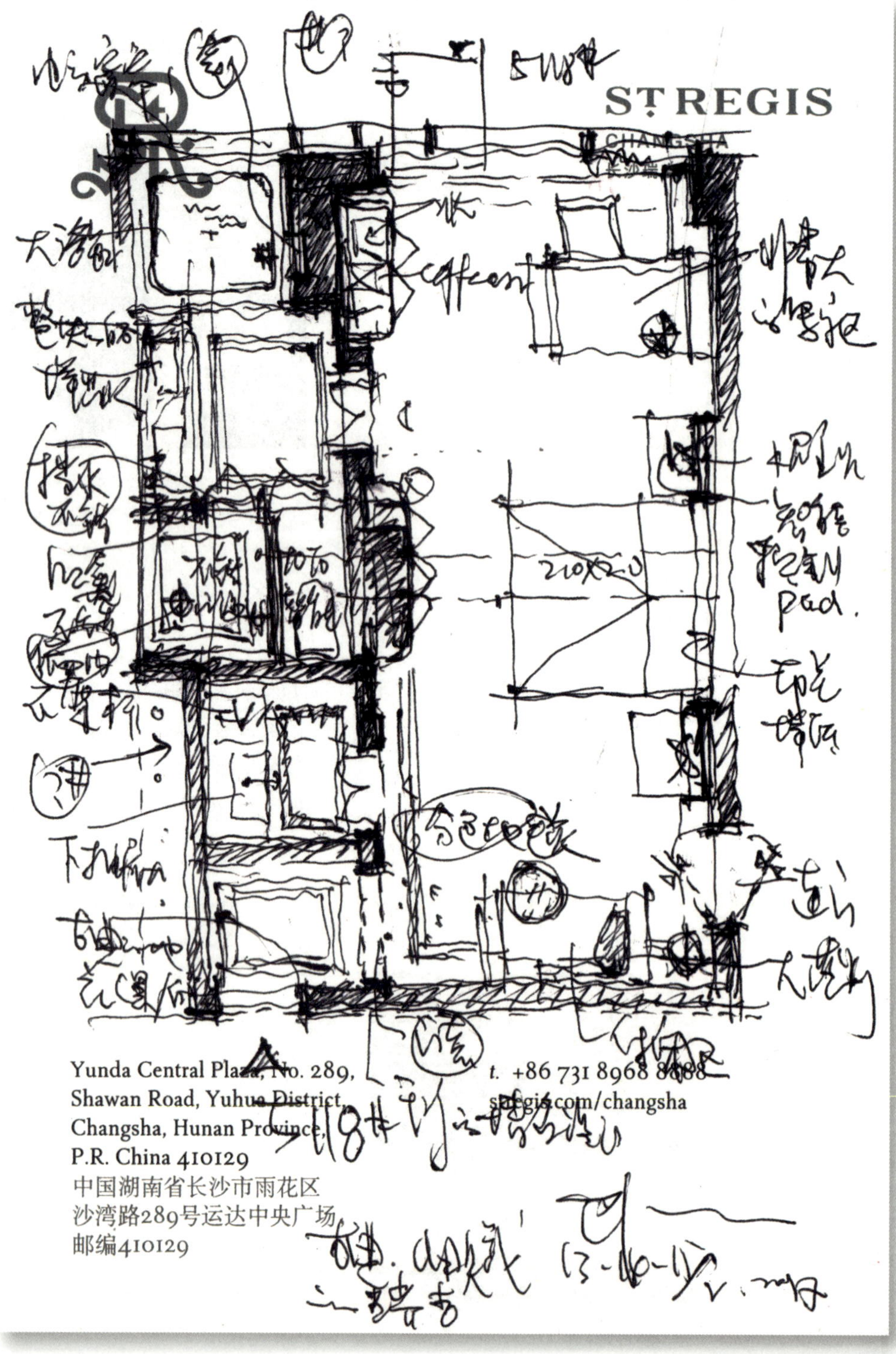

ST REGIS

CHANGSHA

长沙瑞吉酒店

王家总、常总、方总、姚总们：

入住这些地段的瑞吉酒店还是很值得的，虽然地段是不够好，但结构设施的酒店住宅都是最好的，包括造园服务的团队也是最好的，人之住宅是在前几年的设计生活的，家庭的生活也是当下最好的景观酒店。回到这个瑞吉酒店，从今天的建筑空间到日常流线，仪式感都是非常有前瞻性的，可谓是全球中了最好的酒店之一，它的品牌还有它价值。

1680元一天的基本房费，600元之一住的品质是值得。

丁明

①

Yunda Central Plaza, No. 289,
Shawan Road, Yuhua District,
Changsha, Hunan Province,
P.R. China 410129
中国湖南省长沙市雨花区
沙湾路289号运达中央广场，
邮编410129

t. +86 731 8968 8888
stregis.com/changsha

ST REGIS
CHANGSHA
长沙瑞吉酒店

瑞吉酒店客房一向以注重特色和功能的空间[illegible]

[illegible]开放式[illegible]、大空间[illegible]特征,

印象[illegible](深圳瑞吉 2012)。深圳的瑞吉[illegible]

[illegible]

[illegible]。长沙虽然是一线城市,但瑞吉酒店

[illegible]

55m²[illegible],1.5m的开[illegible]

[illegible]

[illegible]区 [illegible]一[illegible]

Yunda Central Plaza, No. 289,
Shawan Road, Yuhua District,
Changsha, Hunan Province,
P.R. China 410129
中国湖南省长沙市雨花区
沙湾路289号运达中央广场,
邮编410129

t. +86 731 8968 8888
stregis.com/changsha

[illegible]

[illegible]

ST. REGIS
CHANGSHA
长沙瑞吉酒店

沿门前花之入口那个端庄之对称，电梯在正也对中轴以大字（请至大酒店之客房都不是那样做到这一点，你注意到了吗？）地毯颜色之搭配和门墙布背后的各类韵味，深棕色之大面积铜器与漆之瓷器和金属墙面板之手法，我很喜欢它之轻盈、大气，又不失豪华的感觉。

又仁兄您好吧！何况您能大胆在每一处有细节之设计的深意非常，平面布局因地制宜，所以酒店在细节更有个性。写于TOTO 老板仪

Yunda Central Plaza, No. 289,
Shawan Road, Yuhua District,
Changsha, Hunan Province,
P.R. China 410129
中国湖南省长沙市雨花区
沙湾路289号运达中央广场，
邮编410129

t. +86 731 8968 8888
stregis.com/changsha

ST REGIS
CHANGSHA
长沙瑞吉酒店

二、你们应否托我方便的所权手机（包括
手机可能是参考的认字不如手机字）[illegible]
[illegible]
[illegible]
[illegible]
[illegible]
[illegible]
[illegible]

Yunda Central Plaza, No. 289,
Shawan Road, Yuhua District,
Changsha, Hunan Province,
P.R. China 410129
中国湖南省长沙市雨花区
沙湾路289号运达中央广场，
邮编410129

t. +86 731 8968 8888
stregis.com/changsha

[illegible]

入住运达地产开发的瑞吉酒店还是值得的，首先地段是不错的，地产商开发的项目定位都是最好的，包括选用服务的团队也是最顶级的，住宅是老前辈高文安先生设计的，商场的设计也是当下长沙最耀眼的。

It is worthwhile to stay in the Regis Hotel developed by Yunda Real Estate. First of all, it is well located. All the projects were well positioned. It employs the top service team. The residential section was designed by senior designer Mr Gao Wen'an. The design of the shopping plaza is also the most brilliant in Changsha now.

回到这个瑞吉酒店，从公共区的建筑空间到回房间的流线，仪式感都是非常有前瞻性的，可谓完全提升了长沙酒店业的品位。还有定价，1680元一天的基本房费，600多元一位自助餐定价，有胆量。

Return to this Regis Hotel. From the architectural spaces of the public area to the streamlines of the guestrooms, the sense of ceremony is full of foresight. It is safe to say that it has completely enhanced the taste of the hotel industry in Changsha. Then the pricing: 1680 yuan a day, and 600+ yuan buffet—it is bold.

长沙瑞吉酒店 *Stregis, Changsha, China*

瑞吉的酒店客房一向注重特色和功能的完整性。之前拉萨的开放式洗手间，大气而有地域特征，印象挺深的（详见《住哪？2》），深圳的瑞古与三亚的都非常注重入门的礼仪感与窗外景观的照顾。长沙虽不是一线城市，但瑞吉酒店诚恳认真地把客房按照最高的标准去打造，55m²超白金标准，1.5的开间更是让平面布局充满经典的尊贵韵味，袋式入口前厅，隐私性极高，休闲区、睡眠区、写字区一体化的处理，令空间的完整性达到最高。步入式的衣帽间与最奢华的洗手间双掩门入口形成端庄的对称，电视柜正中正对2米大床（许多品牌的客房都不一定能做到这一点，你注意到了吗？）。地域特色的描花私订墙布背幅略显粉味，深木色的大面积钢琴漆的装饰，相信会引来两极的争议：或很喜欢它的稳重、大气，又或感觉超级老气。

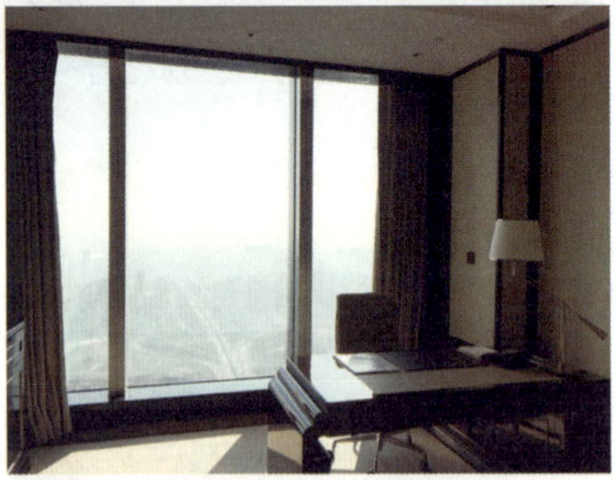

The hotel consistently attaches importance to character and functional wholeness. Previously the open bathroom in Lhasa, tasteful and regionally characteristic, impressed me deeply (see Where to Stay? 2). The Regis hotels in Shenzhen and Sanya emphasize the ceremoniousness of the entrance and the view outside the window. Though Changsha is not a first-tier city, Regis sincerely forges its guestrooms to the top standards: the $55m^2$ *size is up to the extra white gold standard; the 1.5 m width gives the plane layout a classic and noble air;the bag-shaped entrance anteroom provides high-degree privacy; the integrated treatment of the recreation area,the sleeping area and the writing area allows the utmost wholeness of the spaces;the walk-in cloakroom and the entrance of the double hinged doors of the most luxury bathroom form a stately symmetry; the TV bench stands right opposite the 2m bed (the guestrooms of many brands fail to manage to do it — have you noticed that?); the regionally characteristic flowery wallpaper is slightly pinkish;and the large area of dark wooden piano lacquer decoration is set to trigger polarized arguments:some are quite fond of its solemnness and taste and others feel it extra unfashionable.*

见仁见智吧！倒是对白色大理石单一而有细节的洗手间满意非常，平面布局因地制宜，成熟而有细节，更有小惊喜：TOTO马桶是智能化的，侧有云石托，极方便如厕放手机（现代社会可谓是“爹亲娘亲不如手机亲”）。近1米半见方的大淋浴间（容纳两个人不在话下），可手握花洒或天顶花洒。细致在于挡水石的设计，可谓几乎杜绝了沐浴时水往外溢的老毛病，设计师可说棒极了，不一一陈述。全智能化也是长沙这家瑞吉不遗余力的投入，几乎可以说是目前开业的酒店中领先的，相信投入也是不菲的。

Opinion is divided. The bathroom, made of white marble, simple and elaborate, is quite satisfactory. The plane layout, a clever use of the spaces, is mature, detailed and a bit surprising: the smart TOTO toilet has a marble holder on one side for toilet user to place the mobile phone (in modern times, "the mobile phone is dearer than parents"). The nearly 1.5m ×1.5m shower compartment is big enough for two; you can use the hand-held gondola water faucet or the ceiling water faucet; the delicateness lies in the design of the water holding stone, which almost overcomes the old trouble of overflowing when showering. The designer is excellent and I will not dwell on it here. Full smartness is also an aspect Regis Changsha has spared no effort to invest in. It is almost safe to say that it is leading in all the opened hotels. And I believe the input is also handsome.

土豪，可是要靠实力的哦！

Yes, it takes real wealth to lavish money!

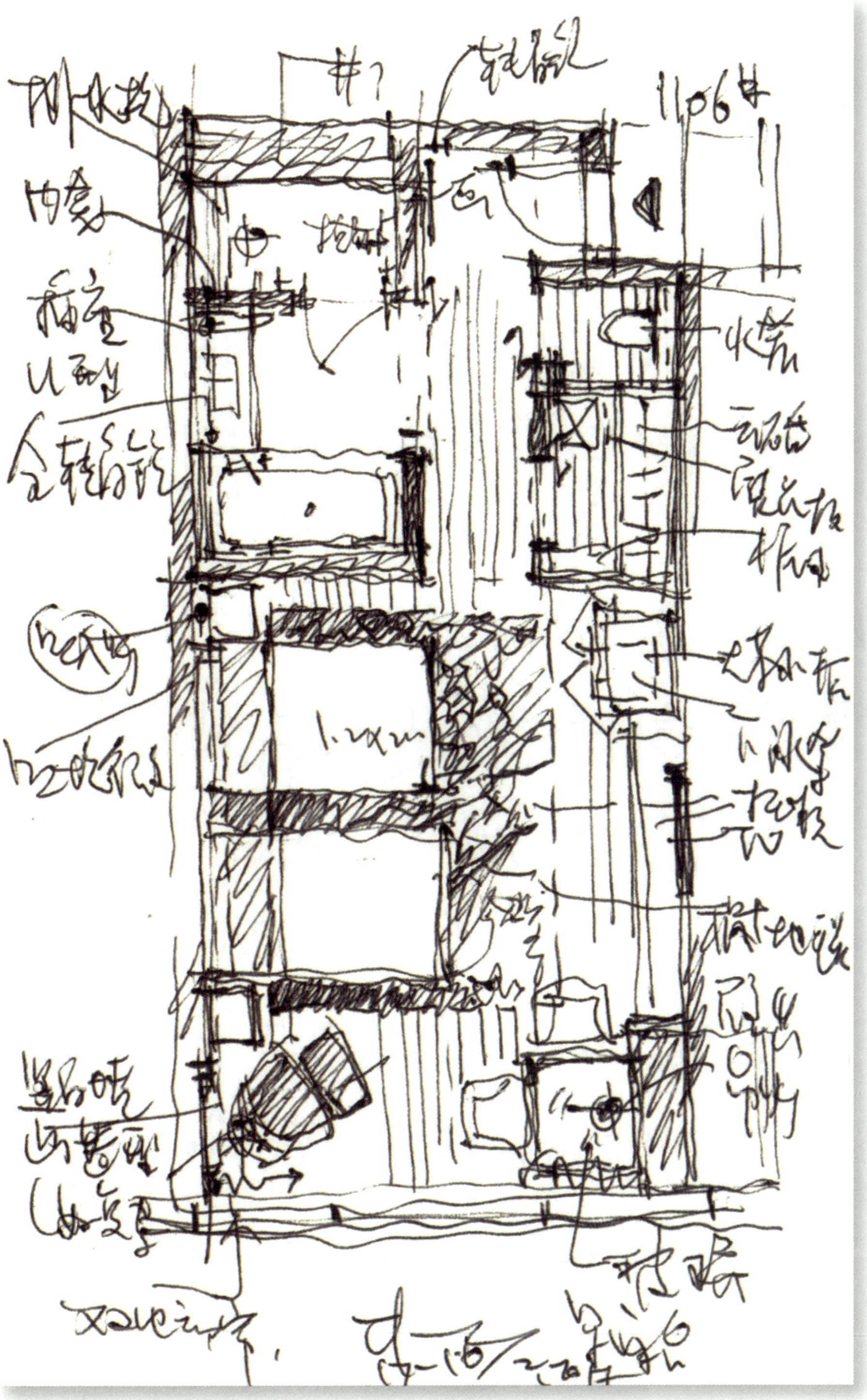

GRAND | HYATT

沈阳君悦酒店

71 ★★★★★

GRAND HYATT SHENYANG CHINA

Address : No. 288A, Qingnian Street, Heping District Shenyang, China, 110004 沈阳和平区 青年大街288号甲

Telephone : +(024) 25121234

Http : //www.grandhyatt.com

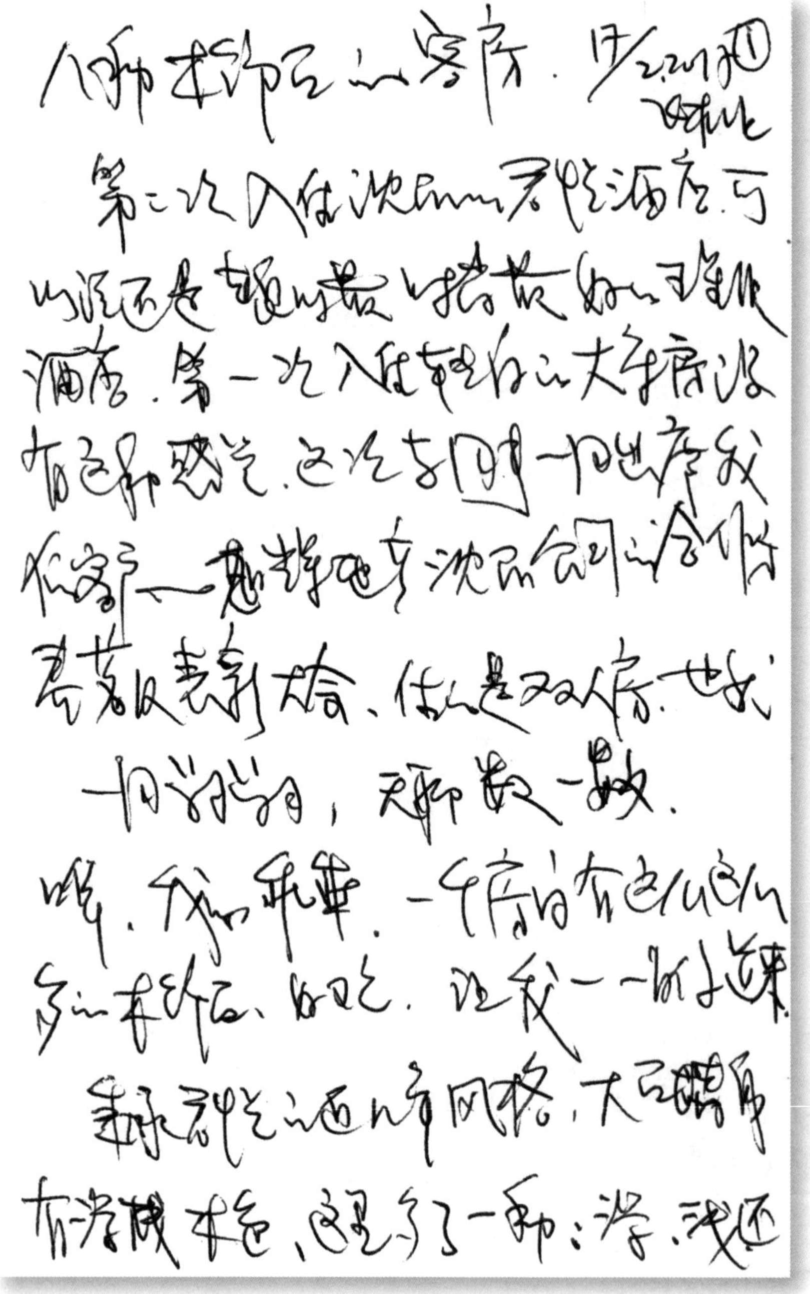

入住书院区的客房。17/2,2017 ①
[illegible]

第二次入住沈阳君悦酒店，可
以说还是[illegible]
酒店。第一次入住郑州的大华店没
有这种感觉。这次去回回一间套房我
们客户一起就住沈阳公司的合作
君悦表示欢迎，住的是双人房，也就
一间普通的，天那里一致。
吧，我和那里，一个房的有区的区的
多的书架，好吃，让我一一体验
新君悦之酒店风格，大堂咖啡
有浓郁书色，透显了一种淡淡，我还

各种衣色的、家具和各种相同和艳色 ②
一个大的设计概念、一个很鲜活、单纯色
一个单纯相同和本色的，艳丽的与偏
红的，也许是设计师以后下层要做的
的指向成为个性气吧。几种了，已经
七种了，还有一种便是我喜欢的。
地毯会本色和我是大和谐式的艳色
术。床垫和布有两种手法的设计，包括
没有个本色的运用还是由设计思想
的，但也是多、变了，也许这个设计
是有设计师体会几种个性的体验。

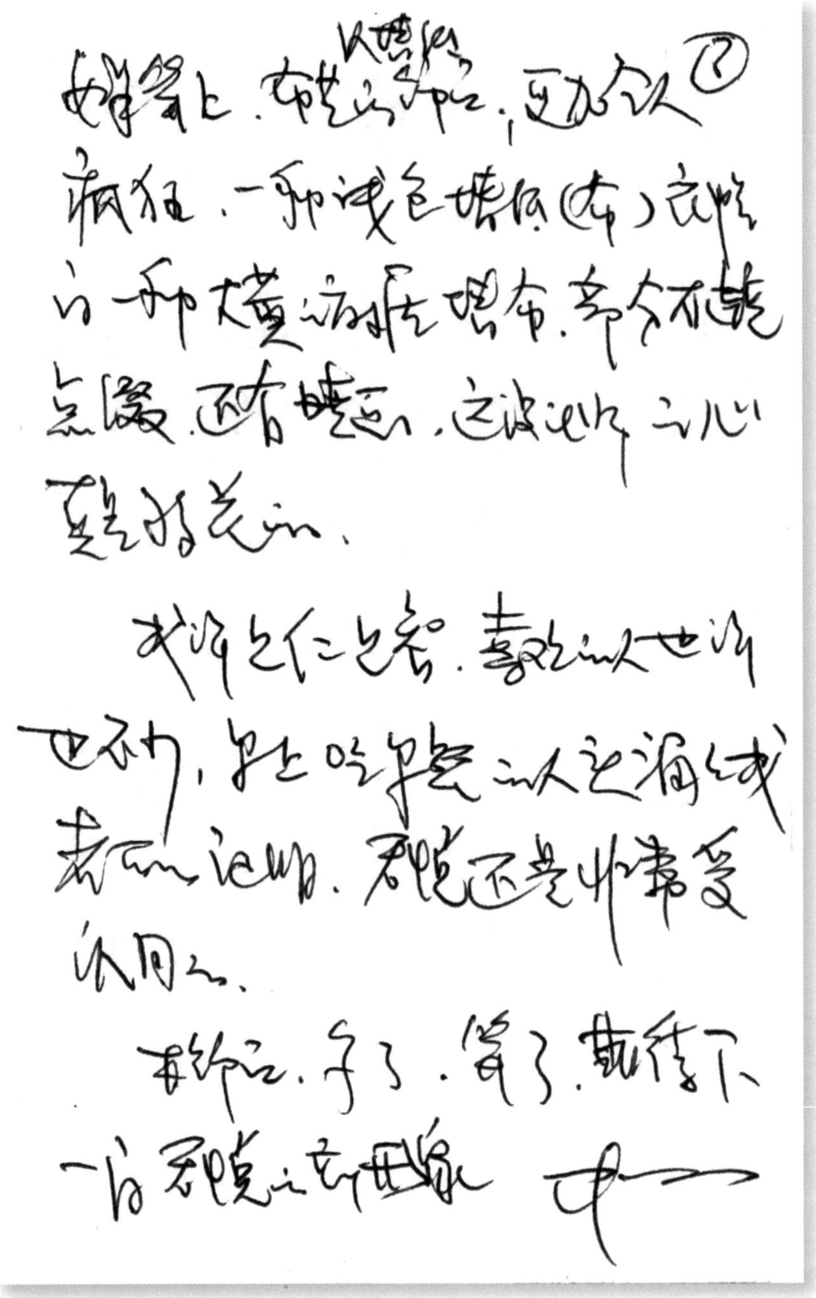

如解笔上，布艺（从墙纸）的印记，更加令人 ⑦
疯狂，一种浅色墙纸（布）之性
的一种大黄满铺墙布，帝令不能
点缀，还有楼道，这就没有一心
连接着花的。

我们之仁之智，喜之的人也们
也不动，但是咱们的人还满足
表示谈明，那就还是收拾受
的用了。

我们之，年了，算了，就给下
一句那就之离开远　中一

第二次入住，沈阳君悦酒店，可以说还是暂时最时尚、最好的五星级酒店，第一次入住转角的大床房没有这种感觉。这次与同事一同出席我们客户——旭辉地产沈阳公司的合作方春茗及表彰大会，住的是双人房，也就一同学习学习，无聊数一数，哗，我的乖乖，一个房间有这么这么多的木饰面，好玩，让我一一道来。

It was the second time I had been in Shenyang Grand Hyatt hotel, which could be rated temporarily as the most fashionable and best 5 star hotel. I had not got such feelings when first staying in the big bed room around the corner. This time, together with my colleagues, I was attending the Spring Festival dinner party and commendation meeting held by the cooperator of CIFI Group Shenyang Branch. As we shared a double room, together we spent time studying and counting ... Wow, there was so many types of wooden veer in this room. It is fun. Let me tell you little by little.

沈阳君悦酒店

Grand Hyatt, Shenyang, China

秉承君悦近几年的风格，大面墙身有深浅木色，这里多了一种：深、浅还有中灰色的。家具床头柜两种艳色：一个是大山纹的桃木，一个是喷漆，黄灰色。一个茶水柜两种木饰面：艳黄的与偏红的，也许是设计师让这个质量一般般的柜子成为焦点吧。几种了？已经七种了，还有一种反而是我喜欢的，地面全木地板，是大杂烩式的杂色木，厚重而有耐人寻味的感觉，应当说整个木饰面的运用还是颇费思量的，但也显多，繁了，也许这个趣味只有设计师多住几次才能体验。

Sticking to Grand Hyatt's style in recent years, the wall was covered in dark and light wooden colors, and there was an extra one here: mid-gray. The furniture and night table came in two bright colors: one was the color of grained mahogany and the other spray-paint gray. A tea cupboard had two types of veneer: bright yellow and red. Perhaps the designer had intended to turn this mediocre cupboard into the focus. How many types now? Seven already. There was still another type, and the one I took a fancy to. The floor was covered in a jumble of mottled gray, massive and inspiring. To be fair, a lot of thought had gone into the use of this veneer. However, it was overdone. Maybe only after being here several times more can the fun be appreciated by the designer.

如果算上布艺以及墙纸的饰面，更加令人疯狂，一种浅色墙纸（布），衣帽间一种大黄的麻质墙布，部分灰镜点缀，还有喷画，这设计师的心真是够花的。

Counting in the veneer on the cloth art and wallpaper, it proved more crazy. One was light gray wallpaper, and the other bright yellow linen wallpaper on the cloakroom. Some parts were dotted with gray mirror and spray paint. The designer was really full of ideas!

沈阳君悦酒店
Grand Hyatt, Shenyang, China

或许见仁见智，喜欢的人也许也不少，早上吃早餐的人头涌涌，或许可以证明君悦还是非常受认同的。

Opinions differ. Maybe there was no lack of people who liked it. In the morning the throngs of dinners attested to the popularity of Grand Hyatt.

木饰面多了，算了，期待下一间君悦的新形象。

Too much veneer here. Forget about it. I am looking forward to a new look in the next Grand Hyatt guestroom.

METROPOLO
JINJIANG HOTELS

锦江都城经典新亚外滩酒店

★★★★★

72

JINJIANG METROPOLO HOTEL CLASSIQ SHANGHAI ROCK BUND

Address : No. 422, Tiantong Road,
Shanghai China, 200085
中国上海市
天潼路422号 200085

Telephone : +(024) 25121234

Http : //www.jjmph.com

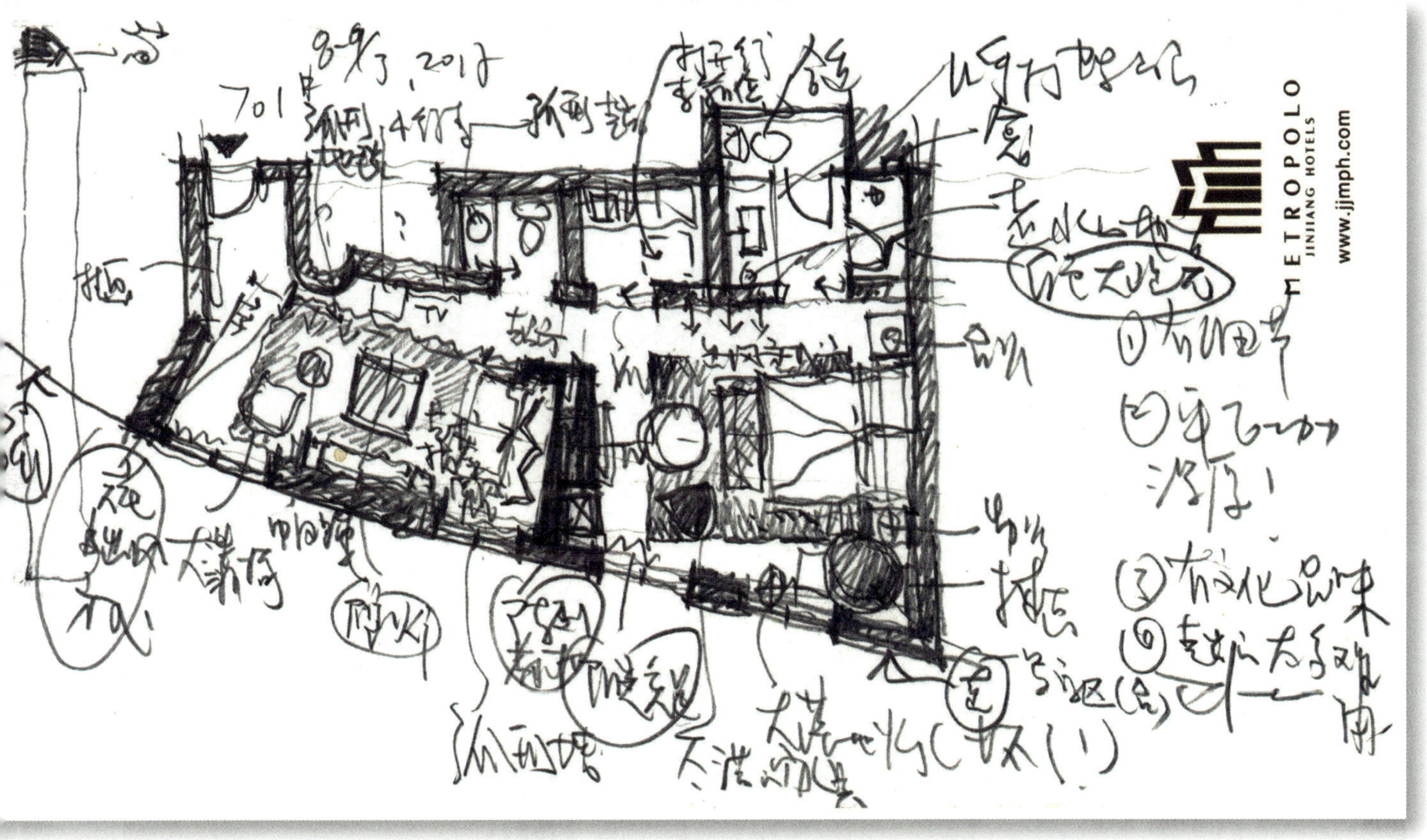

"旧"是一种味道。 9/3, 2017

去上海参加吃"海派上海"，然后造访锦江都城的酒店，参观了酒店修缮的"魔都人"之家、入住外滩地区上的旧新亚大酒店改造之后的这间。

循着旧新亚的味道来到外墙的看板、包括的窗、花饰与装饰艺术的玻璃、合城市节奏的新颖之感，特别值得一提是浅黄色玻璃的入口设计。太精彩了！非常佩服当时设计师们的创意与坚持，传统一直跟着做下去，旧新亚的新酒店，还有当年的西餐，令这座城市的"海派上海"味十足而能展现历史文化的底蕴，令这本身的个性、作品，特点大方得体，完满收获，一步一脚印！

 ①

一进入这个修缮后的、改后的，仿佛到了我们、尽管
细部拐角处、屋顶砖墙、廊柱、雕花也
受到呵护，被改装被设计成"这种
似家"的定位还是相当到位的。我曾经
说过"酒店会的不如家，家如酒店胜
似酒店"，用中国建筑来形容这里
可说是非常恰当的：客房布置得好，有
玉树临风，连地毯都彰显着民国的风
韵，绝对、典雅，门窗的造型，可谓
风情更加完善更趋完美，打开的、半开
的门窗、柏油路、走廊、灯光的感
觉，炽烈亲切，让这个小建筑变得
更加温暖、亲切。

"旧"，应当是有味道
的，是一种有生活气息
的味道。

②

"Old" is a Flavor
“旧”是一种味道

去上海参观“设计上海”，继续选择锦江都城的酒店，咨询了酒店管理方“虞美人”的高见，入住外滩边上的旧新亚酒店改造的这家。

I went to Shanghai to visit "design Shanghai" and went on to choose the Metropolo Hotel. I consulted the hotel manager "Yu Beauty" and checked into the renovated old New Asia Hotel on the bund.

循着旧新亚的味道进行外墙的翻新，包括门窗，最惊喜的莫过于通透的玻璃给城市带来的新鲜气息，特别是黑铁浅黄色玻璃的主入口门斗，太精彩了！非常佩服业主与设计师的创意与勇气，值得一直跟着住下去。旧酒瓶装了新酒，还相当的醇。公共区域的设计“海”味十足而又继承了历史文化的诚意，分区布局、灯光、饰品、挂画大方得体，充满惊喜，一步一景。

Followed the design style of the old New Asia for external wall renovation, including the doors and windows the most surprise is transparent glass brings fresh flavor to the city, especially the main entrance cross-bars of black iron with light yellow glass, so wonderful! I admire the creativity and courage of the hotel owner and the designer. Old wine is bottled with new wine, but still quite an alcohol. The design of the "Shanghai style" in the public area has the sincerity of inheriting the history and culture. The layout, lighting, decorations and paintings of the area are graceful and graceful, full of surprise, one scenery on each step.

锦江都城经典新亚外滩酒店
Jingjiang Metropolo Hotel Classq, Bund Circle, China

锦江都城经典新亚外滩酒店 *Jingjiang Metropolo Hotel Classq, Bund Circle, China*

一进去这个房间，哗，难到我了，尽端弧形拐角处，复杂啊，退房前，硬着头皮动手画，越画越觉得“都城”这种“似家”的定位还是挺高手的。我曾经说过“酒店像家不如家，家如酒店胜似酒店”，用小家碧玉来形容这里确是非常恰当的，客厅有壁炉子，有五折屏风，连地毯都颇费心思地用弧形，饱满、丰富，阳光明媚；卧室配套更加完善，更趋完美，行李间、紧凑的洗手间、小休闲区、写字区，满满的感觉，灯光柔和，让这个旧建筑室内充满温暖，洁净。

While entered the room, wow, it's difficult to me. The corner end arc is so complex. Before checking out, I braced myself to draw the plan of this room. The more I draw, the more I feel how proficient of such "home style" position for Metropolo Hotel. I once said that "the hotel is like home but not home, home is like hotel but better than hotel" . It's very appropriate to describe this hotel room as a beauty from a common family. There is fireplace, five folds screen and even arc carpet in considerable design in the living room, which is full, rich and sunny. The bedroom complement is more improved and perfect. There is luggage room, compact bathroom, small leisure area, writing area. It feels full, with downy lamp light, let this room of old building warm and clean.

“旧”，应当是有味道的，是一种有生活气息的味道。

"Old", should be to have flavor, is a flavor with vitality.

武汉关谷凯悦酒店

★★★★★

73

HYATT REGENCY WUHAN OPTICS VALLEY CHINA

Address : No.1077,Wuhan Hongshan District Luoyu Road near Yujia Lake Road
中国武汉市洪山区
珞瑜路1077号，
近喻家湖路

Telephone : +400 920 5538

Http ://www.hyatt-regency-wuhan.com

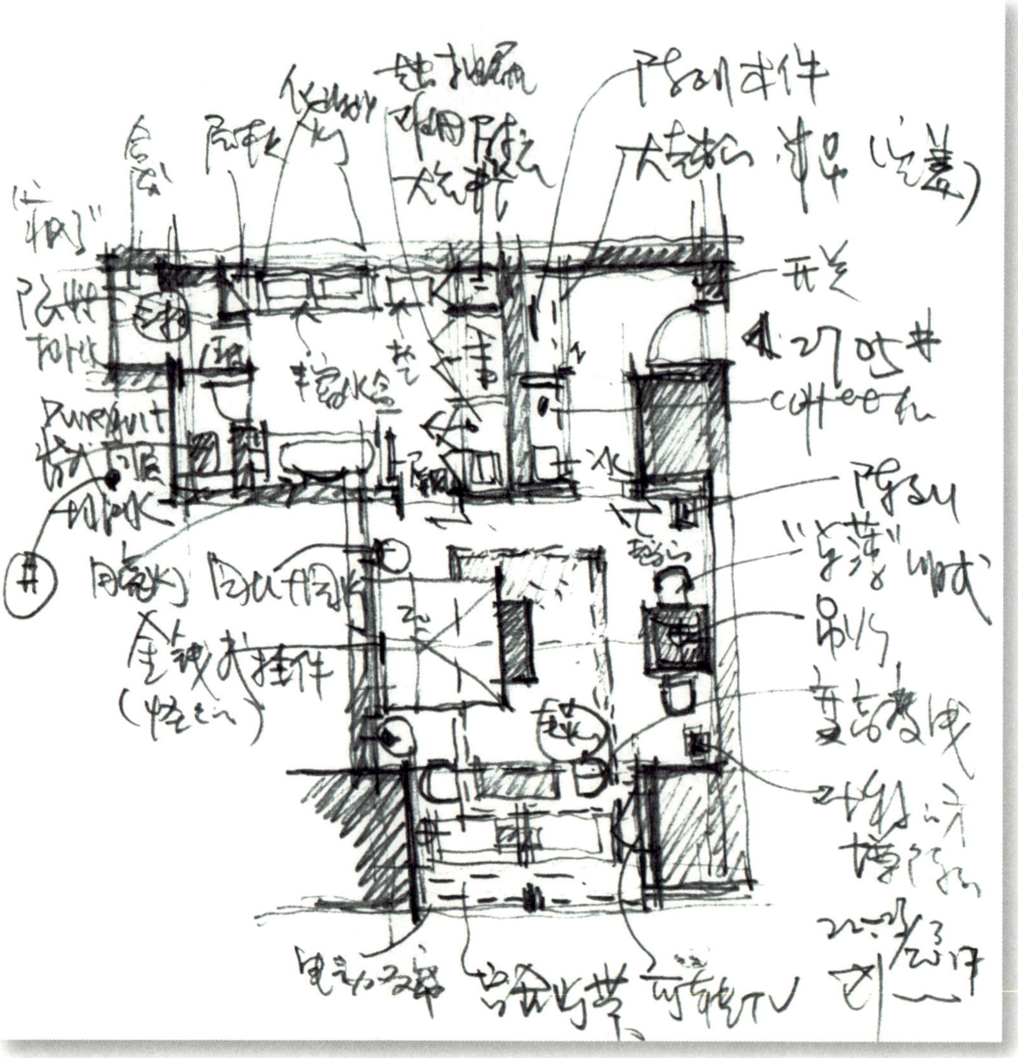

致纪元 23/3

赏·醉樱花

"烟花三月下扬州"，武汉樱花从什么时候开始，却很名的一次樱花，也许知道的人不多，没有想到"樱花成为红人"这么快就传开。三月樱花季节虽说只有三天，但是依然吸引着东湖的樱花全国凑上热闹。关注这几年游人

呼、夜看樱花，不是了，人山人海，甚至都不可躲避的，也不一定有"武大"的诗作，但让我在这里看到，第一次看到樱花还是让人陶醉的！感谢我们的相聚。

B①

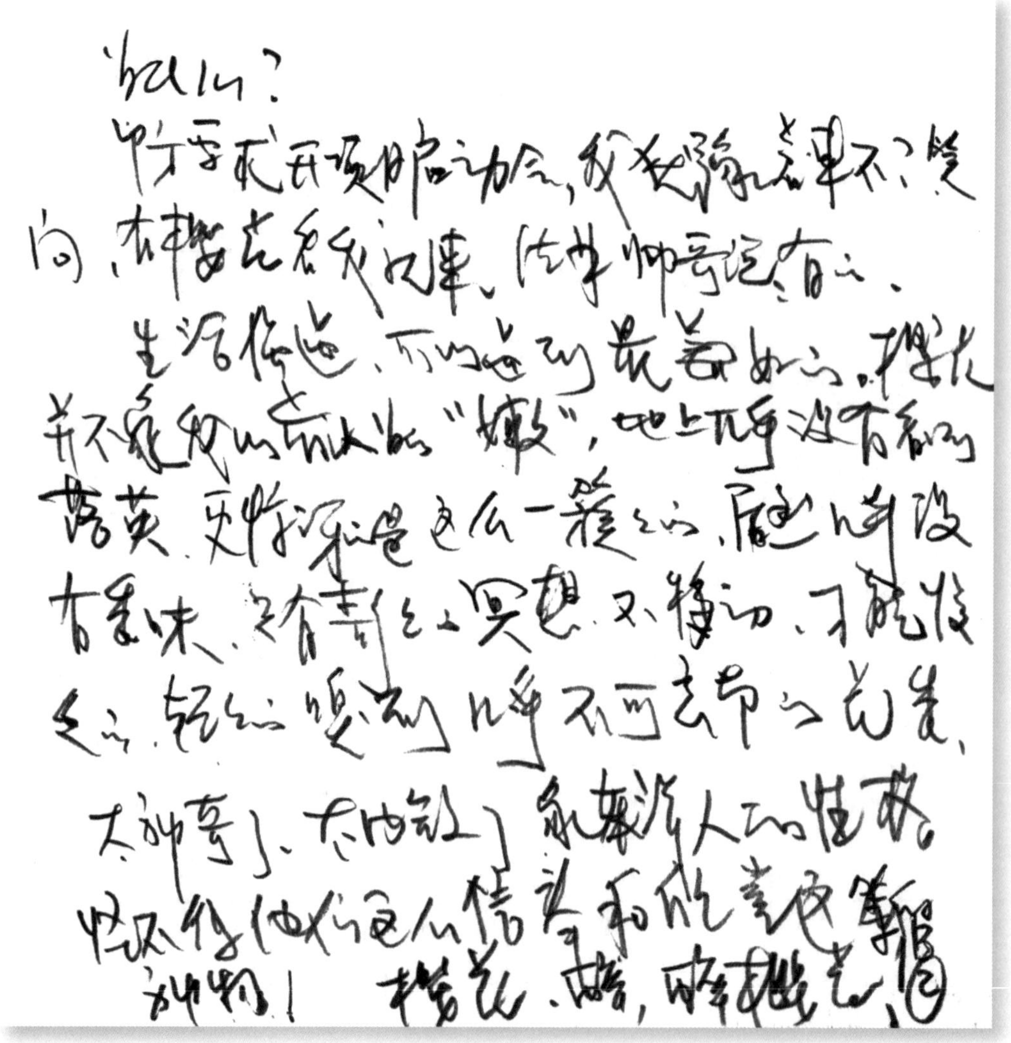

好么？
[illegible]学术[illegible]启动会，我[illegible]
[illegible]樱花[illegible]
生活[illegible]，可以看到[illegible]。樱花
并不[illegible]我们[illegible]的"[illegible]"，也[illegible]没有[illegible]
落英，[illegible]一般[illegible]，[illegible]
有意味，[illegible]思想、[illegible]
[illegible]不可[illegible]节[illegible]
太神奇了，太[illegible]了[illegible]人的[illegible]
[illegible]他们[illegible]信念和[illegible]
[illegible]！ 樱花、[illegible]，[illegible]

Most Intoxicating Cherry Blossoms

最醉樱花

"烟花三月下扬州"，武汉来过几十次，却未及看过一次樱花，也许之前宣传不多，没有趁"武大"樱花成为"红人"之前去光顾，结果，三月樱花季节要预约三天，只能灰溜溜地去东湖樱花公园凑凑热闹。

"In flowering March I come to Yangzhou", as the poem line goes. Having been to Wuhan dozens of times, I had never had the the time to see the cherry blossoms. Maybe it was because it lacked publicity. I had not visited the sakura of Wuhan University before It became popular. As three-day-advance reservation was required for the Sakura Festival, disappointed, I had to go to the East Lake Cherry Blossom Park for the fun of the season.

哗！夜看樱花，太美了，小小的激动，关键的是几乎没人，虽不可与日本的比，也不一定有"武大"的气派，但让我在这里看到，第一次看到樱花还是让人陶醉的！感谢我们的项目甲方。

Wow, the blossoms looked so beautiful in the nightlight. I was a little excited. The point was that there was almost nobody around. Though the flowers could not compare with those in Japan, nor did they have the grandeur as those in Wuhan University, I was still grateful to Party A of our project, who allowed me to see the intoxicating flowers for the first time!

为什么?

Why?

甲方要求开项目启动会，我犹豫着来不来？笑问，有樱花看我就来，结果帅哥说：有的。

When Party A requested me to take part in the project launching meeting, I, still indecisive, responded jokingly that I would go if there were cherry blossoms to see. Then the handsome guy replied: "Yes, there is."

生活偶遇，可以遇到最美好的。樱花并不像我以前认为的“嫩”，地上几乎没有看到落英，更惊讶的是这么一簇簇的，居然几乎没有香味，只有静静的冥想，不移动，才能慢慢地、轻轻地嗅到几乎不可忘却的花香，太神奇了、太内敛了，像东洋人的性格，怪不得他们这么信奉和欣赏这等神物。

You may meet the best in real life change meetings. Cherry blossoms were not so "infirm" as I had thought before as there was hardly any found fallen on the ground. More surprisingly, near these clusters of flower, I could barely smell their sweetness. Only when lost in quiet meditation, motionless, slowly and gently, was I able to smell its aroma. It is so magic and reserved, suggestive of the character of Japanese people. No wonder they worship and admire the flower like a holy thing.

樱花；醉；醉樱花。

The cherry blossom, you are intoxicating; I am intoxicated.

武汉关谷凯悦酒店

Hyatt Regency Wuhan Optics Valley, Wuhan, China

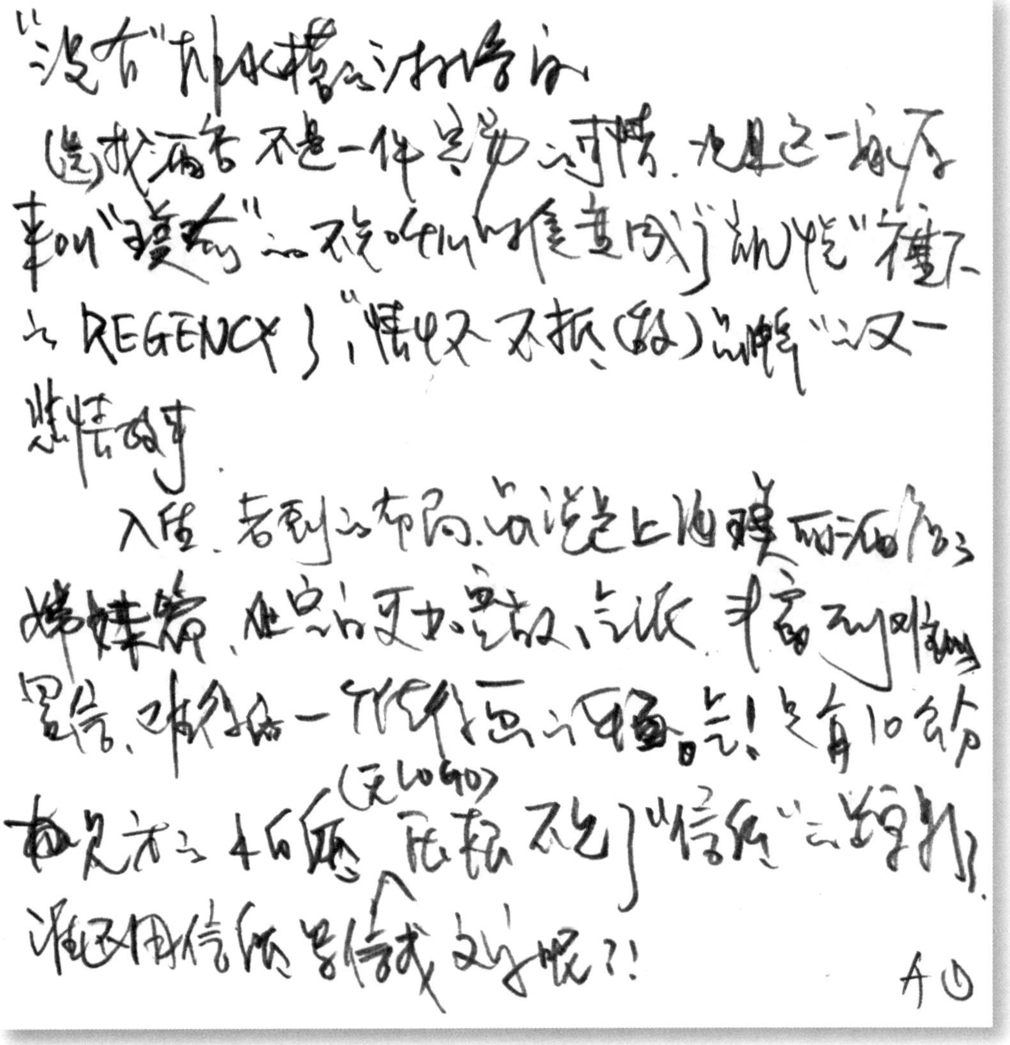
"没有"那么楼之讨论会话
(造)成酒店不是一件容易之事情。尤其这一城市
来的"璞丽"之不说什么作品已成了就像"种不
之REGENCY了，"情也又不抵(触)之触"之又一
悲情故事
入住，看到之布局，如说是上海璞丽酒店之
姊妹篇，但只是更加豪放、气派。再看到酒店
客信，难得给一个很低级之精品店！是有10余分
都是大之4的(无logo)纸张不见了"信纸"之踪影了。
谁还用信纸写信或文字呢？！
A ①

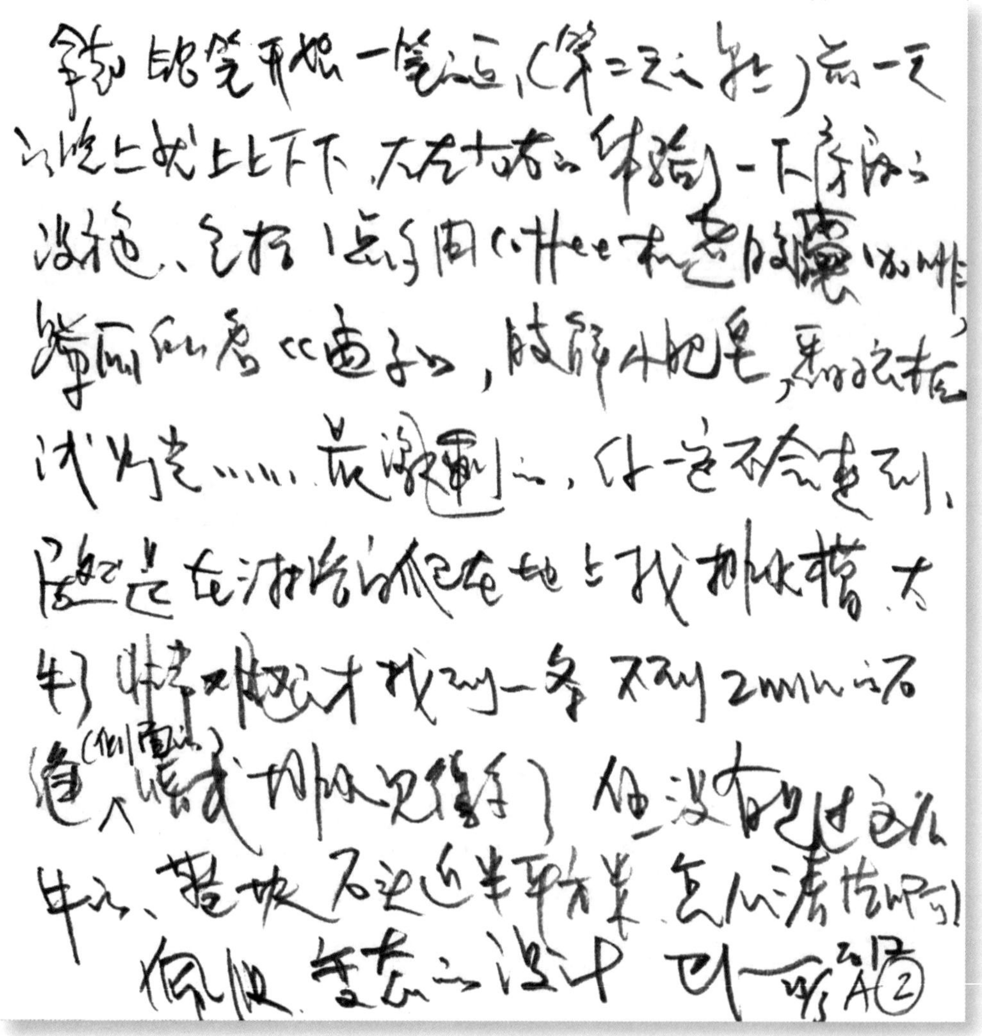
钟到昭觉开始一笔之后（第二天之早上）在一天之路上就上上下下，左左右右之体验到一下彝族之[illegible]、包括[illegible]……[illegible]"画子"，[illegible]……最深刻之，什么不会忘到。居然是在满路的爬在地上找那块[illegible]，太[illegible]了，[illegible]才找到一条不到2mm之石[illegible]（[illegible]）[illegible]，但没有[illegible]……[illegible]

Shower Compartment with no Drainage Tray

没有排水槽的淋浴间

(选)找酒店不是一件容易的事情，况且这一家原来叫“璞瑜”的不知什么时候变成了“凯悦”旗下的REGENCY了，“情怀不抵（敌）品牌”的又一悲情故事。

Selecting a hotels is not so easy. More so as this hotel, formerly named Puyu, had changed into a Regency under Hyatt-another story of “emotion cannot beat brand”.

入住，老到的布局，虽说是上海璞丽酒店的姐妹篇，但空间更加宽敞、气派，丰富到难以置信，是难得的一个值得画的平面。气！只有10厘米见方的小白纸（无LOGO），压根不见了“信纸”的踪影子，谁还用信纸写信或文字呢？！

Then I checked in. Skillfully laid out, the hotel, though a cousin of Shanghai Puli Hotel and Spa, was more spacious and grand, incredibly rich—a rare plane worth drawing. Damn! There were only 10cm×10cm paper sheets (without logo) available, no trace of “letter paper”. Who is still using writing paper to write letters or simply write on? !

拿起铅笔开始一笔笔地画（第二天的早上），前一天的晚上，就上上下下，左左右右地体验了一下房间的设施，包括一点多用咖啡机煮胶囊咖啡，蹲厕所看《孟子》，肢解小肥皂，翻衣柜、试灯光……最刺激的，你一定不会想到，居然是在沐浴间趴在地上找排水槽。太牛了，非常艰难才找到一条，不到2mm的石缝（侧面的）。暗式排水见得多了，但没有 见过这么牛的，整块石头近半平方米，怎么清洁啊！

With the pen in hand, I began to draw carefully. The next day morning or the night before I had going up and down the room experiencing the equipment—using the coffee machine to boil the capsule coffee at a little bit past one in the morning, reading Mencius while sitting on the toilet, dismantling the small soap, ransacking the wardrobe, testing the lighting … The most provoking, I bet, you cannot imagine—I found myself kneeling on the ground looking for the drainage tray. Fabulous—with great difficulty I finally found a slit in the lateral side of the stone, less than 2mm wide. I had seen lots of hidden drainage trays, but never one so fabulous like this. The whole stone was nearly half a square meter. How to clean it!

佩服，变态的设计。

I cannot help admiring such a freakish design.

长沙保利瑜璟阁公寓

★★★

74

POLY Y J RESIDENCE CHANGSHA CHINA

Address : No.9,Changsha Shishu Road,
Poly International Plaza
湖南省长沙市书院路9号
保利国际广场B1栋
Telephone : +(0731)8305 0988

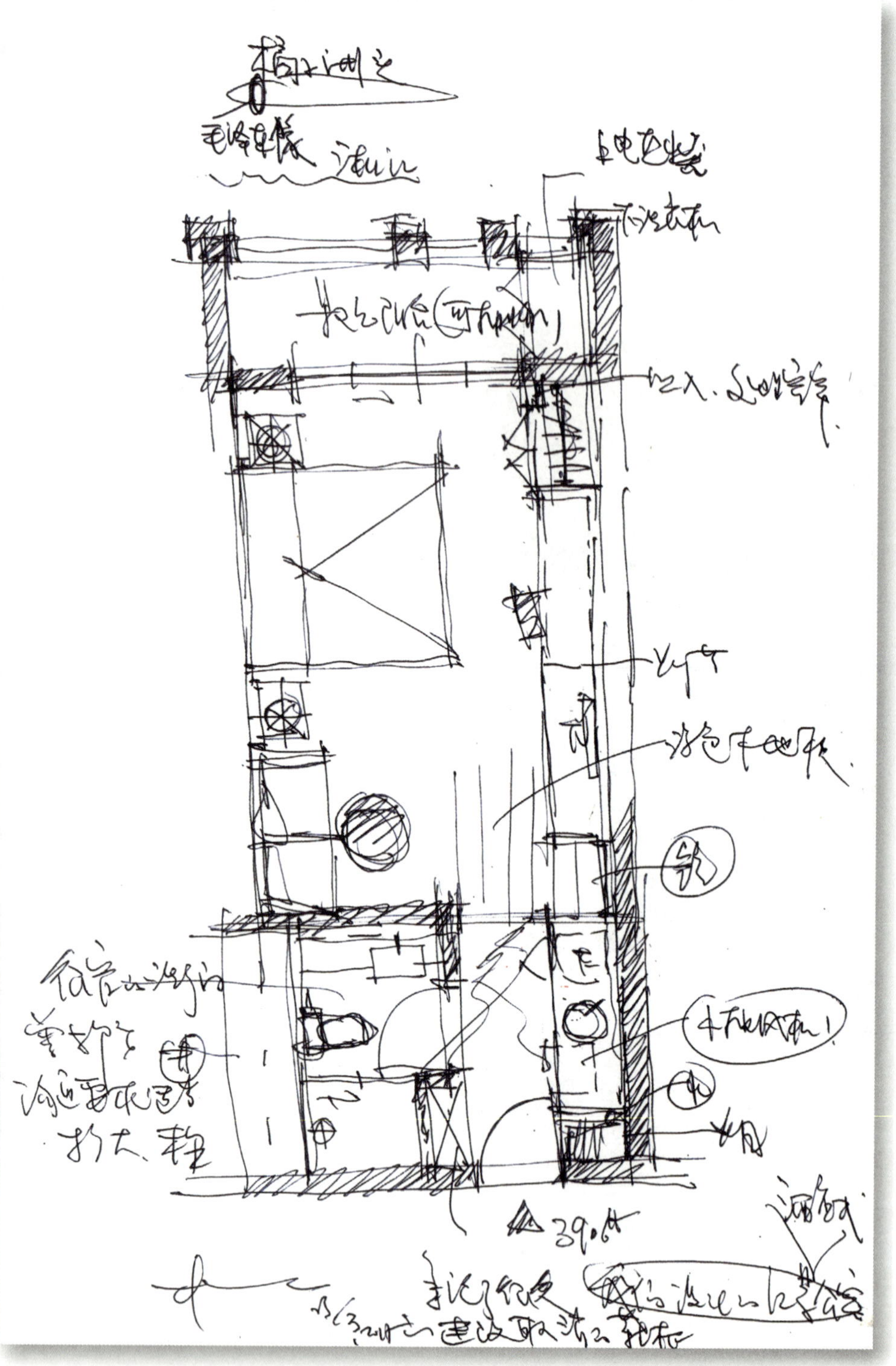

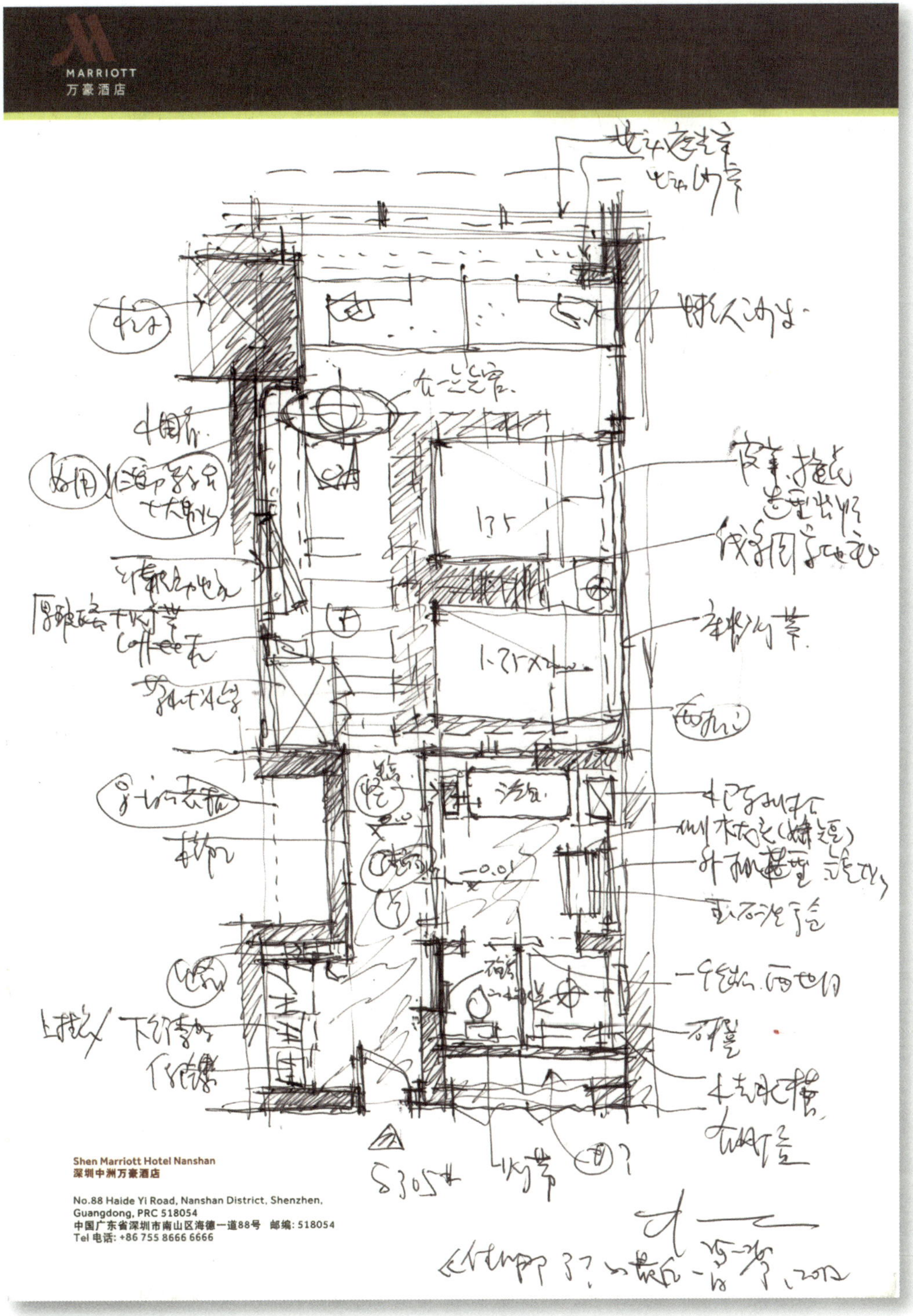

Living • Marriott

深圳中洲万豪酒店

75 ★★★★★

SHEN MARRIOTT HOTEL NANSHAN SHENZHEN CHINA

Address : No. 88, Haide Yi Road, Nanshan District, Shenzhen Guangdong, PRC 518054
中国广东省深圳市南山区
海德一道88号

Telephone : +(86) 755 8666 6666

Http : //www.marriott.com.cn

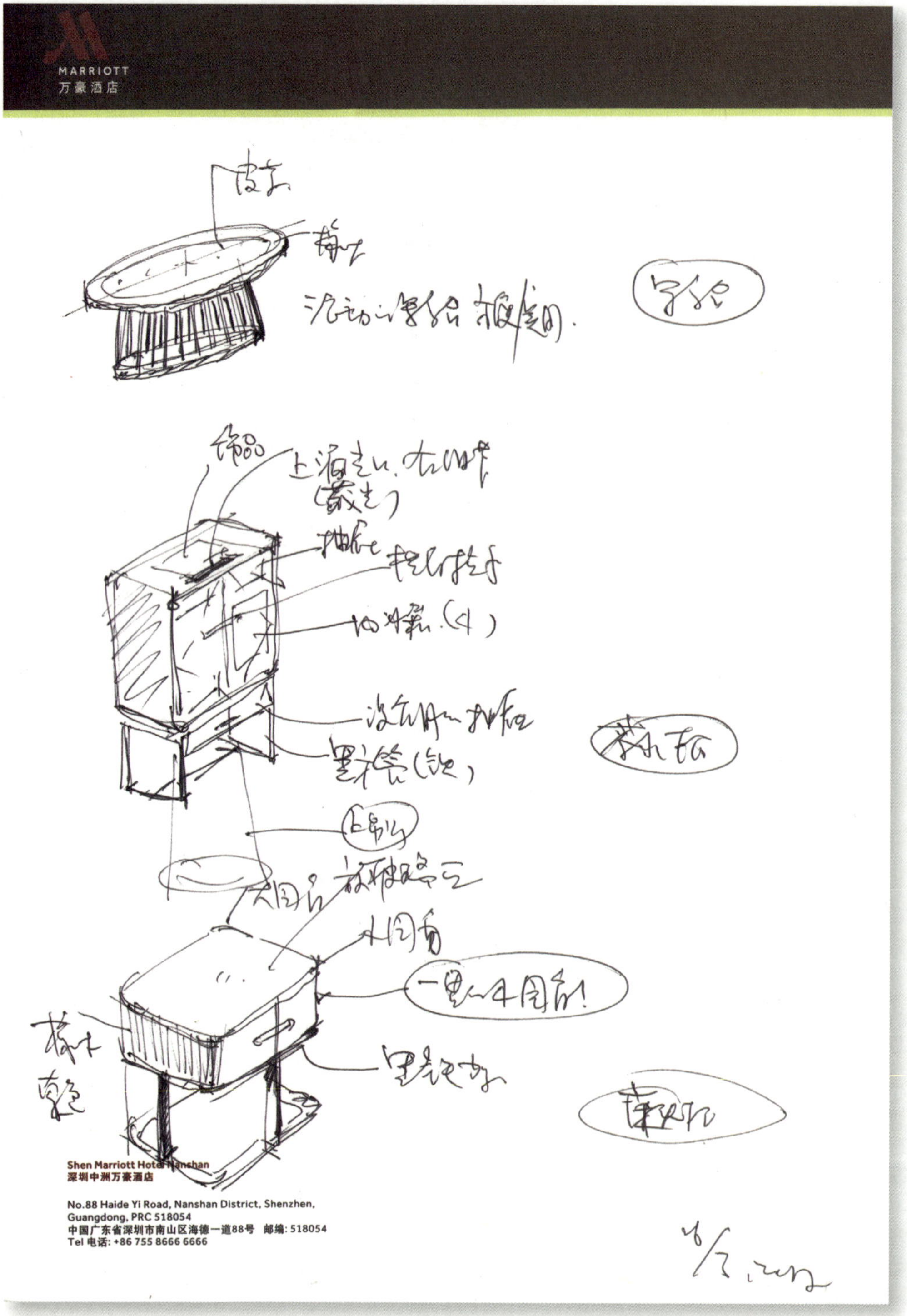
MARRIOTT
万豪酒店
Shen Marriott Hotel Nanshan
深圳中洲万豪酒店
No.88 Haide Yi Road, Nanshan District, Shenzhen, Guangdong, PRC 518054
中国广东省深圳市南山区海德一道88号 邮编：518054
Tel 电话：+86 755 8666 6666

"乾坤园"之设计

周末入住深圳、中洲万豪。之前也住过世界各地的各类酒店

全国各地的酒店，北京的香格里拉，深圳的瑞吉酒店，还是

感到他们不断完善自己的经验与创新。这一次更是一个惊喜。

酒店是集体设计之结晶。从大堂之玄关进入到

艺术品（装置）之选择与布置，恰到好处。近期代表的也是很有

特点。所以大胆猜测是陆的设计，是人、家具、陈设

融合在空间之中，是新时期世界各地建筑的一角。

中餐之体验更有意思，令人流连，布局、空间、氛围、细节、

一步一景之手法，吸引了当今阶层之人们，精致之制作，自由舒适

之，完全达到了深层次的一揽之美。相对于深圳其他

入住之是乾坤酒店，这里可谓是"乾坤之作"，值得一住与

学习研究，特别是陈设、小品布置（细节）。

客房是另有意境的设计，也是展示本地文化之艺术，

开始第一期是我之设计"藏宝"之标本是润

题为全面（是由同行们值得学习之作品）

①

MARRIOTT
万豪酒店

[illegible]

这的酒店，总体来说还是[illegible]的，可以"大家不[illegible]"的说，[illegible]"[illegible]"的酒店！

一次不错的住！

[illegible]

26/3, 2012

Shen Marriott Hotel Nanshan
深圳中洲万豪酒店

No.88 Haide Yi Road, Nanshan District, Shenzhen,
Guangdong, PRC 518054
中国广东省深圳市南山区海德一道88号 邮编: 518054
Tel 电话: +86 755 8666 6666

深圳中洲万豪酒店
Shen Marriott Hotel, Nanshan Shenzhen, China

"Self Surpassing" Design
"超越自己"的设计

周末入住深圳，中洲万豪，之前也住过这家国内知名设计公司设计的酒店：北京三里屯洲际、深圳瑞吉酒店，还是能看到他们不断总结自己经验的影子。这一次更是一个经典。

One weekend I checked into Shenzhen Zhongzhou Marriott. Before I had patronized other hotels designed by this nationally famous design company: Beijing Sanlitun International and Shenzhen St.Regis, giving the impression of constantly learning from their its past. This time it was especially a classical example.

深夜，是体验灯光的好时机。从大堂的夜环境到对重点艺术品（装置）的点缀灯光，恰到好处，让视线柔和地关注在点上。白天以大玻璃幕墙的阳光为主角，让人、家具、陈设沐浴在晨光中，色彩的叠加也成为完美画面的一角。

Midnight is the best time to experience the lighting. From the nightly environment of the hall to the ornate lighting designed for the key works of art (devices), it was proper to perfection, casting soft light on the intended points. During the day, the sunlight on the big glass curtain wall played the leading role, bathing the guests, furniture and furnishings in the sun. The overlapped colors also became a corner of the perfect picture.

早餐的体验更为这家公司添分，丰富、层叠、知性、细腻、一步一景的手法，吸引着用餐的人们。多变化的制作区，自由而有序，平面规划的深厚功力一览无遗。相对之前所入住的其他几家酒店，这里可谓其"巅峰之作"，值得一住与学习研究：特别是陈设，非常有范（国际范）。

The breakfasting experience was another plus of this company. The interior, rich, overlapping, reasonable, fine, varying all the way, appealed to the diners. The food making area, full of variation, free and orderly, showed fully the superb skill of the graphic designer. Compared with several other hotels I had patronized, this one could be called the pick worthy of a stay and study: the furnishings were especially exemplary (internationally).

客房是待的时间最长的地方，也是最有体验感的地方。开门的第一感觉就是“雅”：浅灰黄的橡木基调贯通全房（虽然因为几个供应商的供应，木色五花八门）灰色皮革，米白墙纸，麻布沙发……细节非常多。回到平面布局上，基于奢侈的投入，平面细化了的功能用更多的层次造型去实现。整体效果还是非常能达到完美打动人的程度的，不愧为全球排名前三位的酒店方面的设计公司，不断进步。

The guestroom is where the guests stays the longest and where they feel the most of the hotel. What greeted me on opening the door was “elegance”: the basic tone of light gray yellow oak ran through the room (for all the assortment of the colors as the woods were from several different suppliers): the gray leather, off-white wallpaper, linen sofa, and untold details. Back to the plane layout, due to the luxury input, the detailed functions were realized through many more tiers. On the whole, it touched people to perfection. It was really the work of a worthy design company rated top three globally in hotel design, which is constantly making progress.

这家酒店，整体来说还是非常棒的。可以“大言不惭”地说，确实“超越自己”的设计了。

All taken into consideration, this hotel was fantastic. To put it without humility, it had really “surpassed itself” in design.

一次不错的住！

A nice hotel stay!

成果

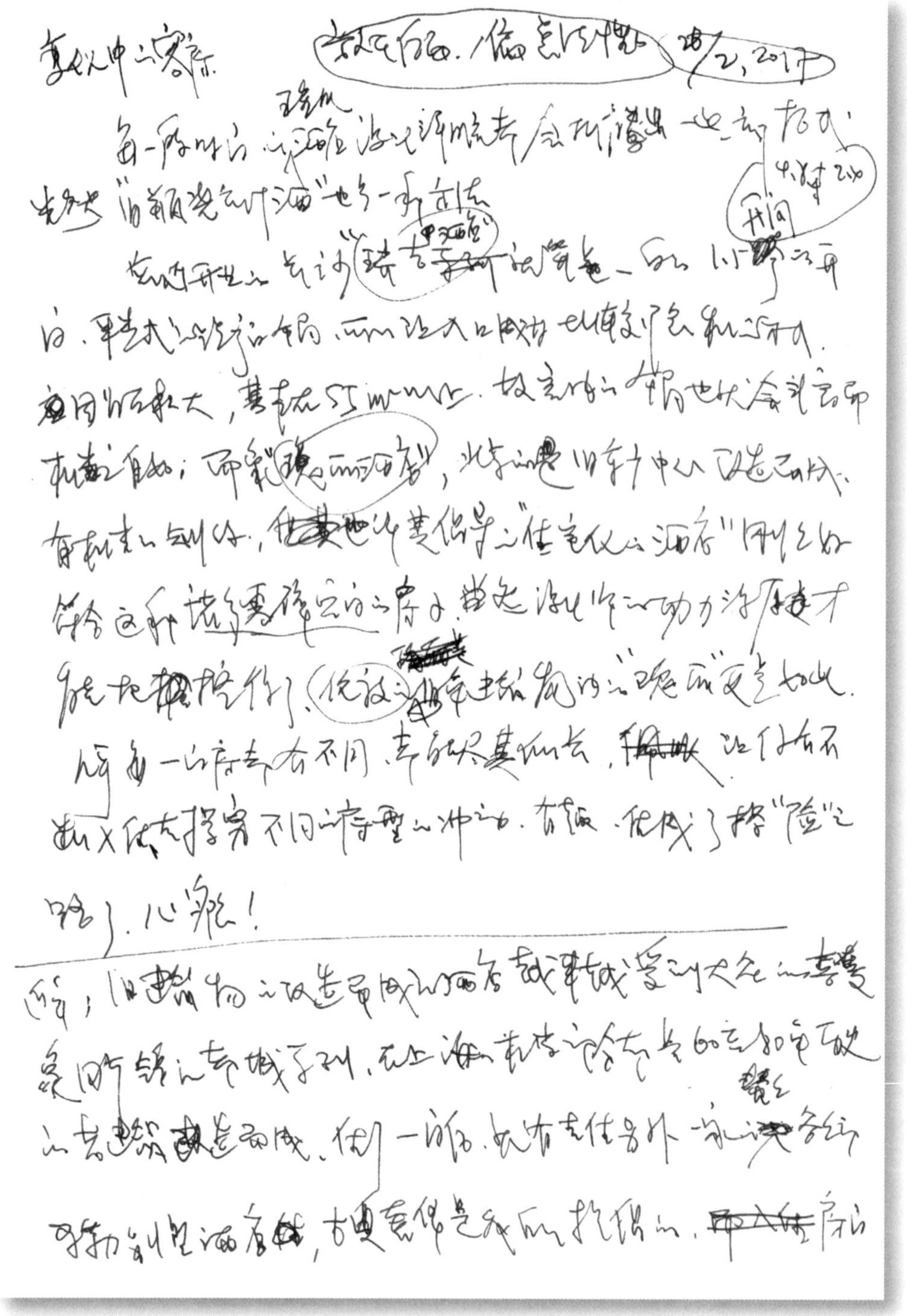

通常的小小曲折的一些构图的（特别是文学艺术等形式的不同

注意节奏的处理“对比”）很容易引导进入而达成“会说”的

说法。去把他们看不到的，感动人们，从而各生的作用

显露出来）。大作所表，无一不是从体验中得到的。

“大运之气”是另一种主观的酒店（旅馆），也给以特色，适合了偏爱个性

的酒店品牌。个别人认为西和个性不适当。将会使“入乡随俗”

努力追求一些艺术、文化、历史、当地文化和婚姻（相关的）自然植入，很

好地融合。当然加入新艺术也不错。洪的的居住酒店

可说是其中的代表。大的旅馆，房的开设多的到大的面积

合宜，而且给人大体上相当。在非常旅行和植入地域文化，总是还

不够。

楼的方式也在向小楼发展了，当然也存在清楚下。你在此的

楼的房的不如一般，但也不能太大，也没有更多要求，不能太

深，太现实有种情况下是的，也会使得得到于的好的，是

新的，多样的可以达到一个奥秘，但然后也是很难

的。虽然还是一定满意，特别是设计的太浓的地域区］

最是好的人的感受之上的。喜欢不喜欢，你给他还有

不同的“光彩”（位在何）好像的激情追求！——

Ever Changing Guestrooms

变化中的客房

每一段时间的五星级酒店设计潮流都会折腾出一些新招式，当然“旧瓶换新酒”也是一种方法。

The trend for five-star hotel design shows new moves at a time. Of course, "old bottles for new wine" is also a way.

新开业的长沙“瑞吉酒店”就算是。一向1.5开间，4.8米+2.4米的开间，采光式的洗手间布局，可以让入口成为比较隐私的方式，因为面积大，基本在55平方米以上。故室内的布局也就丰富而相当自如；而像“瑰丽酒店”，北京的是旧京广中心改造而成，有相当的制约，也许其倡导的“住宅化的酒店”刚刚好符合这种诸多零碎空间的房子，当然设计师的功力深厚才能把控得了。伦敦的旧建筑物内的“瑰丽”更是如此，几乎每一间房都有不同，都能尽其所长，让你有不断入住去探究不同的房型的冲动。有趣，住成了探“险”之路了，心瘾！

Newly opened Changsha St. Regis Hotel is an example. The guest room has 1.5 bays, 4.8m+2.4m frontal width and lighting toilet layout, making the entrance private. The area is large, which is basically more than 55m^2. The layout of the interior is also rich and comfortable; for example, Rosewood Hotel in Beijing is transformed from the old Jing Guang Center, which has considerable constraints. The hotel advocates the "residential hotel", and rooms with many fragmentary spaces can meet this. Of course, the designer's skill is magnificent. Rosewood Hotel in London's old buildings is also like this. Almost every room is different and can do its best, making guests want to explore the different types of room. It's interesting that living becomes a exploration, which is addictive.

近年，旧建筑物改造而成的酒店越来越受到大众的喜爱，像国产锦江都城系列，在上海有相当部分都是20世纪60~80年代历史的老建筑改造而成，住了一间后，就有去住另外一家的蠢蠢欲动。马勒别墅酒店，古典奢华是我所提倡的，房间通常因为旧建筑的结构局限（特别由于类文物等级，不让设计轻易去“破坏”）。往往非常迂回和造成“应该”的浪费，高高低低在所难免，马勒就是，所以服务生的作用就显露出来了，大件行李，女士是很难搬到房间的。

In recent years, hotels transformed from old buildings become more and more popular. For example, in Shanghai, lots of domestic Metropolo Jinjiang hotels are transformed from old buildings in 1960s to 1980s. When guests live one, they will definitely want to try another one. For Hotel Villa Rein, I like its classical luxury, but the guest room is influenced by the structural limitations of the old building (especially the cultural relics do not let the designer easily "destroy"). In particular, the rooms are very roundabout and cause some waste. It is inevitable that the rooms are at different levels. At Hotel Villa Rein, the role of waiter is revealed, they must carry large luggage to the room for ladies.

“大道至简”是另一种五星级酒店主流，也许比较适合于偏商务性的酒店品牌，个人认为这种个性不显著的慢慢也 “入乡随俗”，努力通过加入一些“艺术、文化、人文”，与当地文化联姻，相对的合理投入，得到新的混合体，旧套路加新艺术也不错，沈阳的君悦酒店可说是其中的代表，大品牌，房间开间足够做到左右的厕浴分离，而至简的大设计方向，亦非常奢侈和植入地域文化，感觉还可以。

"Making it easy" is another mainstream for five-star hotels, which is perhaps more suitable for business hotel brand. Personally I think this kind of slow and not significant influence observes the customs of the place, by adding some "arts, culture and humanities" to connect with local culture. Relative reasonable investment results in a new mixture, and the old routine plus new arts gets good results. Shenyang Grand Hyatt Hotel is one of the representatives. For such a big brand, it has a guest room big enough to separate the toilet and shower. With the general design concept of simplicity, the designer add luxurious elements and implants in regional culture, for me it's good.

横向方式也在市场竞争日益激烈的情况下，偶有出现。横向的房间不好布置，面积不能太大，进深也有要求，不能太深。当然某种情况下最好在建筑设计阶段就计划好，像静安的香格里拉可谓是一个典范； 但短板也是很明显的，居住感觉还不一定满意，特别是洗手间太靠近睡眠区了，最起码心理感觉上是的。喜欢不喜欢，你住住就知道了。

In the increasingly fierce competition, horizontal way occasionally appears in the market. Horizontal room is not good for layout. The area can not be too big, and there are requirements for the depth, which should not be too deep. Of course, in some cases the best way is to make plan in the architectural design stage,Jing'an Shangri-La Hotel is a good example. But its shortcoming is also very obvious. The feeling of living is not necessarily satisfied, especially the toilet is too close to the sleep area, for at least psychological feeling. Like it or not, you can try it.

不同的“花样”让住有了持续的激情与追求！

Different "patterns" make living with a sustained passion and pursuit.

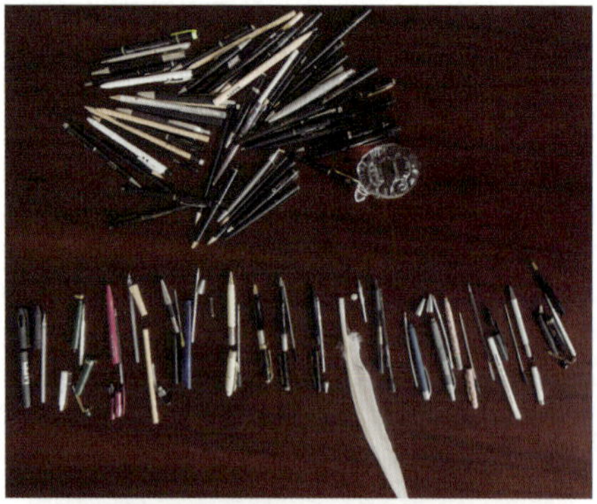

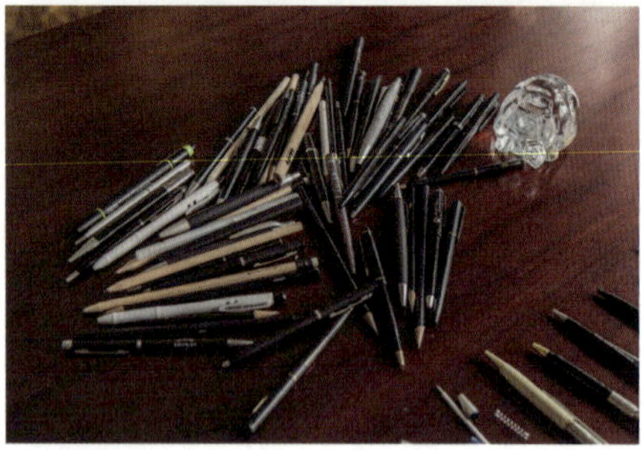

笔：藏而"解"之

往后，收藏笔成了习惯，也不断随手拿起来用作日常之用，体验了这些年，有一些很流畅，出色耐用；而部分很粗糙，很次，当然是与品牌以及耗材品的搭配有直接关系，于是——

肢解，研究一下它们内部构造、用材、组合方式，还是有些有趣好玩的！

记得有一次在中央二台讲到生产"圆珠笔"的设备居然没有一台是中国制造与研发的，不禁自嘲一笑，看似简单的一支笔，包含着相当"吹毛求疵"的技术含量，当然以我这个设计师是绝对不能理解个中的奥妙与含金量的，但一支笔从零配件到外观设计都需要很精密的设计和生产流程的。比较之下，从各品牌所收集来看，一步步"肢解"再装配好一支笔，与一个抢救患者挽回的第二生命一样，应当是十分有趣、喜悦、兴奋而满意的。

也许，一支笔就会让你爱上一个品牌，留在你的手里！情感留在心里！爱抚于书写的过程中。

古有"庖丁解牛"，今有我——解笔。

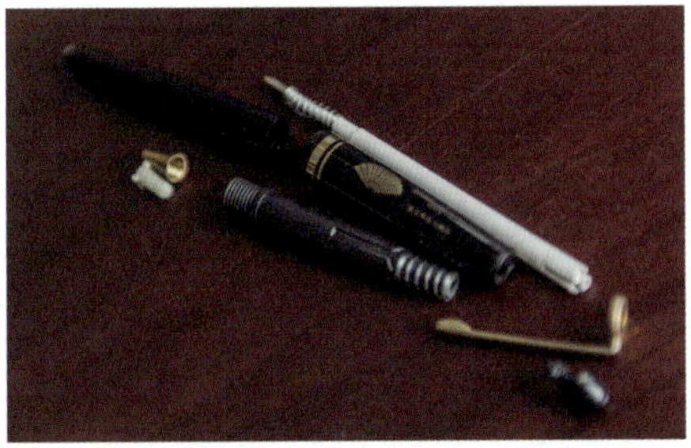

The Habit of Collecting and Disassembling Pens

笔：藏而“解”之

住店，收藏笔成了习惯，也不断随手拿起来用作日常之用，体验了这些年，有一些很流畅，亦很耐用，而部分很粗糙，很次。这当然与品牌对常耗品的投入有直接关系，于是——

While traveling around, I have formed the habit of collecting pens. Now and then they have come in handy for my daily use. It turns out that over the years some of them write smoothly and last well, while some others are quite crude and of poor quality. Of course, it has something to do with what a brand puts into consumer goods.

“肢解”，研究一下它们的内部构造、用材、组合方式，还是挺有趣好玩的！

Then I began to disassemble them in order to study the internal structure, material and composition. And it has turned out to be a lot of fun!

记得有一次看中央二台讲到全世界生产“圆珠笔”的设备居然没有一台是中国制造与研发的，不禁目瞪口呆，看似简单的一支笔，包含着相当“吹毛求疵”的技术含量，但一支笔从笔芯再到装配件再到外观设计都是需要很精密的设计和生产流程的，比较之下从各品牌酒店收集来看，一步步“肢解”再“装配”好一支笔，与一个枪手熟悉装拆自己的第二生命——枪一样，应当是十分有趣、喜欢、娴熟而经常的。

I still remember the time I was stunned by the fact revealed by CCTV II that none of the ball-pen making equipment in the world is produced or developed by China. A pen, simple as it looks, contains rigorous techniques, requiring exact design and production processes, from the refill to the assembly and to the exterior design. As for these pens I have collected from various brand hotels, it is naturally with fun, love, ease and regularity that I have dismantled and reassembled each of them step by step, just like a marksman knowing his gun like the back of his hand assemble and disassembling his second life — the gun.

也许，一支笔就会让你爱上一个品牌，温度留于你的手里！情感留在心里！爱呈现于带给你的书写过程中。

Maybe a pen can make you fall in love with a brand as it leaves warmth in your hand and affection in your heart, and represents itself as a form of love as you write with it.

古有“庖丁解牛”，今有我——解笔。

In ancient China, there was the famous butcher who was good at dismembering cattle; now I am good at disassembling pens.

People to Gratitude

感谢的人

每次出书会惊动很多人，这一次《住哪？3》也不例外，有帮忙整理乱七八糟的原稿，收集、核准酒店的各种资讯，初步编辑、排版，协助翻译及核对，联系出版事宜的一批同事们；有支持我在此“耗费”大量时间的朋友、家人；有一起去体验和讨论酒店的同事们、同行们及旅伴们（哈哈，当然还有女伴）；有费了不少眼神为书写序的人：资深的“大内高手”，酒店设计及管理的翘楚；更有乐意让我们折腾的中国建筑工业出版社的朋友，每一次的合作，都会令我受益良多。

Every time I write a book, I disturb a lot of people. At this time, Where to Live ? III is no exception. My colleagues help me to organize a mess of the original script, collect and check the various information of hotels,edit, set type, assist in translation and collation and contact the publisher. My families and friends support me to "consume" lots of time to the book. My colleagues and travel companions (of course female companions) experience and discuss hotels with me. I would like to also give my sincere gratitude to people who spent lots of wisdom to write preface for the book. They are senior professionals and elites in the field of hotel design and management. Meanwhile, to friends from China Architecture & Building Press who are tortured by me, every time our cooperation makes me benefit a lot.

谢谢！
我身边的人！

Thank you.
For being always with me!

2017年7月18日
July 18, 2017